FROM CELL TO ORGANISM

Readings from
**SCIENTIFIC
AMERICAN**

FROM CELL
TO ORGANISM

With Introductions by

DONALD KENNEDY

STANFORD UNIVERSITY

W. H. FREEMAN AND COMPANY
SAN FRANCISCO AND LONDON

Each of the SCIENTIFIC AMERICAN articles in *From Cell to Organism* is available as a separate Offprint at twenty cents each. For a complete listing of approximately 550 articles now available as Offprints, write to W. H. Freeman and Company, 660 Market Street, San Francisco, California 94104.

Library of Congress Catalog Card Number: 66-30156

Preface

The cell, as a "least common denominator" of living things, is an orderly and understandable unit: it is well integrated and self-sufficient, and it displays the same basic set of structures and abilities everywhere one finds it. When combined into multicellular organisms, however, cells form highly complex interdependent systems that show many "new" properties. This book is an attempt to explore why this is so.

From Cell to Organism—like its predecessor *The Living Cell*—is a collection of articles originally published in SCIENTIFIC AMERICAN. The earlier volume focused on the "universals" of living organization; it dealt mainly with the modern insights that have developed from the Cell Theory, and it showed how fully justified biologists have been in their hope that the basic mechanisms of processing energy, replication, and chemical synthesis would be nearly as ubiquitous as the structural entity of the cell itself. The title of this new collection reflects a different kind of emphasis, and a much more difficult task: that of showing how *cellular* properties may be transformed into the integrated activities of complex multicellular organisms. Our new information about the molecular basis of cellular events, marvelous though it is, will be disappointing indeed if it cannot be used to nourish discoveries about communication between cells, which is the hallmark of the complex organism. It is this challenge to the "New Biology" that the title *From Cell to Organism* reflects.

Because this challenge is still largely unmet, many of the answers provided in the following articles are incomplete, and many are necessarily couched in language less precise and less conclusive than that used to define cellular energetics, recombination in bacteriophage, or the structure of a protein molecule. There is a reason for this: the problems are much more complicated. They are also extremely important, and—because their complexity often prohibits a straightforward solution—they have been approached in diverse and ingenious ways. Before we can hope to solve these problems, we must deal with them as they are and obtain dismayingly complex answers in order to ask further questions. For this reason, much of the so-called "physiology" of whole organisms is descriptive because it states the general performance characteristic of a system. Such a description is an essential prelude to an ultimate molecular analysis; the description in itself, moreover, may contain the answers to important questions about the whole organism in its environment.

The theme of this book is the organism. An organism is different from a mere aggregation of cells because its constituent cells interact in ways that are cooperative as well as competitive. There is evidence that the movements, differentiation, metabolic activities, and other behavior of certain cells are systematically affected by other cells. In short, the biology of organisms is in great part the biology of intercellular communication. Most of the articles in this book are concerned, at least indirectly, with this intercellular communication.

Unlike *The Living Cell,* this volume contains several articles of relatively early vintage. For example, George Wald's elegant account of vision, "Eye and Camera," was published in 1950. Such articles remain current because their authors purposely covered the broad development of an area of research. Furthermore, the recent advances in "organism" biology have not been as explosive as those in molecular and cellular biology; progress has unfolded at a deliberate pace, as though awaiting the application of new insights from the molecular level.

Material that supplements some of the older articles has been incorporated in the introductions. I have added selected references to the bibliographies that appear at the end of the book, and have brought some of the biographical sketches up to date. Jeanne Kennedy prepared the index.

DONALD KENNEDY

Contents

V. NERVOUS INTEGRATION

VI. BEHAVIOR

Note on cross-references: Cross-references within the articles are of three kinds. A reference to an article included in this book is noted by the title of the article and the page on which it begins; a reference to an article that is available as an offprint but is not included here is noted by the article's title and offprint number; a reference to a SCIENTIFIC AMERICAN article that is not available as an offprint is noted by the title of the article and the month and year of its publication.

Part I

DEVELOPMENTAL INTEGRATION

I
Developmental Integration

INTRODUCTION Long ago in the course of evolution, multicellular organization began. It probably originated in several independent instances among the plants, and at least twice among the animals. We may never know whether it resulted from the accidental coherence of prospectively independent cells following division from some secondary aggregation mechanism, or even from the subdivision of a multinucleate single cell. Whatever its origin, the new scheme clearly had its advantages; and though a wide range of successful unicellular organisms still exists today, the dominance of complex, well integrated multicellular plants and animals is an easily recognized fact.

All of these multicellular forms repeat, in their own life cycles, the transition from single to multicellular organization. In most forms, this transiton begins with a succession of divisions of a fertilized egg, followed by a program of individual growth, migration, association, and differentiation of the resulting cells to construct the organism designated in its genetic blueprint. This is a demanding project for the individual cell: at various times it may be called upon to grow to a certain length and then to stop, to move among other cells and then to take up association with a specific group, or to specialize according to the dictates of messages from its immediate environment. All these feats require of cells a common competence—the ability to communicate with other cells.

This ability appears to be the very basis of organization of the community that makes up an organism. It is shown even in some examples of primitive cellular association, examples so obscurely related to the more familiar world of multicellular organisms that biologists hesitate to put them on the same continuum. Yet they display, in simple and easily analyzed form, the same kind of communication and differentiation that underlies the development of all organisms. Such an example is described by John Tyler Bonner in "Differentiation in Social Amoebae." In social amoebae multicellularity is established by secondary aggregation, which is very different from the embryonic development of most higher plants and animals. The individual amoebae are attracted by a chemical, acrasin, and aggregation results after the feeding stage has ended. The cells form a compact mass in which some differentiation between front and rear is evident. The mass responds to surgical separation by re-establishing the original ratio of differentiated cell types. The capacity to regulate this ratio testifies that information exchange occurs among individual cells in the mass.

In the development of animal embryos, the definitive final form is largely the result of the movements of cells from one place to another. Such movements cause the process of gastrulation as well as the formation of many internal organs. The question of what guides and orients the cells in these journeys is perplexing and almost completely unanswered. Equally difficult is the question of what *stops* them: what influences persuade the migrant that a particular environment is the appropriate place to terminate the journey? There is some evidence that associated molecules unique to the species (perhaps like antigens in function) are capable of providing the appropriate cues for other cells. The tissue cells of different vertebrate embryos separate out according to tissue type when they are cultured, suggesting that a cell may communicate its state of differentiation as well as its genetic background.

Cell populations in a particular place receive from their neighbors a barrage of instructions, which help to determine what they will become. Such interactions are crucial in vertebrate development and have been the subject of intensive study since the early part of this century. George W. Gray provides a stimulating and eloquent history of this work in "The Organizer." Since 1957, when his article appeared, there has been some progress in the attempt to establish a chemical basis for the phenomenon of embryonic induction. It has been shown that one tissue may influence the differentiation of another across thin pieces of Millepore filter but not across thicker pieces. This finding suggests that materials associated directly with the cell surface are responsible for transmitting the inductive message. In other sorts of experimental systems, chemicals diffusing over a longer range can affect the differentiation of other cells, without the requirement for close contact. The basic strategy of inductive interaction, however, has been firmly established by the work described in Gray's article.

Although the identity of the signals that may guide the differentiation of cells during embryonic development is unknown, it is clear that the cells do become different from one another. Since each receives the same set of genetic instructions, it is asked *how* they become different. In "The Embryological Origin of Muscle," Irwin R. Konigsberg illustrates some of the new techniques available to attack this very old problem. Clones of muscle cells may be grown in tissue culture from a single ancestral cell; Konigsberg has demonstrated that such clones do indeed differentiate independently, developing the capacity to synthesize the contractile proteins actin and myosin and arranging them in a characteristic way. Though the original question is still unanswered, techniques like those described in this article have provided the bases for possible new directions of inquiry.

The developmental machinery of plants is very different in basic organization from that of animals: size is constantly increasing; growth never stops at a certain point. Indeed, growth is the plant's equivalent of behavior. It is growth that establishes and then appropriately alters the plant's primary orientation in the environment, and hence its relationship to such critical factors as dissolved nutrients and sunlight. Growth proceeds by the rapid division of cells at a growing point or *meristem* at the tip of root or shoot. Cells left behind this active apex elongate, thrusting the tip ahead; then they differentiate *in situ* either into elements for the upward or downward conduction of fluids, or into supporting cells, or into constantly dividing units that provide for increase in girth. The branching linear form of the plant is a direct outcome of this apical growth pattern. Because plant cells are enclosed in cellulose walls which restrict their movement, cellular migrations play almost no part in morphogenesis.

It is not surprising, in view of the crucial significance of light to autotrophic organisms, that their developmental system is closely geared to the timing and the direction of illumination. In "Light and Plant Development," W. L. Butler and Robert J. Downs discuss one aspect of this photic control of growth processes: the seasonal timing of morphogenetic events in accordance with variations in the length of day, a phenomenon known as *photoperiodism*. The clock used by plants to measure changes in the length of day apparently consists of an interconvertible pair of light-sensitive pigments. The authors present evidence that this substance, phytochrome, in one of its two forms functions as an enzyme that catalyzes or prevents various specific growth responses, such as germination, flowering, or the ripening of fruit.

SPHERICAL MASSES OF SPORES of the social amoeba *Dicty-ostelium discoideum* are held aloft by stalks composed of other amoebae of the same species. When the spores are dispersed, each can liberate a new amoeba. The stalks are about half an inch high.

1

DIFFERENTIATION IN SOCIAL AMOEBAE

JOHN TYLER BONNER   December 1959

Recently I was asked to talk to two visiting Russian university rectors (both biologists) about the curious organisms known as slime molds. Communication through the interpreter was somewhat difficult, but my visitors obviously neither knew nor really cared what slime molds were. Then, without anticipating the effect, I wrote on the blackboard the words "social amoebae," a title I had used for an article about these same organisms some years ago [see "The Social Amoebae," by John Tyler Bonner; SCIENTIFIC AMERICAN, June, 1949]. The Russians were electrified with delight and curiosity. I described how individual amoebae can come together under certain conditions to form a multicellular organism, the cells moving into their appropriate places in the organism and differentiating to divide the labor of reproduction. Soon both of my guests were beaming, evidently pleased that even one-celled animals could be so sophisticated as to form collectives.

Of course there are other reasons why slime molds hold the interest of biologists. The transformation of free-living, apparently identical amoebae into differentiated cells, members of a larger organism, presents some of the same questions as the differentiation of embryonic cells into specialized tissues. In the budding embryo, moreover, cells go through "morphogenetic movements" which seemingly parcel them out to their assigned positions in the emergent organism. The only difference is that the simplicity of the slime molds provides excellent material for experiments.

The slime-mold amoebae, inhabitants of the soil, do their feeding as separate, independent individuals. Flowing about on their irregular courses they engulf bacteria, in the manner of our own amoeboid white blood cells. At this stage they reproduce simply by dividing in two. Once they have cleared the food away, wherever they are fairly dense, the amoebae suddenly flow together to central collection points. There the cells, numbering anywhere from 10 to 500,000, heap upward in a little tower which, at least in the species *Dictyostelium discoideum*, settles over on its side and crawls about as a tiny, glistening, bullet-shaped slug, .1 to two millimeters long. This slug has a distinct front and hind end (the pointed end is at the front) and leaves a trail of slime as it moves. It is remarkably sensitive to light and heat; it will move toward a weak source of heat or a light as faint as the dial of a luminous wrist watch. As the slug migrates, the cells in the front third begin to look different from the cells in the two thirds at the rear. The changes are the early signs of differentiation; eventually all the hind cells turn into spores—the seeds for the next generation—and all the front cells cooperate to make a slender, tapering stalk that thrusts the mass of spores up into the air.

To accomplish this transformation the slug first points its tip upward and stands on end. The uppermost front cells swell with water like a bit of froth and become encased in a cellulose cylinder which is to form the stalk. As new front cells arrive at the frothy tip of the stalk they add themselves to its lengthening structure and push it downward through the mass of hind-end cells below. When this process, like a fountain in reverse, has brought the stalk into contact with the surface, the continued upward migration of pre-stalk cells heightens the stalk lifting the presumptive spore cells up into the air. Each amoeba in the spore mass now encases itself in cellulose and becomes a spore. The end result is a delicate tapering shaft capped by a spherical mass of spores. When the spores are dispersed (by water or by contact with some passing creature such as an insect or a worm), each can split open to liberate a tiny new amoeba.

What mechanism brings the independent slime-mold amoebae together in a mass? More than a decade ago we found that they are attracted by the gradient of a substance which they themselves produce. In our early experiments we were unable to obtain cell-free preparations of this substance (which we named acrasin); cells actively secreting it were always necessary to start an aggregation. Later B. M. Shaffer of the University of Cambridge got around this barrier in an ingenious experiment. He took water that had been near acrasin-producing cells (but was itself free of cells) and applied it to the side of a small agar block placed on top of some amoebae. The amoebae momentarily streamed toward the side where the concentration of acrasin was higher. Shaffer found that the water must be used immediately after it is collected in order to achieve this effect, and that it must be applied repeatedly. He therefore concluded that acrasin loses its potency rapidly at room temperature. The loss of potency, he showed, is caused by enzymes that are secreted by the amoebae along with acrasin; when he filtered the fluid through a cellophane membrane to hold back the large enzyme molecules, he was able to secure a stable preparation of acrasin. Presumably the enzymes serve to clear the environment of the substance and so enhance the establishment of a gradient in the concentration of acrasin when it is next secreted. Maurice Sussman and his

AGGREGATING AMOEBAE of *Dictyostelium discoideum* move in thin streams toward central collection points. Each of the centers comprises thousands of cells. This photograph and facing one were made by Kenneth B. Raper of the University of Wisconsin.

co-workers at Brandeis University in Waltham, Mass., have confirmed Shaffer's work and are now attempting the difficult task of fractionating and purifying acrasin, steps leading toward its identification.

Meanwhile Barbara Wright of the National Institutes of Health in Bethesda dropped a bombshell. She discovered that urine from a pregnant woman could attract the amoebae under an agar block just as acrasin does. The active components of the urine turned out to be steroid sex hormones. This does not necessarily mean that acrasin is such a steroid. Animal embryologists were thrown off the track for years when they found that locally applied steroids induce the further development of early embryos. Only after much painful confusion did it become clear that steroids do not act directly on the embryo, but stimulate the normal induction substance. We must therefore consider the possibility that the steroids act in a similarly indirect manner on the amoebae. The purification of acrasin will, we hope, soon settle the question.

From observations of the cells during aggregation, Shaffer has come to the interesting conclusion that the many incoming amoebae are not responding to one large gradient of acrasin but to relays of gradients. That is, a central cell will release a puff of acrasin that produces a small gradient in its immediate

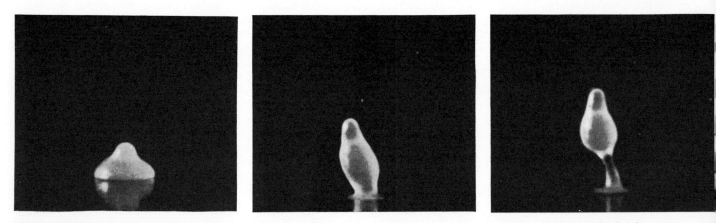

DEVELOPMENT OF THE FRUITING BODY of a slime mold is shown in this series of photographs made at half-hour intervals. At far left the tip cells are starting to form a stalk. In the next two pictures the stalk has pushed down through the mass to the

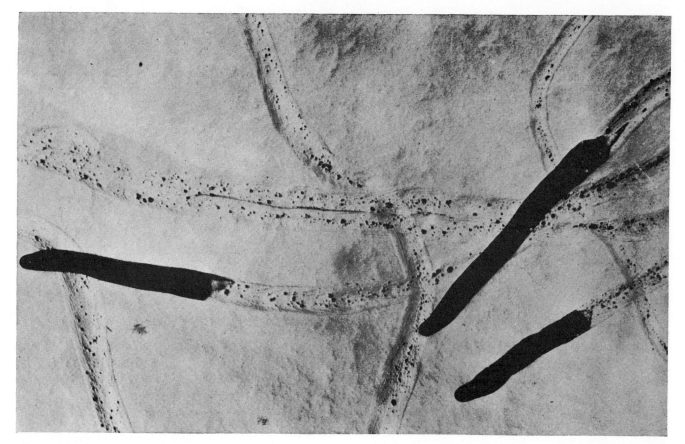

MIGRATING SLUGS of *Dictyostelium discoideum* leave trails of slime behind them as they move. The photographs in this article appear in *The Cellular Slime Molds*, by John Tyler Bonner, and are reproduced with permission of Princeton University Press.

vicinity. The surrounding cells become oriented, and now produce a puff of their own. This new puff orients the cells lying just beyond, and in this way a wave of orientation passes outward. Time-lapse motion pictures show the amoebae moving inward in waves, which could well represent the relay system. If this interpretation is sound, then the rapid breakdown of acrasin by an enzyme plainly serves to clear the slate after each puff in preparation for the next. The cells do not depend entirely on acrasin for orientation; once they are in contact they tend to stick to one another and the pull-tension of one guides the cells that follow. This is a special case of contact guidance, a phenomenon well known in the movements of embryonic cells of higher animals.

After the amoebae have gathered together, what determines their position within the bullet-shaped slug? One might assume that the cells that arrive at the center of the heap automatically become the tip of the slug, and that the last cells to come in from the periphery make up the hind end. If this were the case, chance alone would determine whether a cell is to become a front-end cell and enter into the formation of the stalk, or a hind-end cell and become a spore. If, on the other hand, the cells rearrange themselves as they organize into a slug, then it is conceiv-

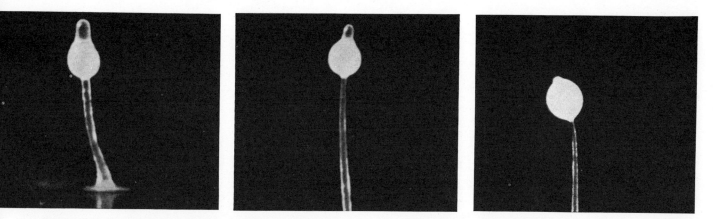

surface and is starting to lift the cell mass. In the fourth picture the spores have formed their cellulose coats, making the ball more opaque. In the last two pictures the spore mass moves up to the very top of the stalk, as the stalk itself becomes still longer.

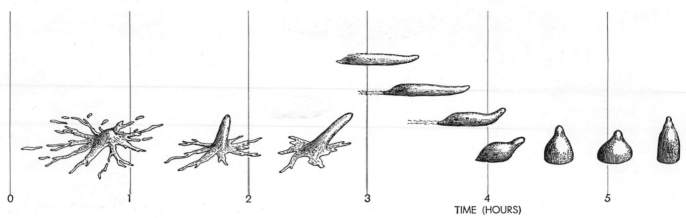

0   1   2   3   4   5

TIME (HOURS)

LIFE CYCLE OF A SLIME MOLD, typified by *Dictyostelium discoideum*, involves the aggregation of free-living amoebae into a unified mass (*first three drawings*), then the formation of a slug which moves about for a time (*next four drawings*) and finally

able that the front end might contain selected cells, differing in particular ways from those in the hind end. I am embarrassed to say that in 1944 I presented some evidence to support the idea that their chance position was the determining factor—evidence that, as will soon be clear, was inadequate. It is some comfort, however, that I was able to rectify the error myself.

The first faint hint that the cells do redistribute themselves in the slug stage came when we repeated some experiments first done by Kenneth B. Raper of the University of Wisconsin. We stained some slugs with harmless dyes and then grafted the hind half of a colored slug onto the front half of an unstained slug. The division line remained sharp for a

number of hours, just as Raper had previously observed. But later we noticed that a few stained cells were moving forward into the uncolored part of the slug. In the reverse graft, with the front end stained, a similar small group of colored cells gradually migrated toward the rear end of the slug. Still, the number of cells involved was so small that it could hardly be considered the sign of a major redistribution. Next we tried putting some colored front-end cells in the hind end of an intact slug. The result was a total surprise: now the colored cells rapidly moved to the front end, traveling as a band of color up the length of the slug.

Here was a clear demonstration that the cells do rearrange themselves in the

slug and that there is a difference between the cells at the front and hind ends. The difference between front-end and hind-end cells—whatever its nature—was confirmed in control experiments in which we grafted front-end cells to the front ends of the slugs and hind-end cells to the hind ends; in each case the cells maintained their positions.

It looked as if front-end cells were selected by their speed; the colored cells simply raced from the rear end to the front. When we placed hind-end cells in the front end, they traveled to the rear, outpaced by the faster-moving cells, which again assumed their forward positions. We tried to select fast cells and slow cells over a series of genera-

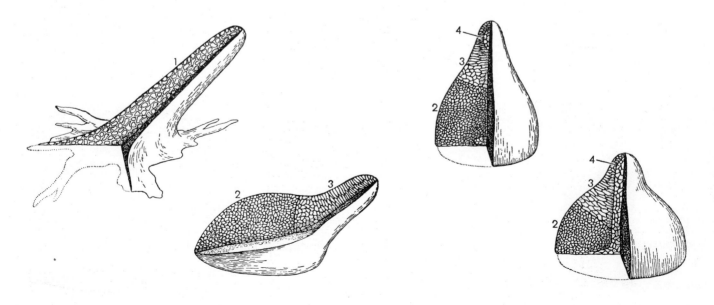

CUTAWAY DRAWINGS of five stages show how the cells change. At the end of aggregation all cells appear the same (1), but in the slug they are of two types (2 and 3). The cells near the tip (3) gradually turn into stalk cells (4) and move down inside the

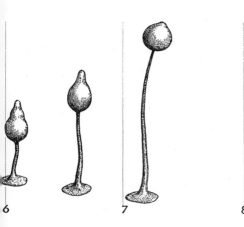

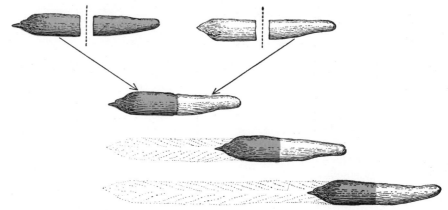

6   7   8

the development of a fruiting body (*last six drawings*). Times are only approximate.

tions to see if speed was a hereditary trait, but after selection the cultures showed no differences from one another or from the parent stock.

Quite by accident a new bit of evidence turned up in an experiment designed for totally independent reasons. Instead of using the fully formed slug we stained amoebae colonies in the process of aggregation and made grafts at this stage by removing the center of the stained group and replacing it with a colorless center, or vice versa. In either case the resulting slug was always uniformly colored, indicating a rapid reassortment of the cells during the formation of the slug.

The evidence for a rearrangement of

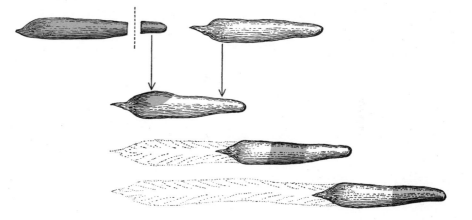

GRAFTED SLUG composed of the hind end of a stained slug and the front end of an unstained slug retains a sharp line of demarcation between the parts even after several hours.

COLORED TIP taken from a stained slug can be inserted into the hind part of an intact slug. The colored cells then move forward as a band until they again are at the front tip.

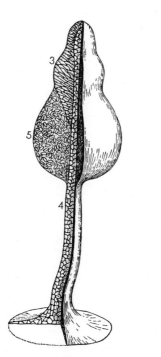

mass. The others (2) become spores (5) as the growing stalk lifts them into the air.

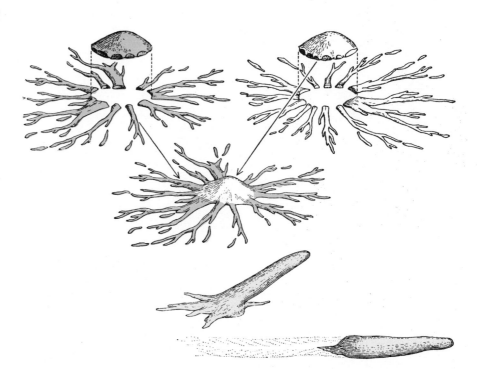

COLORED AGGREGATE in which the center has been replaced by a colorless center produces a uniformly colored slug, indicating that the cells are rearranged as the slug forms. The experiment illustrated in these drawings was originally performed by Kenneth B. Raper.

cells was becoming impressive, but I felt uneasy about the reliability of tests with dyes because such tests had led me into my earlier error. We needed to confirm our results by a different method.

At about this time M. F. Filosa, who was working in our laboratory on his doctoral dissertation, discovered that many of our amoeba cultures contained more than one genetic type. By isolating and cultivating single cells of each type he was able to obtain pure strains that displayed various recognizable abnormalities—in the way they aggregated, in

the shape of their slugs or in the form of their spore masses [*see the illustration below*]. The discovery of these strains furnished natural "markers" for identifying and following cells.

Of course there remained one technical problem: How could the individual cells be identified? Fortunately Raper had shown some time earlier that each fragment of a slug that has been cut into pieces will form a midget fruiting body. Spores derived from the several fragments can then be cultured individually. The amoeba from each spore will give

rise to many daughter amoebae which can be scored for mutant or normal characteristics as they proceed to form slugs and fruiting bodies.

In one experiment we started with a culture of cells in the free-living feeding stage, into which was mixed 10 to 15 per cent of mutant cells. If we were to find a higher concentration of one type of cell in one part of the resulting slug, then we could conclude that there had been a rearrangement. We allowed the cells to form a slug and cut it up into three parts. Upon culturing the individ-

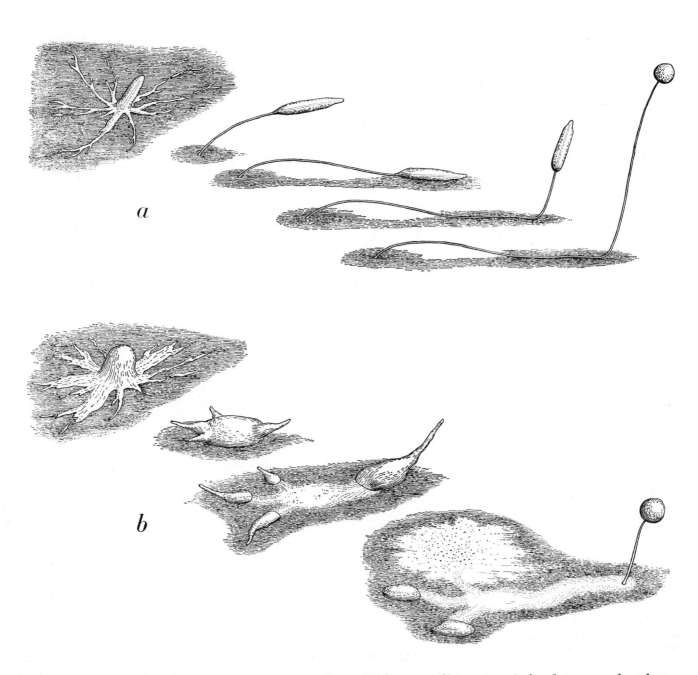

NORMAL AND MUTANT STRAINS of *Dictyostelium mucoroides* are contrasted in these drawings. The normal form (*a*) aggregates in thin streams, and its slug remains anchored by a thin stalk. The "MV" mutant (*b*) aggregates in broad streams and produces a starfish-like slug which then breaks up into smaller slugs. The stalk of the mutant is usually shorter than that of normal strain.

ual spores produced by each part, we found that the hind third had 36 per cent mutant cells, the middle third 6 per cent and the front third 1 per cent. Nothing could be more clear-cut; obviously the cells sort themselves out in a way that brings the normal cells to the front end of the slug. In another experiment, with a larger percentage of mutant cells in the mixture, hind and middle fractions contained 91 per cent mutant cells, and the front end only 66 per cent. Further experiments, including some with other species of slime mold, all led to the same conclusion. During the process of slug formation some cells are more likely to reach the front end than others, and the position of a cell in the slug does not merely depend upon its chance position before aggregation.

One must assume that certain cells move to the front because they travel the fastest, while the other, slower cells are left behind in the rear end of the slug. Considering the different fates of the front and rear cells, however, it is natural to wonder whether there are any other discernible differences between the front and hind cells. Size is one of the easiest qualities to measure, and comparison of spores from the front and rear portions showed that cells of the front segment are larger. From this it might be concluded that the fastest cells are the largest. But size is related to many other factors; some evidence indicates that cells in the front end divide less frequently than those in the hind regions, and this could affect their size. The possibility of a correlation between size and speed can only be settled by further experiment and observation.

But one fact is inescapable. The cells that tend to go forward are not identical with those that lag behind. Do the differences ultimately determine which cells become stalk cells and which will be spores? The most obvious deduction is that among feeding amoebae roughly a third are presumptive stalk cells, and the rest are predestined to be spores. This interpretation is clearly false, however, because then it would be impossible to explain how a single fragment of a cut-up slug can produce a perfect miniature fruiting body. The cells in the hind piece, which would normally yield spores, recover from the surgery that isolates them from the large slug, and one third of these presumptive spore cells proceed to form the midget stalk. This remarkable accommodation to a new situation is also exhibited by many types of cells in embryos and in

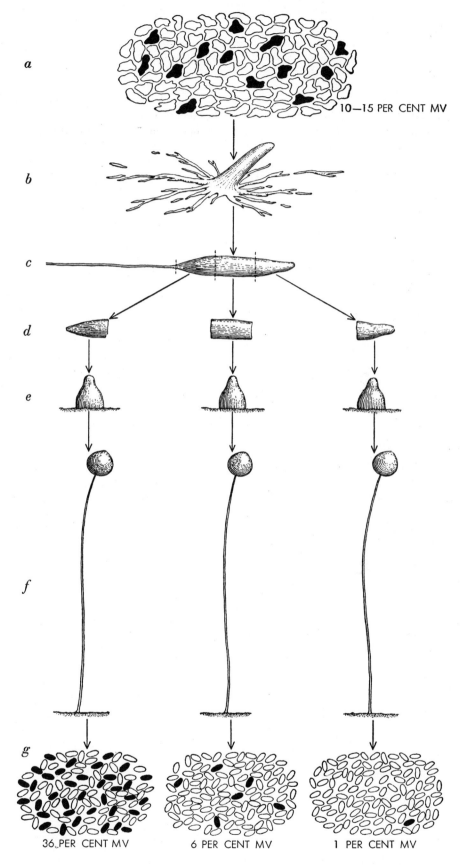

REDISTRIBUTION OF CELLS was proved in an experiment in which MV mutant cells (*black*) were randomly mixed with normal cells at feeding stage (*a*). The cells aggregated (*b*), and the resulting slug (*c*) was cut into three parts (*d*). Each part produced a fruiting body (*e* and *f*). Spores of each were then identified (*g*) by culturing them separately. The concentration of mutant cells was markedly higher in spores from the hind part of the slug.

animals capable of regenerating limbs and organs.

A more reasonable way to explain the relation between sorting-out and differentiation is to visualize the aggregating amoebae as having all shades of variation in characteristics between the extremes found at the ends of the slug. As they form a slug the cells place themselves in such an order that from the rear to the front they display a gradual increase in speed, in size and perhaps in other properties not yet measured. Thus each fragment of a cut-up slug retains a small gradient of these properties. It is conceivable that the gradient, set up in the process of cell rearrangement, actually controls the chain of events that leads the front cells to form a stalk and the hind cells to become spores. For the present, however, this is only conjecture.

At this point let me emphasize that the sorting-out process is not unique to slime molds. Recently A. A. Moscona of the University of Chicago and others have found that if the tissues of various embryos or simple animals are separated into individual cells, the cells can come together and sort themselves out [see "Tissues from Dissociated Cells," by A. A. Moscona; SCIENTIFIC AMERICAN, May, 1959]. For instance, if separate single pre-cartilage cells are mixed with pre-muscle cells, the cartilage cells will aggregate into a ball and ultimately form a central mass of cartilage surrounded by a layer of muscle. By marking the cells in a most ingenious way Moscona showed that there was no transformation of pre-cartilage cells into muscle cells or vice versa; each cell retained its original identity but moved to a characteristic location. In animals, then, sort-ing-out appears to be a general phenomenon when the cells are artificially dissociated. Since the movement in slime molds is part of their normal development, this raises the challenging question whether such sorting-out occurs in the normal development of animal embryos as well.

One must concede that slime-mold amoebae do profit by collectivization: the aggregate can do things the individuals cannot accomplish alone. In the amoebae's society, however, all are not created equal; some rise to the top and others lag behind. And then there is this distressing moral: Those that go forward with such zest to reach the fore are rewarded with sacrifice and destruction as stalk cells. It is the laggards that they lift into the air which survive to propagate the next generation.

2

"THE ORGANIZER"

GEORGE W. GRAY
November 1957

It's a very odd thing—
As odd as can be—
That whatever Miss T. eats
Turns into Miss T.
—Walter de la Mare, *Peacock Pie*

Individuality is the hallmark of life. In the realm of physics and chemistry an investigator must deal with crowds; he can rarely if ever single out one atom or one molecule for study. But a biologist can focus on a single cell, on the nucleus of the cell, on the individual strands of material that make up the nucleus, even, indirectly, on the activity of a single gene. And so he learns that not only is Miss T. unfailingly able to convert steak and potatoes into the unique pattern of the tall, angular, blond woman that is Miss T., but every one of the billions of cells that make up her body carries the individual design that marks it as exclusively her own. The cells are not a crowd but members of an organized community, each serving a special function according to a pre-established plan.

How this organization is brought about is the central problem of biology. If man is ever to understand what life is, he must solve the mystery of how a living thing takes inanimate material and builds it into a germ cell, and how this one cell, after fertilization by merger with another cell, divides into two, and then each into two more, and so on

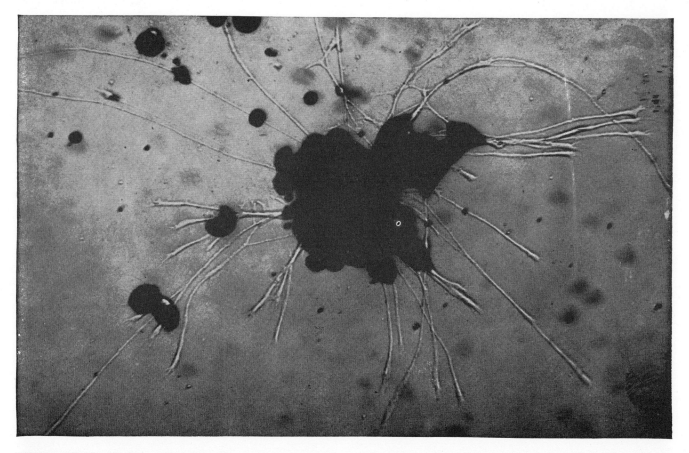

NERVE CELLS with their typical long fibers were unorganized ectoderm eight days before this photomicrograph was made by M. C. Niu of the Rockefeller Institute. The change was induced by fluid taken from a culture of embryonic "organizer" cells.

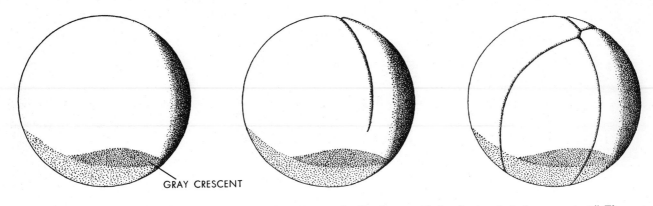

GRAY CRESCENT

DIVISION OF THE EGG of the salamander is depicted in this series of somewhat schematic drawings. The first drawing shows the fertilized egg, with its characteristic "gray crescent." The second drawing shows the first cleavage of the egg, which is perpendicular

through a succession of 40 to 50 cell generations until a human being is born. Everything that is now expressed in the 25 million million cells of the newborn baby was precisely blueprinted in the original germ cell. Not only the architect's plan, but the machinery for building according to the plan was carried in that seed of life not much bigger than the point of a pin.

But how? By the operation of what laws is a single cell able to multiply into such different structures as skin cells, bone cells, muscle cells, blood cells, brain cells and all the rest—and at the same time marshal this wide diversity into a closely coordinated and smoothly working whole?

Epigenesis *v.* Preformation

Embryology dates back to the shadowy dawn of Greek medicine. Two thousand years before there was any knowledge of the biological cell, physi-

cians observed the differences between organs and began to speculate on how the organs were formed. The question was raised by one of the Hippocratic writers, and quite early in history two concepts arose.

Aristotle argued that the mother contributes the substance and the father the structure of their offspring. He pictured the male's semen as the moving element which organized the substance provided by the female, just as an artist "imparts shape and form to his material." From observations of animals, but mainly of the developing chick in the egg, Aristotle deduced that the first organ to emerge was the heart. He said:

"Either [the organs] are formed simultaneously—heart, lung, liver, eye, and the rest of them—or successively, as we read in the poems ascribed to Orpheus, where he says that the process by which an animal is formed resembles the plaiting of a net. As for the simultaneous formation of the parts, our senses plainly

tell us that this does not occur; some of the parts are clearly to be seen in the embryo while others are not."

Epicurus held another view. Believing that matter is everything ("there are only atoms and the void"), he contended that both parents contribute material and that a child must be completely formed from conception, though in miniature. The Roman rhetorician Seneca later epitomized the idea in these words:

"In the seed are enclosed all the parts of the body of the man that shall be formed. The infant in his mother's womb hath the roots of the beard and hair that he shall wear some day. In the little mass, likewise, are all the lineaments of the body and all that which posterity shall discover in him."

Here are two strikingly contrasting ideas. In Aristotle's view there was a gradual emergence of form from undifferentiated material—a process which has come to be called epigenesis (from

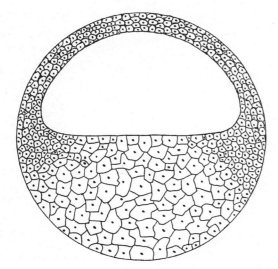

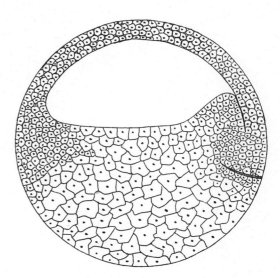

FORMATION OF THE GASTRULA of the frog, which differs somewhat from that of the salamander, is depicted in cross section. The first drawing shows the blastula, a hollow ball partly filled with yolk cells. The second drawing shows the cells beginning to fold

to the gray crescent. The third drawing shows the egg divided into four cells; the fourth, into eight cells; the fifth, into 16; the sixth, into 32. The region just below the gray crescent later gives rise to the blastopore (*see drawings at the bottom of these two pages*).

the Greek, meaning "ensue upon"). The other view, called preformation, asserted the presence of a full-structured organization from the beginning, so that the development of the embryo was simply an enlargement of what already existed in small scale.

It is interesting that in Renaissance Europe it was the scholastics who accepted Aristotle's theory, while medical men and other biologists turned increasingly to the embryology of the Epicureans. By the 17th century the sway of the preformationists was almost unchallenged, and the concept was pushed to the most absurd extremes. If the embryonic germ is a complete body, it was argued, then the germ must contain all the organs and parts, including the seed of the next generation, and that seed in turn the seed of the next, and on and on, like a series of Chinese puzzle boxes. Mother Eve, it was said, carried in her body the forms of all the people to be born.

Meanwhile the power of the glass lens had been discovered. The pioneering Dutch microscopist Anton van Leeuwenhoek focused his "optik glass" on a drop of human semen and saw the spermatozoa, which he named "animalcules." Others took up the new instrument, and among them was an ardent preformationist who thought he saw in each animalcule the form of a tiny human being, complete with head, body, hands and feet! This observation led to a great schism among preformationists. For it suggested that Adam, rather than Eve, contained all mankind. Many forsook Eve to espouse the new dogma, and they were called animalculists; those who remained loyal to Mother Eve were ovists.

Observations and Experiments

The first breath of fresh air came from Germany. At the University of Halle, Kaspar Friedrich Wolff watched the de-velopment of the tip of a growing plant through his microscope and made careful drawings of what he saw. Shoot after shoot showed only homogeneous tissue. There was no sign in this tissue of the leaves, flowers and other organs which later emerged from the shoots. Wolff noticed, moreover, that when the specialized parts did begin to form, each appeared first as an almost imperceptible prominence or swelling in the undifferentiated tissue.

Wolff next trained his microscope on the developing chick in the egg to see what he could learn of animal tissue. He found that the intestine gradually formed from tissue which at the beginning showed no rudiment of the organ that was to come. And so with other organs. Neither in a plant nor in an animal could he see any trace of preformation, and he concluded that in both the developmental process was epigenesis. Wolff's 18th-century observations inaugurated a rational approach to embry-

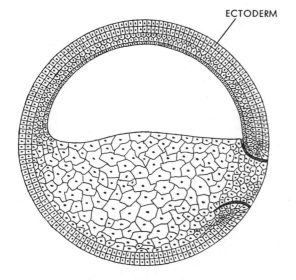

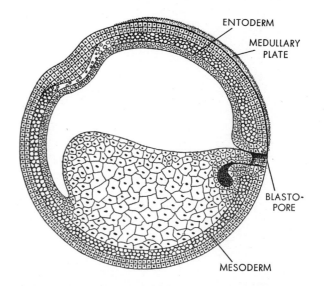

inward to form the gastrula. The third and fourth drawings show the formation of the blastopore and the three layers of the gastrula: the ectoderm, the mesoderm and the entoderm. The medullary plate, from which springs the nervous system, grows out of the ectoderm.

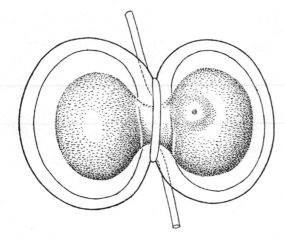

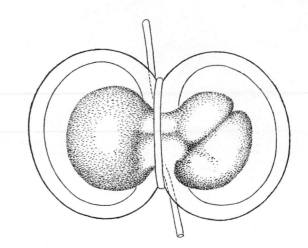

CLASSIC EXPERIMENT performed by Hans Spemann involved tying the salamander egg into two halves across the gray crescent.

The nucleus of the egg (*marking at right in drawing at left*) was confined to one half. At first only the half containing the nucleus

ology, and other advances paved the way: improvement of the microscope, the firm establishment of the cellular theory, the invention of the microtome, the discovery of evolution, of the gene, of proteins and of nucleic acids. Up to the latter part of the 19th century, however, the epigeneticists and the preformationists continued their war of words without direct experimental test, and embryology remained almost entirely a descriptive science. "Take 20 or more eggs," an experimenter of the fourth century B.C. had instructed, "and let them be incubated by two or more hens. Then, each day from the second to that of hatching remove an egg, break it and examine it." This preoccupation with what we may call the natural history of the embryo, paying little attention to causal relationships, continued to domi-

nate embryological research until almost the turn of the century.

In the 1880s the speculations of a German zoologist, August Weismann, precipitated a significant investigation. He developed a germ-plasm theory which pictured the nucleus of the fertilized egg as a mosaic in which "primordia" (starting points of the organs and tissues) "stand side by side, separate from each other like the stones of a mosaic, and develop independently, although in perfect harmony with one another, into the finished organism." If this were the case, then an embryo at the two-cell stage would have one half of the individual in each cell. Wilhelm Roux, an anatomist at the University of Breslau, decided to test Weismann's hypothesis. He figured that if he removed one of the two cells at this stage,

the incubation of the remaining cell would provide the needed test. And so in 1888 Roux performed a historic experiment. Taking the two-cell embryo of a frog, he killed one of the cells with a hot needle and let the other develop in its natural water medium. The result was a half-tadpole. This seemed to demonstrate that each cell carried half of the machinery for constructing the frog, and the experiment was hailed as proof of the mosaic theory of preformation.

But there were doubting Thomases. They pointed out that the killed cell had remained attached to the living one and might conceivably have influenced its development. Various attempts were then made to separate embryonic cells, and in 1891 this was accomplished by Hans Driesch, working in a laboratory in Naples. Driesch shook two-cell em-

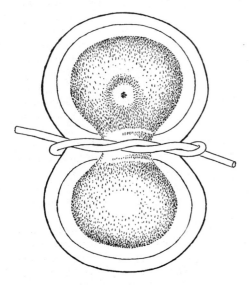

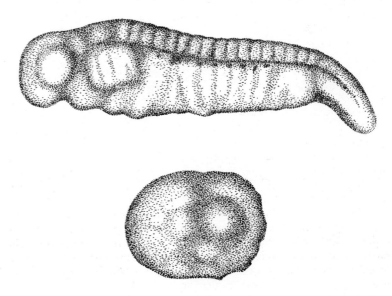

SECOND EXPERIMENT by Spemann was to tie the egg parallel to the gray crescent. The half of the egg with the gray crescent de-

veloped into a normal embryo (*upper right*), but the other half produced only an unorganized "belly-piece" (*lower right*).

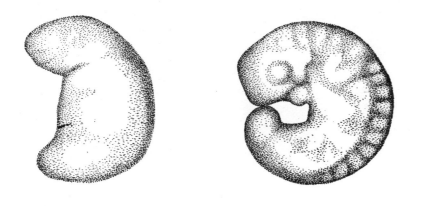

divided (*drawing second from left*). Later the other half began to divide. Eventually both halves gave rise to normal embryos, one younger than the other (*drawings at right*).

bryos of sea urchins in a vial of sea water and succeeded in disjoining the two cells without injury. Under incubation each cell developed into a whole sea-urchin larva. In later experiments he separated a four-cell embryo into its components, then an eight-cell, and finally one that had reached the 16-cell stage—and from each he obtained on incubation a complete animal.

This spectacular emergence of the whole from a fragment was so contrary to the popular dogma of the day that it aroused embryologists. There was a rush to the laboratories, and investigators who had been content merely to watch life unfold now became ardent experimenters. By the turn of the century a whole new school of laboratory workers were engaged on problems which the conflicting results of Roux and Driesch had posed. Among them were Hans Spemann, a 31-year-old zoologist at the University of Würzburg, and Ross G. Harrison, a 30-year-old associate professor of anatomy at the Johns Hopkins University. These two became the leaders, builders and teachers of the modern science of developmental biology.

Spemann and Harrison

Both men were superb experimenters. They knew how to ask the right questions of nature. Employing techniques which in some cases had been pioneered by others without definitive results, they had the imagination to strip the problem to its simplest terms, bend the experimental procedures to the new approach and thereby frame a question in such a way that the subject could respond. Spemann used to say that he regarded the embryo as "a conversational partner who must be permitted to answer in his own language"; Harrison had the same attitude. As one reads the papers of these two masters of research, Harrison appears to be more matter-of-fact, more objective, a greater realist, while Spemann seems more philosophically minded, more concerned with symbols, a searcher for wholeness. They never collaborated in investigation but were warm personal friends, frequently consulting each other on speculations and experimental results. Harrison often spent part of his summer vacation visiting Freiburg, where Spemann was professor from 1919 to his retirement in 1935.

Harrison is most widely known, perhaps, for his invention of the tissue-culture technique. It was devised to tackle a problem in embryology, namely, the origin of the peripheral nerves that connect the brain with the end-organs of touch, taste, smell and the rest. The connection is through the spinal column, of course, but how are the lines laid down during development? Some investigators believed that the nerves grew out from the neural tube, the primitive spinal cord of the embryo. Others argued that the fibers originated in the organs and grew toward the cord.

Harrison conceived the idea of trying to culture a minute fleck of tissue from an embryonic neural tube in the nutrient fluid to which it was accustomed, and watching to see whether nerves could grow out of it. He devised the "hanging drop" technique, putting the bit of tissue in a drop of fluid which was then suspended from a glass cover slip laid over a hollowed-out well in the center of a glass slide [*see drawing on page 21*]. The tissue was a particle of neural tube cut out of a frog embryo, and the fluid was a drop of the animal's lymph.

Harrison watched the bit of tissue almost continuously under a microscope, and soon results began to show. After a few hours delicate protuberances began to bud out from the tissue. They grew into filaments which extended into the clotted texture of the lymph. Then, in the same way, Harrison cultured bits of non-nervous tissue—a piece of embryonic muscle, a particle of intestinal wall, other fragments of early organs. All of them grew, but none sent out any nerve fibers. Thus he demonstrated that nerve development is a growth outward from the neural tube, not inward from the organs.

The hanging drop became one of the most powerful tools of experimental biology. In 1917 the Nobel prize committee of the Swedish Karolinska Institute chose Harrison for the Nobel award in physiology and medicine, but for some reason the Institute decided not to give the prize that year, and Harrison never received this honor. But he has had many others, and only last year the Academy of the Lincei in Italy, the world's oldest learned society, sought him out, at the age of 86, to give him its Antonio Feltrinelli prize of $8,000.

Spemann, too, was the recipient of numerous decorations. In 1935, six years before his death, the committee selected him for the Nobel prize and this time the Karolinska Institute decided to give it. The citation said that the award was for "his discovery of the organizer effect in embryonic development."

The Organization of a Newt

Spemann's discovery was made in experiments with newts, a variety of salamander. These small, lizard-like amphibians originate from eggs whose habitat is water, so it is a simple matter to keep the temperature and other conditions favorable for incubation and nurture. Comparative studies have shown that the development of the salamander egg parallels closely that of other backboned creatures, including man.

A peculiarity of the salamander's egg is that about half of its surface is dark-colored, the other half light or colorless. Immediately after fertilization a small, crescent-shaped segment of the boundary region between the light and dark areas takes a grayish hue. This so-called "gray crescent" is the first visible manifestation of profound changes occurring within the egg. It appears just before the self-duplication of the fertilized cell.

Spemann, like many other biologists of his day, was fascinated by the prob-

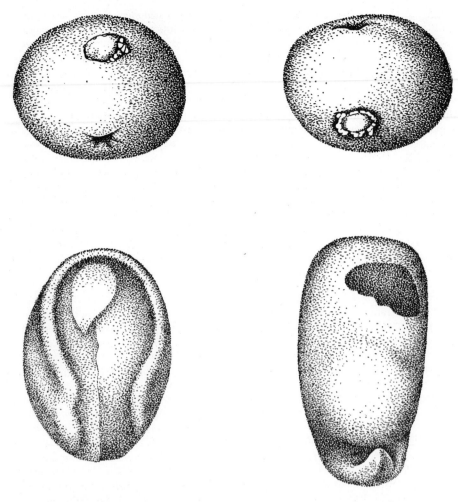

TISSUE WAS TRANSPLANTED by Spemann from a region above the blastopore of a salamander egg (where it would normally grow into belly skin) to a region below the blastopore of another egg (where it would normally grow into nerve tissue), and vice versa (*drawings at top*). Later the cells which would have become skin tissue became nerve tissue (*lower left*), and the cells which would have become nerve tissue became skin (*lower right*).

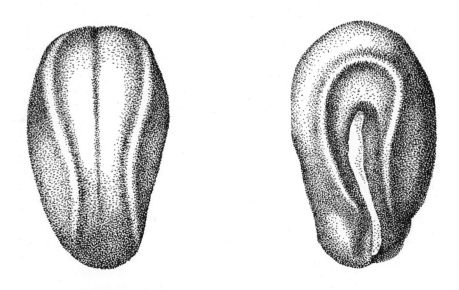

TISSUE WAS REMOVED from the lip of the blastopore of a colorless salamander gastrula and implanted in the ectoderm of a pigmented gastrula. The blastopore tissue of the colorless gastrula induced the pigmented gastrula to make a second medullary plate. At left is the medullary plate of a pigmented embryo; at right, the induced medullary plate on the other side of the embryo. The light area in the induced plate is the implanted tissue.

lem of testing Weismann's mosaic theory of preformation. Suppose, instead of separating the two-cell embryo into its halves as Driesch had done, the egg were simply constricted across its middle just before it cleaved? Spemann procured a strand of baby hair, made a slip noose, looped it over the cleaving egg and drew the strand tight, leaving only a slender bridge of protoplasm between the two halves. The nucleus of the cell was sequestered in one half [*see drawing at top of page 18*]. It began to split and draw apart in two spindles, duplicating itself in the process known as mitosis. The cell on its side of the constricting noose thus divided into two cells, then into four. Meanwhile the cell on the other side of the noose remained just a single cell. Soon, however, it too began to divide, after mitosis had occurred on the other side in a cell close enough to the pinched waist to send nuclear material through the narrow bridge. The net result was two embryos, the one on the side with the original nucleus developing first, but the other finally catching up. Eventually two complete salamander larvae developed. Spemann repeated this experiment many times, once getting 32 cells on one side before nuclear material slipped through and started development on the other.

Suppose the cleaving egg were pinched in a direction at right angles to the first—that is, parallel to the gray crescent instead of across it [*see drawing at the bottom of page 18*]. Would this make a difference? Spemann made the experiment and found the result indeed quite different. The half without the nucleus eventually divided and subdivided, but its final product this time was only an unorganized mass which Spemann called *Bauchstück*—belly-piece. This belly-piece contained liver cells, lung cells, intestinal cells and other abdominal material—but it had no axial skeleton, no nervous system, no unifying pattern.

Why? Why should the unnucleated half of the egg now produce only a crowd of miscellaneous cells? Spemann decided that the distinguishing difference between the experiments must be the fact that in the first, each half of the egg had part of the gray crescent, whereas in the second, one half had most or all of the gray crescent, while the other—the half that failed to develop—had little or none of it.

What was so significant about the gray crescent? Spemann considered its "geographical" relationship to the known facts about development of an embryo. After the embryo has developed to the form of a hollow sphere, a dimple begins to form on its surface, and this deepens into a crater called the blastopore. The cells around the lip of the crater suddenly begin to slide in over the brink, as if pushed by an invisible hand. At the end of this process, known as gastrulation, the salamander embryo looks like a rubber ball with half of it pushed in. Now the cells on the concave outer surface, called ectoderm, will become skin, brain, nerves, ears, eye lenses and other sensory organs. A layer of the cells that have migrated to the inside, called entoderm, will develop into lungs, liver, stomach, intestines and certain other abdominal organs. Another infolded layer of cells, the mesoderm, is destined to produce bones, cartilage, connective tissue, muscle, the blood and its vessels, and organs of the urogenital system.

The Organizing Lip

How are all these fates fulfilled? Spemann knew that the lens of the eye is formed by a process of "induction." An eye begins as a tiny protuberance budding off from the embryonic brain; the protuberance grows into a vesicle, and then the vesicle apparently induces the skin overlying it to become a transparent lens. Reflecting on his experiments with pinched eggs, Spemann began to suspect that cells associated with the gray crescent might be the primary inducers of the whole chain of development that produces an organized individual. The gray crescent is merely a surface feature of the egg which soon disappears as the cells multiply. But the blastopore forms just below the area the crescent occupied. For other good and sufficient reasons, Spemann focused his attention on the lip of the blastopore and addressed a series of questions to it.

How would it be if one took a bit of tissue from the specific area of the embryo that is destined to form belly skin and transplanted it to the area behind the blastoporal lip that is to form brain? Spemann did that, using two specimens. From one embryo he cut a microscopic patch from the presumptive flank; from the other he sliced a fragment of equal size from the area that he knew would form brain if left undisturbed. Then he exchanged the two bits, transplanting the presumptive flank tissue into the wound left in the brain area and the presumptive brain into the cut surface of the flank. The grafts grew, the two embryos developed, and lo!—the transplanted flank turned into brain and the transplanted brain into belly skin.

After repeating this and similar experiments many times, and always getting the same answer, Spemann took up a new line of questioning. Suppose, he said, the lip itself were cut out and implanted in another embryo?

The two-story zoology building at the University of Freiburg where Spemann worked was teeming with students, and among them was Hilde Proescholdt. It was her good fortune to be looking for a research project to fulfill the requirements for her Ph.D. Spemann outlined his proposed experiment to the young woman, appointed her his assistant to carry it through, and thus she shared in the great discovery published two years later under their joint authorship.

For this experiment Spemann used two varieties of salamander—one dark-hued, the other colorless. The blastoporal lip was cut from the colorless embryo and grafted into the belly of the pigmented individual. The latter now had two blastoporal lips—its own on its topside, and the implanted lip from the colorless salamander on its underside. Thus there were two centers of organization in a single embryo. And what happened?

Gastrulation occurred at each place: that is, ectodermal cells migrated over each lip into the mesodermal layer beneath. Eventually there developed two axial systems, each complete with a spinal column, head, trunk, legs and tail —two baby salamanders joined like Siamese twins! The extra salamander had whole sheets of tissue made of dark

cells, as well as organs and parts built of colorless cells. This showed that the lip transplanted from the colorless embryo had extended its organizing power to multiplying cells of the host embryo.

Here, said Spemann, in the cells that flow around the blastoporal lip, is the primary center of induction. He named this region "the organizer." He did not, however, suppose that it controlled the whole process of development: he saw a succession of organizers at work, one taking up where another left off, each having its part in the sequence which began with the migration of a sheet of epidermal cells over the lip of the blastopore and ended with the birth of a coherent, sentient being.

The New Questions

Spemann's work is one of the great landmarks of biological research. It exemplifies, in the simplicity and directness of its approach, how the mind of a master investigator works. And his experiments and interpretations brought a sense of unity to the science of embryology, leading it sufficiently out of the trees to see the forest and to ask more intelligent questions.

How the organizer and suborganizers work is still a mystery. At first it was thought that the movement of the cells over the lip generated dynamic effects which induced differentiation. But a simple experiment by A. Marx, one of Spemann's students, demonstrated that the primary induction occurred when

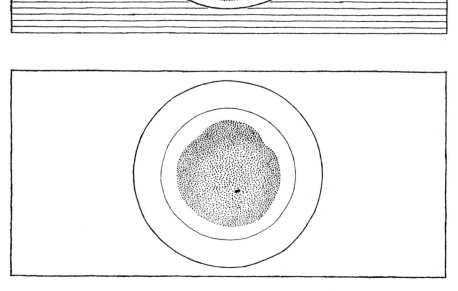

"HANGING-DROP" TECHNIQUE was devised by Ross G. Harrison to grow bits of embryonic tissue *in vitro*. He grew frog tissue in a drop of lymph on the bottom of a microscope cover slip. At the top, the drop and tissue are shown from the side; at the bottom, from above.

there was no movement. Next followed tests in which organizer tissue was subjected to crushing, freezing, heating and other injuries, and then implanted into gastrulating embryos—whereupon it induced the emergence of a central nervous system just as undamaged organizer tissue had done. So the conclusion was that induction must be a chemical effect, and a search began for the potent chemical or chemicals released by the organizer cells.

Johannes Holtfreter, who served his apprenticeship under Spemann, found that even after the lip tissue was killed in alcohol it was able to induce development. The same discovery was independently made by C. H. Waddington at the University of Cambridge. Holtfreter went on to try other salamander tissues and then tested both living and dead tissues of other animals—mouse kidney, liver and brain and extracts from chick embryos. Each induced the embryo to form a neural tube. Various efforts were made to analyze these alien substances and blastoporal lip tissue chemically. But all the analyses have been inconclusive, and the situation was further confused when Waddington, Joseph Needham of Cambridge and Jean Brachet of Brussels discovered the astonishing fact that even the synthetic dye methylene blue induced formation of nerve tissue when injected into a gastrulating embryo.

With this medley of causes and effects in the record, and more seeming paradoxes being added almost every month, it would seem that the era of experiment has changed into an era of perplexity. Perhaps that is another way of saying that the embryo is more versatile and complex than anybody dreamed. The situation can be likened to that which arose in biology when it was discovered that an unfertilized egg could be made to develop into an animal by pricking its membrane with a needle; indeed, that virgin birth ("parthenogenesis") could also be induced by heat, electric shock, ether and many other agents. Apparently all that is needed is some impulse to activate the egg. Similarly, while blastoporal lip tissue normally supplies the agent which induces cells to differentiate, the inductive force can also be supplied by various other materials, both living and dead.

Recently Victor C. Twitty and M. C. Niu performed an experiment at Stanford University which demonstrated that the action of the primary organizer is mediated through some diffusible substance which it exudes. They cut a mi-

nute piece of blastoporal lip from the early embryo of a salamander and cultured it in a hanging drop. After several days they introduced a bit of ectoderm, consisting of about 15 cells, into the drop. Although the ectoderm fragment was not in contact with the lip tissue, within 10 days it developed into nerve and pigment tissue. They then changed the experiment, this time using not the lip tissue itself but only the fluid in which it had been cultured for several days. As before, the implanted ectodermal cells gave rise to pigment and nerve tissue. Thus it would appear that the organizer is some secreted substance, and Niu (who last year joined the staff of the Rockefeller Institute for Medical Research) is now trying to isolate the active material. He thinks it may turn out to be not just one substance but a group of nucleoproteins, each specific to the induction of a particular tissue. Indeed, embryologists are fairly sure that "the organizer" is not a single substance but a complex of agents and reactions.

The Organizer and the Genes

All that the organizer does, apparently, is to release capabilities already present in the cell but dormant. You cannot force a cell into an alien pattern. You can change the direction of development, but each cell has only a limited "repertory." Its repertory becomes more and more restricted as development proceeds, and this restriction of potentialities is the very essence of embryonic development. It is a restriction under the influence of the genes which reside in the cell's nucleus.

Oscar E. Schotté, now professor of biology at Amherst College, who studied under Spemann at Freiburg and later at Yale University under Harrison, conducted an experiment which brings out beautifully the influence of the genes on development. His project was to transplant an embryonic bit from a frog to a salamander. These two animals belong to different orders and have striking contrasts in structure. The salamander larva has teeth, and on each side of its head are balancers which aid it in swimming. The tadpole of the frog, on the other hand, has no teeth but horny jaws, and the protuberances on the sides of its head are suckers. From the underside of a frog embryo Schotté took a slice of cells which normally would become flank skin and transplanted it to the prospective head area of a salamander embryo. The salamander developed horny jaws instead of teeth, and the

head suckers of a tadpole instead of balancers!

From this we conclude that while the organizer determines in general what organs are to be formed, the genes control the details of those organs. It is as if the genes in the frog cells said to the salamander: "You tell us to form a mouth, but we don't know how to make your mouth; we can make only a frog mouth." It is a case of the genes of one animal confronting the alien organizer of a different animal. The fact that they are still able to team up to produce an animal rather than a confused collection of cells is an unsolved puzzle.

This much is sure: development is a business of *both* preformation and epigenesis. The blueprint of the individual is carried in the fertilized egg, but the pattern takes form, organ by organ, as it is called into being by the organizer and is shaped in detail by the genes.

Medical Implications

All these studies, with their changing tactics, have meaning for us. The plastic embryonic cells of sea urchins, salamanders, frogs and chicks are, if not brothers to our own cells, at least cousins. What is learned of them applies to all. And there are many practical problems whose solution may hang on the scientist's fuller understanding of what the organizer is. The cruel malformations that occasionally arise during the development of the embryo—such as Siamese twins or babies born without arms or legs—are failures of organization. Cancer is a lawless crowd of unorganized cells. As we gain in knowledge of the laws of organized growth we may get new clues to the nature of the wild growth that produces malignancy.

Recently, at the Rockefeller Institute, Paul A. Weiss obtained striking evidence of the capacity of cells to organize themselves. From the embryo of a chicken he cut bits of skin tissue, of limb-bud cartilage and of tissue destined to become the coating of the eyeball. He treated all these with an enzyme which dissolves or loosens the "glue" holding the cells together, and so got a mixture of completely dissociated cells of three kinds of tissue. Yet in a tissue culture the cells reassembled themselves according to their kind, and the limb-bud cartilage proceeded to form bone, the eye cells to form eyeball coating and the skin cells to form feathers. "These experiments imply," said Weiss, "that a random assortment of cells which have never been part of any adult tissue can set up conditions—a 'field,' I call it—which will cause members of the cell group to move and grow in concert, following the pattern of a feather in one case, of an eye in another and of a bone in still another."

In a second set of experiments Weiss was able to watch organization at the subcellular level. He cut a salamander and then observed the healing of the wound. The wound cavity filled with a mucus-like liquid. While new skin grew over the cut, connective tissue beneath sent fibrils into the mucus. The tiny fibrils at first were a jumble, like a log jam in a river, but presently they began to assume an orderly arrangement, forming alternate layers like cordwood being stacked crisscross. "Two processes were at work," said Weiss. "The underlayer of connective-tissue cells produced the fibrils, organizing them out of mole-cules, while the overlying layer of skin cells organized them into a subcellular construction."

Down to the Molecules

Basically, of course, cell differentiation depends on the operation of mole-cule-forming processes. The molecular building blocks of one tissue (*e.g.,* muscle) differ radically from those of another (*e.g.,* brain). Heinz Herrmann, head of the laboratory of chemical embryology at the University of Colorado Medical School, has called my attention to the fact that investigators of developmental biology have begun to focus on the protein-forming systems of the embryonic cell in their search for the "organizer." Recent experiments in bacteriology are highly suggestive. They show that in a bacterium a new enzyme may suddenly emerge in response to an environmental change. In other words, a new differentiation suddenly appears in one of the molecular building blocks, and as a result the cell acquires a new property, for every feature of its molecular construction of course is controlled by the catalytic action of enzymes.

Herrmann points out that embryology, having arrived at the protein-forming systems as the center of its search, now joins forces with other branches of biology—physiology, immunology, microbiology—which likewise are seeking ultimate answers to their basic questions in the hidden mechanisms by which cells make proteins. Thus the mystery of the development of an organism emerges from its long isolation as a separate study and becomes an integral part of the many-sided inquiry into the nature of life itself.

3

THE EMBRYOLOGICAL ORIGIN OF MUSCLE

IRWIN R. KONIGSBERG   August 1964

In recent years large strides have been made toward an understanding of how living organisms transmit their characteristics from one generation to the next. Much of this knowledge has emerged from experiments with bacteria, which, being single-celled organisms, convey identical genetic instructions to all their descendants (except for an occasional mutant). It is now quite clear that the functions of all cells, from bacteria to the cells of man, are governed by the hereditary blueprint embodied in the molecules of deoxyribonucleic acid (DNA). It is also generally assumed that in a many-celled organism such as man the DNA present in the fertilized egg is reproduced in exactly the same form in all the cells descended from the egg. Yet these cells, unlike bacteria, differentiate into many kinds of cells, namely the cells of such specialized tissues as skin, nerve and muscle. This raises the question: If all the cells have the same blueprint, how can they differentiate? Or, to put the question another way: How is the same blueprint translated differently in all the cells descended from the fertilized egg?

These questions introduce a complication into the study of the mechanism of heredity, and it was to avoid this complication that geneticists chose bacteria for their experiments. Such experiments, however, have provided uncomplicated models of cell behavior that can now be tested with the cells of multicellular organisms. They have also given other biologists an appreciation for the advantages of working with pure populations of a single type of cell.

In the earliest stages of embryonic development all the cells of the embryo exhibit certain synchronous changes. In later stages the cells progressively di-

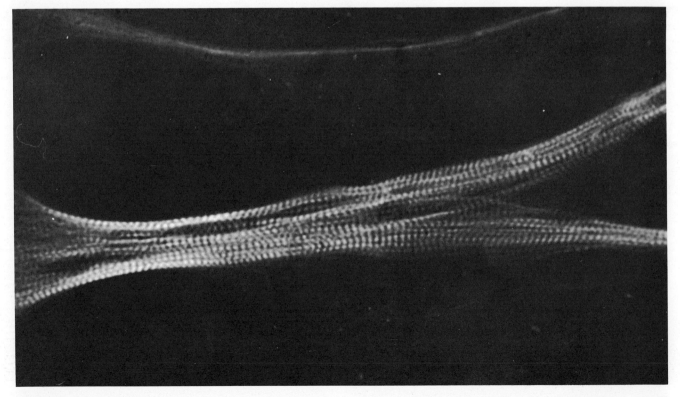

DIFFERENTIATED MUSCLE was grown in culture from a single chick-embryo cell isolated 13 days earlier. Magnification of 500 diameters in polarization microscope reveals characteristic striated pattern of skeletal muscle in the colony of which this is a small part. The experiment demonstrates that single embryonic cells will multiply and differentiate in culture under appropriate conditions.

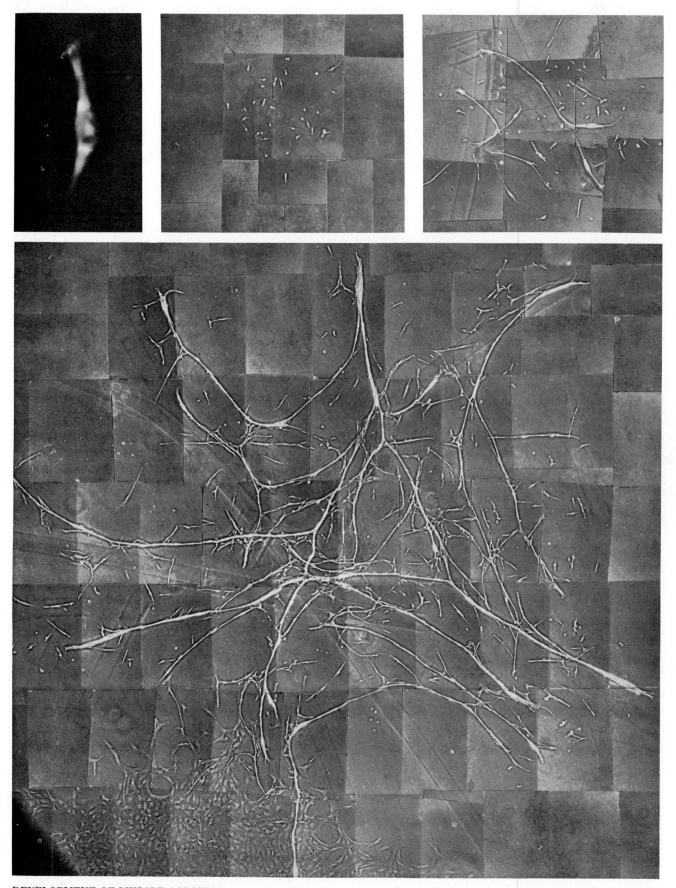

DEVELOPMENT OF MUSCLE COLONY from a single cell proceeds rapidly. The single cell, magnified some 850 diameters, is at top left. Three days later a number of cells have appeared (*top center*). By the sixth day several long, multinuclear cells have formed, presumably by cell fusion (*top right*). Scattered mononuclear cells are also seen. The large mosaic is made of photomicrographs taken on 13th day. Magnification is about 45 diameters. An invading colony of cells of unidentified type is at bottom edge.

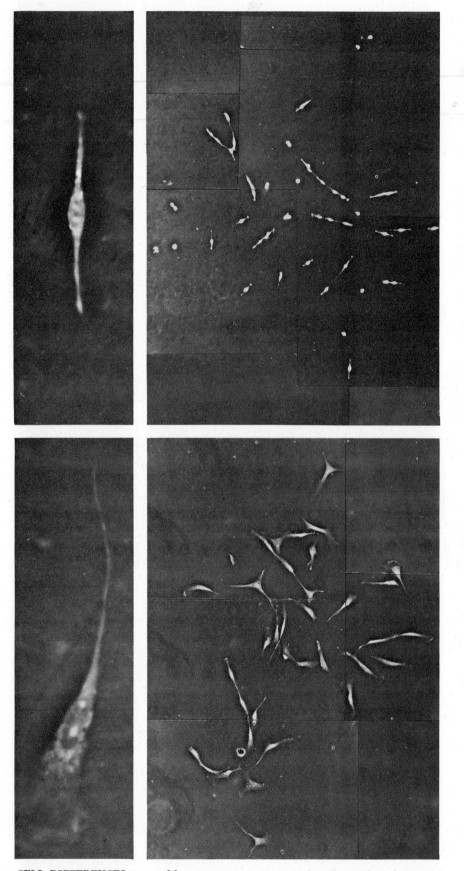

CELL DIFFERENCES are readily apparent as soon as single cells attach to bottom of Petri dish and flatten out. Embryonic muscle cell (*top left*) is distinguished by its spindle shape. "Fibroblastic" cell (*bottom left*) is more irregular. Colonies of living cells, photographed on the fourth day (*right*), show same differences. Magnification is some 740 diameters in photomicrographs at left and approximately 95 diameters in those at right.

verge in their characteristics. At this point the student of differentiation must focus his attention on specific types of cells, but until recently he has had to compromise by studying particular areas of the embryo. Now he can isolate a single embryonic cell and grow a pure population of cells descended from it.

The technique of culturing animal cells from a single progenitor, called cloning, was developed largely by the efforts of the late Wilton R. Earle of the National Institutes of Health and Theodore T. Puck of the University of Colorado School of Medicine and their respective associates. The cells usually grown in such cultures, although originally isolated from animal tissues, have become adapted to the culture environment and have lost those properties that characterized them in the original tissue. Accordingly they are not suited to studies of normal differentiation. For the purpose of such studies it seemed reasonable to apply the cloning technique to newly isolated embryonic cells, and at the Department of Embryology of the Carnegie Institution of Washington we undertook to do this.

It was first necessary to determine if the methods we would have to use would allow differentiation to occur at all. Cells can be liberated from living animal tissue by controlled digestion with an enzyme; moreover, the enzyme treatment does not in itself prevent differentiation, provided that the cells are allowed or encouraged to re-form a compact mass. Our first question was: Would cells freed by such procedures be capable of differentiating when they were grown not in a compact mass but dispersed on a glass surface?

In undertaking to answer the question we used cells of skeletal muscle (muscle of the type that is involved in voluntary movement); the specialized features of such cells can be easily identified both microscopically and biochemically. Cell suspensions were prepared from developing leg muscle dissected from a 12-day-old chick embryo. Equal numbers of cells were pipetted into each of a series of Petri dishes containing a liquid nutrient medium. The cells settled separately on the bottom of the dish and attached themselves to the glass. Incubated at body temperature, they multiplied and in three to four days yielded a population so dense that it formed a continuous layer, one cell thick, completely covering the bottom of the dish.

As the sheet of cells became continu-

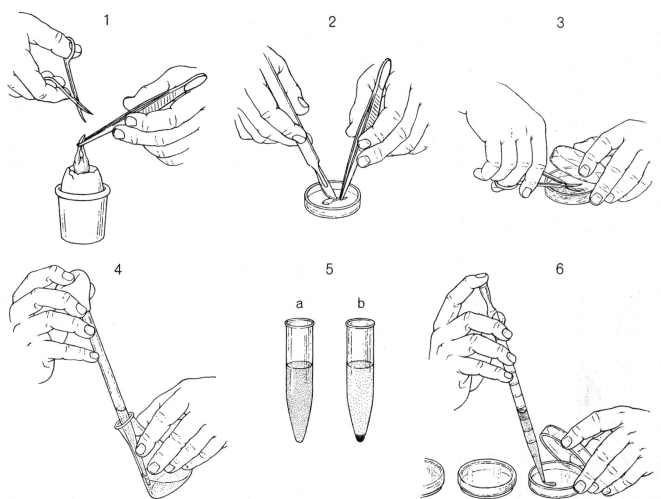

ISOLATION OF CELLS begins with removal of chick embryo from the egg (1). Muscle tissue is dissected from bone (2) and minced (3). The minced muscle, suspended in a dilute solution of the enzyme trypsin, is transferred in pipette to a small flask (4), where the cells separate during a brief incubation. Solution is placed in a tube (5a) and centrifuged, which packs cells into a pellet at tip of tube (5b). After resuspension in liquid medium the cells are counted and then inoculated into Petri dishes (6).

ous, large numbers of long, ribbon-like cells appeared. This new type of cell was identified as a muscle cell on the basis of three criteria. First, each cell contained many nuclei. (A considerable body of evidence has accumulated in recent years that indicates that muscle cells form by the successive fusion of individual cells.) Second, such cells were observed to undergo a series of rapid contractions spontaneously. Third and most conclusive was the presence of the cross-striated pattern typical of skeletal and heart muscle. This pattern, which reflects the ordered periodic arrangement of the muscle proteins actin and myosin, is not present in any other type of cell.

Such experiments demonstrated that embryonic muscle cells, even though cultured as randomly distributed individuals, were able to continue their differentiation. We were closer to our goal of studying differentiation in pure populations of a single cell type, but we were still dealing with heterogeneous mixtures of cells. Embryonic muscle tissue is composed primarily of two specialized cell types: (1) fibroblasts, which lay down the connective-tissue framework of the muscle, and (2) myoblasts, which become the contractile muscle cells themselves. Our dispersed cultures contained both cell types in intimate association. Could one type multiply and differentiate in the absence of the other?

An approach to this question was provided by the elegant technique developed by Puck and his associates for growing colonies from single animal cells. Following these procedures, we prepared cell suspensions from embryonic muscle tissue as before, but now we pipetted into each Petri dish a relatively small number of cells (200 to 400). Dispersed over the surface of the dish, the cells were isolated from one another by virtue of the large distances that separated them. In the course of a week each viable cell multiplied into a colony large enough to be seen with the naked eye.

These initial trials, however, were disappointing. We could not detect any indication of muscle differentiation in any of the colonies. Later we extended the culture period to 10 to 13 days instead of terminating the experiment after a week. It then became apparent that one deficiency in the design of the earlier experiments had been patience. After the longer culture interval we could detect in some of the colonies (about one in 10) the familiar long, multinucleated cells with the characteristic pattern of cross striations.

The fact that only one in 10 colonies exhibited these characteristics was puzzling. Did this ratio indicate that only one cell in 10 was a future muscle cell, or myoblast, or did it indicate that the progeny of only certain myoblasts could differentiate under the particular culture conditions we used? We began to test modifications of our culture conditions that might be effective in increasing the yield of muscle colonies.

Our earlier experiments, in which muscle cells had differentiated after forming a continuous sheet, had suggested that a relation might exist between the degree of crowding and the initiation of the process of cell fusion, by which the multinucleated muscle cells are formed. This relation might indicate simply that the intimate contact between cells imposed by crowding was a prerequisite for fusion. But certain other experiments, while not entirely excluding the effect of physical crowding, focused our attention on the possibility that the steadily increasing population of cells was changing the culture medium in which they were growing. Simply by reusing the nutrient medium removed from older cultures to initiate new cultures we obtained differentiation 24 hours sooner than in freshly prepared medium. Apparently the composition of the medium had been changed by the metabolic activities of the first population of cells cultured in it, making it more effective in supporting the differentiation of the second population.

When such "conditioned" medium was used in cloning small numbers of cells, we found that it not only supported a luxuriant growth of colonies but also increased fourfold the proportion of differentiated colonies. In essence, when small numbers of cells are pipetted into conditioned medium, the net effect is to simulate the presence of a larger mass of cells. Similar tricks have been used before to promote the growth of single cells. Our results indicate that this one is an effective means of promoting not only growth but also differentiation.

As to exactly what changes have occurred in the medium during the first period of conditioning, we can at present only speculate. Actively metabolizing cells would remove some substances from the medium and contribute others. Harry Eagle and his associates at the Albert Einstein College of Medicine have recently described an interesting example of the relation of the cultured cell to its environment. In examining the nutritional requirements of several strains of animal cells they found that some of the "essential" constituents of the medium required for cell growth

could in fact be synthesized by the same cells. These substances had to be added to the medium, Eagle found, simply because the cells were leaky: they lost the substances to the medium faster than they could replace them by synthesis. When the substances were included in the medium at a high enough level, the cells could maintain an internal concentration of them at the level required for growth.

The same principle may apply to the differentiation of cells. For many years embryologists have recognized that in culture, differentiation was favored by maintaining the cells in a compact mass. Such conditions would tend to minimize the loss of substances from the cells and thus compensate for any leakiness.

Having developed a procedure for obtaining a relatively high yield of muscle colonies, we could turn our attention to a more critical and detailed examination of the pattern of growth and differentiation in the colonies. To obtain this information we used the simplest of experimental designs: we looked. Eighteen hours after inoculating a Petri dish we scanned the bottom of the dish with a phase-contrast microscope. When we located single cells that had not yet divided, we circled their position on the underside of the dish. Then at regular intervals we examined and photographed each of the colonies that developed.

These single cells multiplied at a remarkably rapid rate—at least during the first four days, when it was still feasible to count their descendants. From the photographic records we calculated that the cells must have divided every 12 to 18 hours. By the end of the fourth day

the colonies usually consisted of 50 to 60 individual cells. At some time during the fifth or sixth day short multinuclear cells formed. These were the first muscle fibers. By the end of the second week the colonies measured several millimeters in diameter and consisted of a delicate tracery of interlaced fibers. When they were appropriately stained, they could be distinguished from the disk-shaped colonies of fibroblast cells even with the unaided eye [see bottom illustration].

These studies of the sequence of growth and differentiation in colonies helped to settle a troublesome question. It was of considerable importance to establish rigorously whether or not a colony of muscle cells could actually develop from a single cell. We had not been able to exclude the possibility that our muscle colonies had arisen from small groups of cells; cell suspensions commonly contain a few clumps of from two to six cells. Our sequential observations indicated, however, that a solitary cell could indeed give rise to a muscle colony. As a further check on this point we ran several series of experiments in which the single cell under observation was not only separate from other cells but also physically isolated from them. This was accomplished by placing a small glass cylinder over the cell and sealing the cylinder to the Petri dish with silicone grease. Within the confines of the glass cylinder single cells again multiplied and formed colonies of differentiated muscle.

It often happens that an experiment devised to answer a specific question also provides information of a totally unexpected kind. The sequential studies of colony formation led us to the realiza-

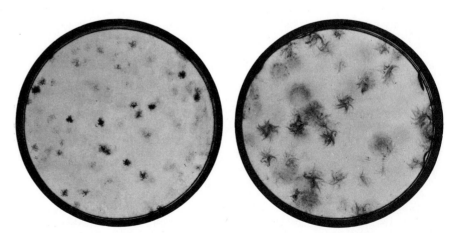

TWO CULTURES, shown actual size, were each inoculated with 200 cells. The culture at left was fixed and stained after seven days, that at right after 13 days. Colonies at right that appear to be interlaced threads are differentiated muscle grown from single cells.

tion that the parent cell of the muscle colony—the myoblast—had certain distinguishing features that made it possible to identify it with a high degree of accuracy. Such cells had a rounded body with two slender, symmetrically arranged processes extending from it [*see illustration on page 26*]. The myoblasts were readily distinguishable from the highly flattened cells, irregular in outline, that gave rise to "fibroblastic" colonies. (Whether or not any of these colonies represent true fibroblasts has yet to be determined.)

We were surprised at how reliably myoblasts could be identified because so many factors can alter the shape of a cell in culture. For this reason shape alone is generally considered to be a poor criterion of cell type. Since we had established that a particular cell gave rise to a muscle colony, however, it was not surprising that the cell happened to have a bipolar shape. In the embryo, at a particular time in its development, the future muscle cell similarly assumes a bipolar shape. Does the change in shape also mark the stage at which the cell acquires the capacity for giving rise to a muscle colony in isolation? This question remains to be investigated by the culture of single cells that have been taken from the embryo at earlier stages.

The fact that the isolated myoblast retains its characteristic morphology is more than just an interesting observation. It provides a marker with which we can predict the developmental fate of a *living* cell. Ordinarily the identifi-

ONE COLONY, magnified 17 diameters, developed in 13 days. It grew from a single cell that had been isolated from neighboring cells by walling it off with a tiny greased ring. This colony is like those in the Petri dish at the right in the illustration on page 28.

cation of embryonic cell types necessitates techniques that kill the cell. This situation was somewhat analogous to the uncertainty principle of the physical sciences. We could say only that a particular cell *was* a myoblast. Cloning makes it possible for us to say that a cell *is* a myoblast because it allows us to test our prediction.

The cloning of embryonic cells is potentially a powerful tool because it permits such predictions to be made. Being able to deal with a homogeneous population of cells whose developmental fate is predictable, we can now ask what factors permit that fate to be realized. Can it be interrupted? Can the course of one type of differentiation be reversed and the cells channeled into another type of differentiation? This last question has long remained unresolved simply because it has not been possible to start with uniform populations of cells of a predictable normal fate. That obstacle, we know now, is not insurmountable.

4

LIGHT AND PLANT DEVELOPMENT

W. L. BUTLER AND
ROBERT J. DOWNS   December 1960

Various kinds of plant germinate, grow, flower and fruit at different times in the year, each in its own season. Thus some plants flower in the spring, others in the summer and still others in the autumn. And in the autumn, trees and shrubs stop growing in apparent anticipation of winter, usually well before the weather turns cold. What is the nature of the clock or calendar that regulates these cycles in the diverse life histories of plants? Some 40 years ago it was discovered that the regulator is the seasonal variation in the length of the day and night. Since this is the one factor in the environment that changes at a constant rate with the change of the seasons, in retrospect the discovery does not seem so surprising. But how do plants detect the change in the ratio of daylight to darkness? The answer to this question is just now becoming clear. It appears that a single light-sensitive pigment, common to all plants, triggers one or another of the crises in plant growth, from the sprouting of the seed to the onset of dormancy, depending upon the plant species. This discovery is a major breakthrough toward a more complete understanding of the life processes of plants, and it places within reach a means for the artificial regulation of these processes.

The pigment has been called phytochrome by the investigators who discovered it at the Plant Industry Station of the U. S. Department of Agriculture in Beltsville, Md. It has been partially isolated, and it has been made to perform in the test tube what seems to be its critical photosensitive reaction: changing back and forth from one of its two forms to the other upon exposure to one or the other of two wavelengths of light that differ by 75 millimicrons. (A millimicron is a ten thousandth of a centimeter.) Phytochrome appears to be chemically active

in one of its forms and inactive in the other. In the tissues of the plant it functions as an enzyme and probably catalyzes a biochemical reaction that is crucial to many metabolic processes.

The first step toward the discovery of phytochrome was taken in the 1920's, when W. W. Garner and H. A. Allard of the Department of Agriculture recognized "photoperiodism" [for more information, see "The Control of Flowering," by Aubrey W. Naylor, Offprint # 113]. They showed that many plants will not flower unless the days are of the right length—some species flowering when the days are short, some when the days are long. Indeed, some plants seem to react not simply to seasonal changes but to changes in the length of the day from one week to the next. Photoperiodism explained why plants of one type, even though planted at different times, always flower together, and why some plants do not fruit or flower in certain latitudes.

Other investigators soon reasoned that if a plant needs a certain length of day to flower, then keeping it in the dark for part of the day would inhibit its flowering. They tried to demonstrate such an effect in the laboratory. Nothing happened: the plants always bloomed in the proper season. The riddle was solved when the reverse experiment was tried, that is, when the night was interrupted with a brief interval of light. Chrysanthemums, poinsettias, soybeans, cockleburs and other plants that flower during the short days and long nights of autumn and early winter remained vegetative (i.e., nonflowering). Moreover, they could be made to bloom out of season, when the night was lengthened by keeping them in the dark at the beginning or the end of a long summer day.

Conversely, a brief interval of light interrupting the long winter night induced flowering in petunias, barley, spinach and other plants that normally bloom in the short-night summer season. Artificial lengthening of the short summer night kept these plants vegetative. Interrupting or prolonging the nighttime darkness correspondingly affected stem growth and other processes as well as flowering in many plants.

It was evident that light must act upon a photoreceptive compound, or compounds, to set some mechanism that runs to completion in darkness. As a first step toward elucidating the chemistry of the process H. A. Borthwick, Marion W. Parker and Sterling B. Hendricks of the Department of Agriculture in 1944 set out to determine the wavelength or color of light that is most effective in inhibiting flowering in long-night plants. They exposed each of a series of Biloxi soybean plants, from which they had stripped all but one leaf, to different wavelengths of light from a large spectrograph [as in middle illustration on opposite page]. Several days after the treatment the plants were examined for the effect of this exposure upon the formation of buds. Red light with a wavelength of 660 millimicrons proved to be by far the most effective inhibitor of flowering. The cocklebur and other long-night plants gave the same response. By plotting on a graph the energy of light required to inhibit flowering at various wavelengths, the investigators obtained the "action spectrum" of the mechanism that inhibits flowering. This showed that to interfere with flowering, much more light energy is required at, for example, 520 or 700 millimicrons than at (or very near) 660 millimicrons [see illustration at bottom of page 36]. The wavelength at which the unknown substance absorbs light

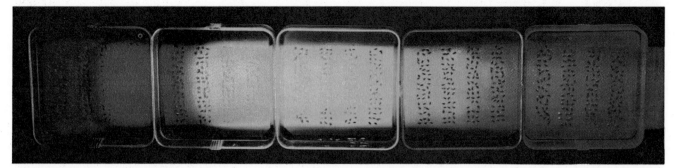

LIGHT FROM SPECTROGRAPH exposes lettuce seeds to different wavelengths. Only 2 per cent of those in far-red light at far left will sprout, while 90 per cent of those in red will germinate. In this photograph only a few of seeds in red region are visible.

CATALPA TREE SEEDLINGS, kept on short days and long nights, are exposed to this spectrum in the middle of the night. Those seen in red will grow. Seedlings in far-red light at far left and all of the others will stop growing and become dormant.

CHRYSANTHEMUM PLANTS at left have had long nights interrupted by period of red light. This divides night into two short dark periods, and plants will not bloom. Chrysanthemums at right also had nights interrupted by red light, but they were irradiated with far-red, as seen here. Far-red light reversed effects of red light, and the plants bloomed just as if nights had been uninterrupted.

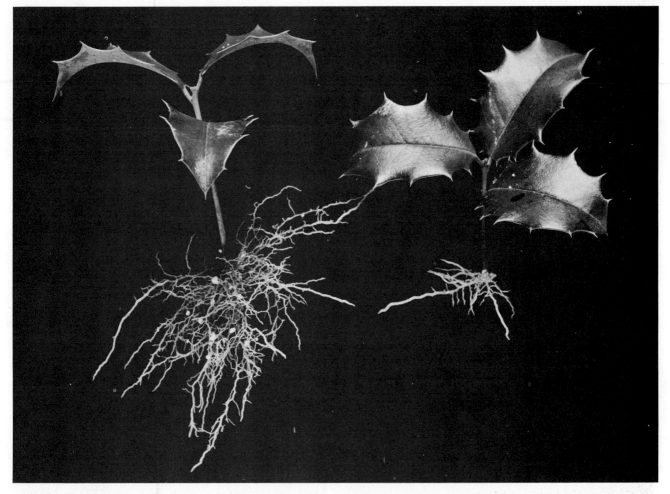

ROOTS OF AMERICAN HOLLY show response to light. Both cuttings were rooted during winter. That at right was kept on natural days and nights. Long nights were interrupted with 30 footcandles of light for plant at left; its roots grew prodigiously.

MATURE APPLES do not turn red if they are kept in the dark while ripening (*right*). Ethyl alcohol collects instead of red pigment. The apples can manufacture anthocyanin, the red pigment, only if they are exposed to light when they are mature (*left*).

most efficiently was thus shown to be 660 millimicrons.

Borthwick and his associates then turned to short-night plants. The same spectrographic experiments yielded exactly the same action spectrum. But in this case the effect of the exposure was to promote—not inhibit—flowering! Since the same wavelength of light caused the greatest response in both cases, the investigators could only conclude that a single photoreceptive substance was involved in these two diametrically opposed responses.

Subsequent experiments implicated the same compound in the control of stem elongation and leaf growth. Recent work has shown that light at 660 millimicrons also acts upon mature apples to turn them red by enabling them to make the pigment anthocyanin. The same red light controls the production of anthocyanin in a number of seedling plants.

With the collaboration of a research group headed by Eben H. Toole, Borthwick and his associates next set out to determine which wavelengths of light trigger germination in those seeds that must be exposed to light in order to grow. Many weed and crop seeds are of this type. The action spectrum for the promotion of seed germination turned out to be essentially the same as that established for other plant responses. Whereas about 20 per cent of the seeds germinated in the dark or when they were exposed to green, blue and other shorter-wavelength colors, more than 90 per cent sprouted after irradiation by red light at 660 millimicrons.

This finding led the two groups of investigators to the study of an entirely different effect of light upon germination. It had been observed in the late 1930's that germination was inhibited when seeds were exposed to the longer wavelengths of far-red light which are invisible to the human eye. That observation was speedily confirmed. In fact, seeds that had been pushed to maximum germinative capacity by exposure to red light failed to germinate when they were subsequently irradiated with far-red light. The plotting of the action spectrum for this effect showed that far-red light at 735 millimicrons wavelength is the most potent in inhibiting the germination of seeds.

Still more interesting was the discovery that the diametrically opposed effects of red and far-red light upon germination are fully reversible. After a series of alternate exposures to light of 660 and 735 millimicrons, the seeds re-

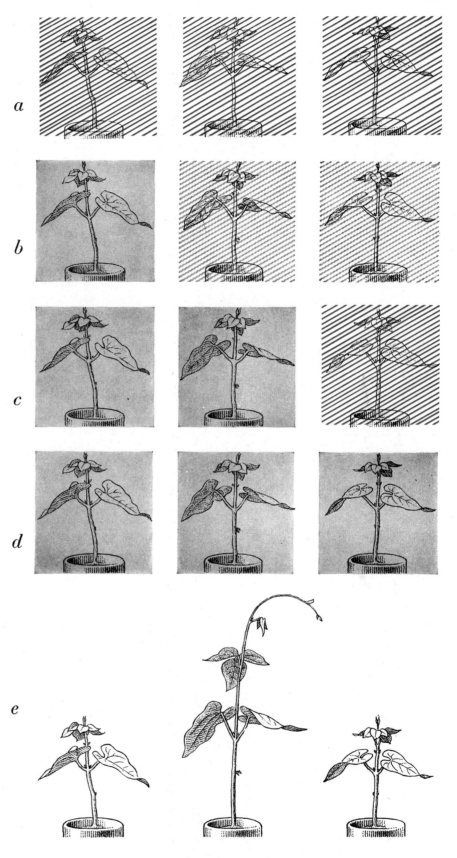

STEM ELONGATION in pinto-bean seedlings is promoted by exposure to far-red light on four successive evenings. All three plants are in red-irradiated condition at end of each day (a). Second and third plants are exposed briefly to far-red light (b); third plant is given dose of red light, which reverses effects of far red (c). Then all three have normal nights (d). Some days later first and third plants are still short, but center plant is tall (e).

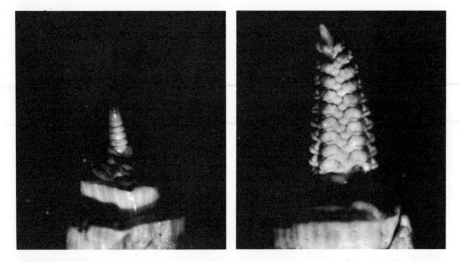

GROWING TIPS OF BARLEY, photographed through dissecting microscope at magnification of 20 diameters, show effects of long and short nights. Tip at left, dissected out from leaves, displays only leaf buds and little growth after being kept on long nights for two weeks. Plant at right had short nights; its tip has grown long and produced many tiny flower buds.

sponded to the light by which they were last irradiated. If it was red, they germinated; if far-red, they remained dormant [*see illustrations at tops of pages 36 and 37*]. Now all of the phenomena of growth and flowering had to be re-examined for the effect of far-red as well as of red light. In each case the experiments demonstrated that irradiation by far-red light reversed the effects obtained by irradiation with red light.

The reversibility of these reactions and the clear definition of their action spectra strongly suggested that a single light-sensitive substance is at work in every case, and that this substance exists in two forms. One form, which was designated phytochrome 660, or P_{660}, absorbs red light in the region of 660 millimicrons. When P_{660} is irradiated at this wavelength, it is transformed into phytochrome 735 (P_{735}), which absorbs far-red light at 735 millimicrons. When P_{735} is irradiated with light at 735 millimicrons, it reverts in turn to the P_{660} form.

In order to substantiate these deductions phytochrome had to be extracted from the plant. This required a method for detecting the presence of the compound other than the responses of a living plant. Since those responses occur at sharply defined wavelengths, there was reason to expect that phytochrome itself would prove to be more opaque, or "dense," to light at the wavelengths of 660 and 735 millimicrons when examined in a spectrophotometer—an instrument that measures the intensity of transmitted light at discrete wavelengths. The transformation of phytochrome from one form to the other would also show up well.

Measuring the very small amounts of phytochrome present in plants was not a simple task. K. H. Norris and one of the authors (Butler), respectively an engineer and a biophysicist in the Department of Agriculture, had been studying the pigment composition of intact plant-tissue, and had developed some sensitive spectrophotometers that measured the absorption of light by leaves and other plant parts. With Hendricks, a chemist, and H. W. Siegelman, a plant physiologist who had been investigating the chemistry of phytochrome, they formed a research group to look for the reversible pigment. Initially the absorption of light by plant parts failed to reveal the presence of phytochrome. The plant tissue, however, contained large amounts of chlorophyll, which absorbs strongly at 675 millimicrons. Apparently the absorption of light by chlorophyll—at a wavelength so near the phytochrome absorption peak of 660 millimicrons—was masking the absorption by phytochrome.

It was no great problem, however, to get around this obstacle. Seedlings can be sprouted in the dark and can grow for a while on the food energy stored in the seed; they do not begin to synthesize chlorophyll until they are exposed to the light. Corn seedlings were accordingly grown for several days in complete darkness. They were then chopped up and exposed to red light to put the phytochrome into the P_{735} form in which it absorbs far-red light. In the spectrophotometer a weak beam of light at 735 millimicrons was projected through the sample, weak light being used so that the phytochrome would not change

form. The absorption spectrum showed that the phytochrome was indeed absorbing light at 735 millimicrons; the same sample passed light at 660 millimicrons. On the other hand, after exposure to relatively bright far-red light the chopped seedlings were found to absorb more light at 660 millimicrons and less light at 735 millimicrons. Repeated demonstration of this reversibility fully confirmed all that had been predicted from the responses of whole, growing plants. No doubt remained that a single compound was responsible for the reversible changes in growing plants.

The spectrophotometer measures the changes in "optical density" with such high sensitivity that it can be used to assay the amount of phytochrome in plant tissue. Thus with the help of this instrument a search was instituted for a plant that would supply phytochrome in sufficient abundance for chemical separation. Certain plant tissues, such as the flesh and seed of the avocado and the head of the cauliflower, showed a relatively high concentration of phytochrome. The cotyledons (the first leaves,

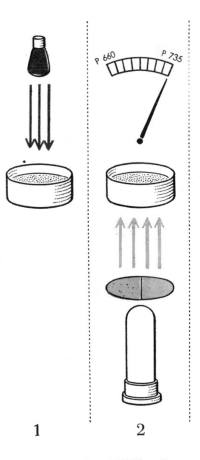

SPECTROPHOTOMETER TESTS, diagrammed here, show effects of red and far-red light on chopped corn seedlings in glass-bottomed dish. Dish is exposed to bright

which feed the seedling) of most legumes synthesize phytochrome, and the concentration reaches its maximum about five days after the seeds start soaking up water. The growing shoot, or hypocotyl, as well as the first leaves of the legumes, contain less phytochrome than the cotyledons. In cabbage and turnip seedlings that have been sprouted in the dark the cotyledons are also a good source of phytochrome. However, five-day-old, dark-grown corn seedlings proved to be the best source, because they develop a high phytochrome content, have large stems and are easy to grow and harvest.

The preliminary and partial chemical isolation of phytochrome was easily accomplished. Corn shoots were ground up in a blender along with water and a mild alkaline buffer, and a clear solution was separated from the solid material by filtration. This extract exhibits exactly the same reversible optical-density changes at 660 and 735 millimicrons as the chopped seedlings themselves. Thorough study of the partially purified material has developed no evidence that

any compound other than phytochrome participates in the photoreaction.

Though phytochrome has not yet been isolated in pure form, the outlook is favorable. The compound shows all the properties of a relatively stable soluble protein, and Hendricks and Siegelman are now using the techniques of protein chemistry to purify it. They have subjected it to dialysis (diffusion through a porous membrane), and it has retained its photochemical activity. Oxidizing and reducing agents do not affect it. Moreover, the photoconversion occurs at zero degrees centigrade as readily as at 35 degrees, as would be expected of a strictly photochemical reaction. Higher temperatures denature the protein and destroy its photochemical activity.

Measurements of the amount of red and far-red light necessary to bring about the conversion show that P_{660} consumes only one third as much energy in being transformed to P_{735} as P_{735} consumes in being changed into P_{660}. In both cases, however, the energy consumption is relatively small, indicating

that the phytochrome absorbs light efficiently. The pigment should turn out to be a blue or blue-green, the colors complementary to red, but the concentration achieved so far has been too low to make the color visible.

Experiments with growing plants had indicated that P_{735} slowly changes back into P_{660} in darkness, whereas red-absorbing P_{660} is stable. Direct measurement of the changes in the form of phytochrome in intact corn seedlings have confirmed these indications. In seedlings that have never been exposed to light, phytochrome occurs entirely in the red-absorbing, or P_{660}, form. These seedlings are exposed briefly to red light to convert the P_{660} to P_{735}. They are then returned to the darkroom and are examined with the spectrophotometer at intervals thereafter. Such measurements show that it takes about four hours at room temperature for the P_{735} to change back into P_{660}.

This conversion in the absence of light is apparently mediated by enzymes. It is markedly retarded by lowering the temperature, and it does not occur at

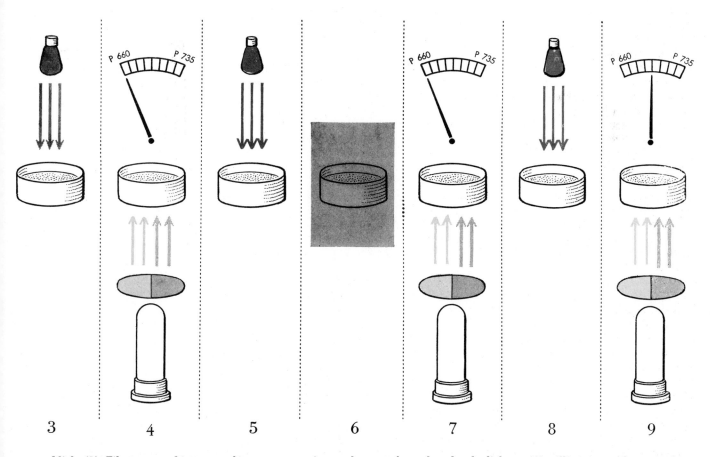

3   4   5   6   7   8   9

red light (1). Filters are used in spectrophotometer to project weak red and far-red light through dish (2); dial shows absorption at wavelength of 735 millimicrons. Now dish is exposed to bright far-red light (3). Next test (4) indicates phytochrome has changed to form that absorbs light at 660 millimicrons. After second exposure to bright red light (5) seedlings are placed in dark for four hours (6), and P_{735} again changes to P_{660} (7). Exposure to red light (8) shows that phytochrome lost half its activity in dark (9).

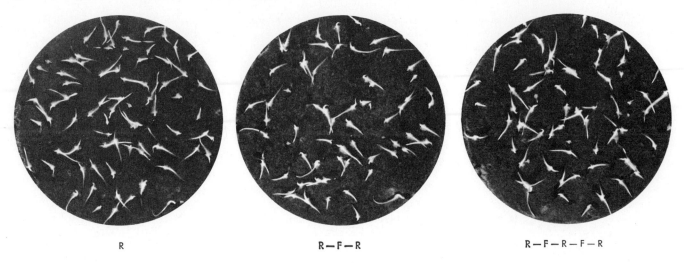

R   R—F—R   R—F—R—F—R

GERMINATION OF LETTUCE SEEDS placed on moist disks of blotting paper is promoted by exposure to red light. "R" indicates an exposure to red light; "F," to far-red light. The last type of light given in the series determines whether the seeds sprout.

all in the absence of oxygen. In the partially purified clear liquid extracts of seedlings, however, P_{735} is stable in the dark, indicating that the dark-conversion enzyme system has been removed. Half the phytochrome activity is lost in intact seedlings during the dark-conversion of P_{735} to P_{660}. After a second illumination with red light, total phytochrome activity declines still more. In continuous light, phytochrome activity is quite low, but is still detectable. It was fortunate that the presence of chlorophyll made it necessary to grow seedlings in darkness for the early experiments. If the seedlings had received even small amounts of light, the unstable P_{735} would have formed and would have soon lost its activity to such an extent that phytochrome might never have been detected.

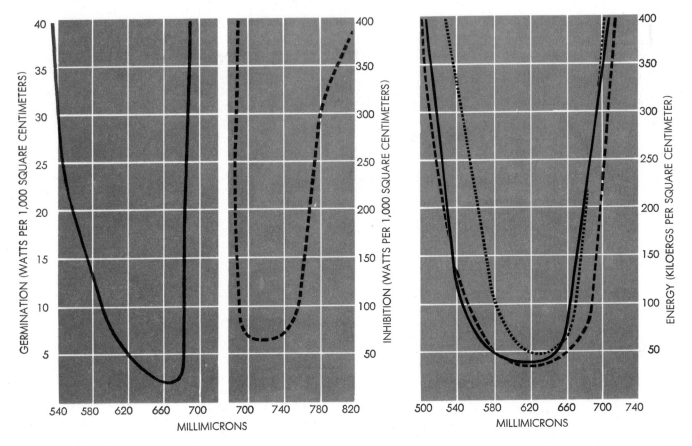

ACTION SPECTRA for the promotion (*left*) and inhibition (*middle*) of germination in lettuce seeds show energy of light required (*vertical scale*) at each wavelength (*horizontal scale*) to produce the desired effect in 50 per cent of the seeds. Curves at right are spectra for the promotion of flowering in barley (*solid line*), and for the inhibition of flowering in soybeans (*long dashes*) and cockleburs (*short dashes*). Barley flowers during short nights, and the other two flower when the nights are long.

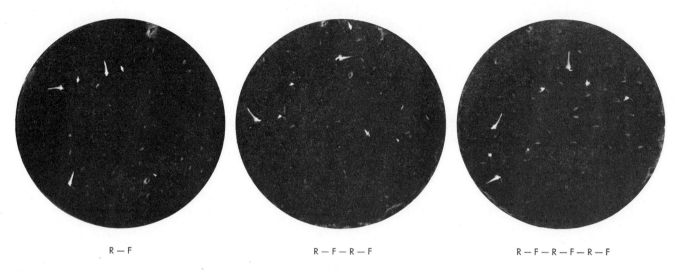

R — F   R — F — R — F   R — F — R — F — R — F

GERMINATION IS INHIBITED by irradiation with far-red light. While the great majority of seeds germinated after final exposure to red light (*opposite page*), practically none of the lettuce seeds irradiated last with far-red light will ever be able to germinate.

How phytochrome exerts its manifold influences on plant growth is still unknown. P_{735} seems to be the active form, while P_{660} appears to be a quiescent form in which the plant can store the potentially active compound. At the close of any period of exposure to light, phytochrome is predominantly in the far-red-absorbing form. The rate at which P_{735} is then carried through the dark conversion back to P_{660} provides the plant with a "clock" for measuring the duration of the dark period. The rate of conversion probably varies from one plant to another, and must depend in part upon such factors as temperature. The effective dark period might be the time required for the complete conversion, or it might be the time in darkness after the conversion is finished. In either case a brief interval of light in the middle of the dark period would cause a plant to respond as though the dark period were short.

Phytochrome is undoubtedly an enzyme—a biological catalyst. Its ability to control so many kinds of plant response in so many different tissues suggests that it catalyzes a critical reaction that is common to many metabolic pathways. Several reactions of this kind are known. One is the reaction that forms the so-called acetyl coenzyme-A compounds. These compounds are essential intermediates in fat utilization and fat synthesis, in cellular respiration and in the synthesis of anthocyanin and sterol compounds. The regulation of the supply of acetyl coenzyme-A compounds would provide an ideal control for growth processes. More than three fourths of all the carbon in a plant is incorporated in this coenzyme at some stage or other.

The extraction and partial purification of phytochrome is the starting point of a major forward movement in the understanding of plant physiology. Further research on this remarkable protein, ubiquitous in the plant world, should answer many questions concerning germination, growth, flowering, dormancy and coloring. It should also provide a means to control all of these plant processes to the great benefit of agriculture.

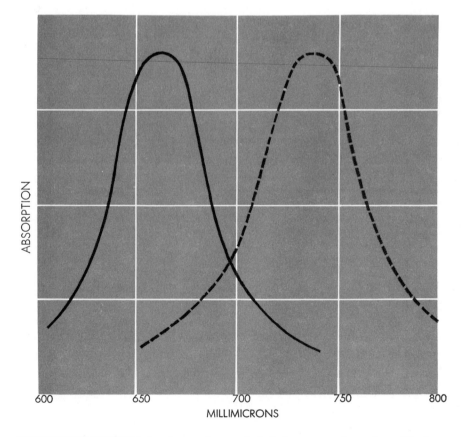

ABSORPTION

600   650   700   750   800
MILLIMICRONS

ABSORPTION SPECTRA for the two forms of phytochrome are shown here. The form known as P_{660} (*solid line*) absorbs the most light at a wavelength of 660 millimicrons, while P_{735} (*broken line*) is far more absorbent, or opaque, to light at a wavelength of 735 millimicrons. The reactions of plants to these wavelengths indicate that P_{735} is the active form.

Part II

EXCHANGE AND TRANSPORT

II
Exchange and Transport

INTRODUCTION

The commerce between the inside of an organism and the world outside is a lively one, involving matter of a variety of sorts (oxygen and carbon dioxide, foodstuffs, waste products) as well as energy. In general, this commerce may be considered as a two-stage process: first, an exchange between organism and environment across the surface boundary of the organism, and second, the transport *within* the organism of the proceeds of exchange. Presumably, the earliest multicellular organisms, like their modern unicellular successors, could handle this commerce easily by the simple process of diffusion, without either the aid of specialized machinery or the expenditure of energy. The reason for this is straightforward: a small three-dimensional object has a high ratio of surface to volume, but as it grows larger its surface increases in proportion to the square of the linear dimension and its volume in proportion to the cube of the linear dimension. Since the organism's needs are generated in proportion to its volume and since matter is exchanged across the surface, the supply problem is obvious.

The importance of size is clearly demonstrated in the following calculation by the great Danish comparative physiologist, August Krogh, who had a special interest in the exchange of respiratory gases. Krogh employed a formula which relates the oxygen tension at the center of a solid to that of an aquatic environment outside, given certain values for diffusion rate, dimensions, and rate of oxygen utilization within the solid. For simplicity, he assumed the object to be a sphere, and he assigned reasonable values (taken from living organisms) for utilization and diffusion rates. He arbitrarily allowed the organism to "survive" when the gradient of oxygen tension reached exactly zero at the very center, and he calculated the external oxygen tension necessary to maintain that gradient in spheres of different diameters. Krogh discovered that a spherical organism 1 cm in diameter would require a preposterous external oxygen tension of 25 atmospheres; in fact, theoretical survival became possible only when the diameter was reduced to 1 mm, the size at which the external oxygen tension required was about that found in nature under the very best conditions.

The evolutionary process has produced a number of devices for overcoming this limitation. One is to expand the surface area greatly, by flattening, folding, or filigreeing the organism in a variety of ways. Another is to make arrangements for moving the solution on the outside, so that the concentration of oxygen or nutrients is constantly renewed. Still another is to divide the surface into separate specialized regions for each different sort of exchange by expanding the area of the surface and equipping many of the specialized regions with a special chemistry so that metabolic energy may be used to drive exchange uphill, that is, against normal diffusion gradients. Finally, internal solutions may be moved in relation to those outside, thus ensuring that a maximal concentration difference will be maintained across the surface and permitting the transport of exchanged material throughout the organism.

All of these strategies have been employed in the evolution of higher plants and animals. Specialized entry points for respiratory gases have evolved: they take the form of gills in aquatic animals, and of lungs in terrestrial animals; in plants, they are the stomata on the undersides of leaves. Animals exchange foodstuffs across complexly folded internal digestive and absorptive epithelia; plants, being autotrophs, do not require a variety of complex organic foods, but they absorb essential mineral nutrients and water through specialized root epidermal epithelia. Both higher plants and higher animals have internal transport systems in which fluids move within tubes; thus *convection*, in which the solute is moved by mass flow of the solvent, replaces simple diffusion. In animals, fluid is transported within multicellular tubes, the blood vessels, and is driven by a hydrostatic pump, the heart; in plants, it moves within a cylindrical column of single cells and is driven by an osmotic pump.

The articles in this section describe some of these specialized systems that transport matter, either between the organism and the environment or within the organism. In "The Kidney," Homer W. Smith recounts the evolution of the remarkable organ that regulates both the body water content of vertebrates and the concentration of a variety of solutes in that water. Though it was published in 1953, this article is still a highly authoritative statement of kidney function. In particular, it communicates the functional variations that can be imposed on a structure during its evolution, and thus enable it to meet a sequence of environmental demands. From their fresh-water origin, vertebrates have moved into other environments—the sea and dry land—in which more water conservation is necessary. The comparative study of the kidney in different groups has provided valuable information not only about this sequence, but also about the functions of the different parts of the kidney. Since 1953, much of renal physiology has been focused on the mechanisms of tubular transport that enable those kidneys possessing loops of Henle to produce a urine hypertonic to the blood. Subsequent research suggests a postscript to Smith's article. As the tubular fluid moves through the loops, which occupy the central medulla of the kidney, salts are actively transported from the tubular fluid across to the extracellular spaces surrounding the nephron. Water follows osmotically, but not in sufficient quantity to secure equilibration; this water leaves the fluid of the medulla with a locally high salt concentration. The urine then passes through the medulla a second time, since the collecting ducts that ultimately connect with the ureter are located there. More water is withdrawn on this trip, so that the contents of the collecting duct are finally made osmotically equal to the locally concentrated extracellular fluid. The mechanism used by the tubular cells to couple metabolic energy to the active pumping of salt, however, is yet to be discovered.

The next two selections, "The Heart" by Carl J. Wiggers and "The Microcirculation of the Blood" by Benjamin W. Zweifach, are concerned with the transport systems of higher animals. Wiggers discusses the properties of the pump; Zweifach, the remarkable system of vessels across which oxygen, carbon dioxide, and foodstuffs are exchanged between blood and cells. Both authors emphasize that the circulatory system does not distribute blood at a constant rate, but rather varies its output in different ways to meet demand. It is useful to think of the circulatory system as a set of pipes, with valves in them, connected to a pump. In such a system, a variety of devices can be used to affect the rate of flow within the pipes: the pump can vary its frequency, or its stroke volume; or the valves in the pipes can open and close to vary the distribution of fluid within them. The circulatory system uses analagous mechanisms to regulate its output: frequency of heartbeat is controlled by the central nervous system, and stroke volume is determined by the degree to which the heart is filled by the venous return. The distribution of fluid among various capillary beds depends upon whether the arterioles serving them are open or constricted. Indeed, as Zweifach points out, the distribution of blood within the vessels must be closely controlled: the total capillary volume is actually larger than the blood volume, and but for this tight control, one could actually bleed to death into his own capillaries.

In the fourth article in this section, "How Sap Moves in Trees," Martin H. Zimmermann discusses the analogous systems in plants that are responsible for the distribution of the organic products of photosynthesis, which are largely sugars, and for the ascent of water and nutrients from roots to leaves. The mechanism controlling upward transport is quite different from that governing downward movement. The traffic in the xylem, which moves sap from roots to leaves, takes place within the cellulose walls of dead cells and is propelled by such physical forces as cohesion and evaporation. The movement of sugars from leaves to other places in phloem occurs in living cells. It is probably caused by an osmotic pump powered by metabolically derived active transport. Because sugar is actively taken up by phloem cells in the leaves, it is very highly concentrated and produces a powerful osmotic pressure that drives water into the phloem at the top. Since the concentration of sugar at the other end of the system is much lower because it is being used more rapidly, there is a strong pressure gradient from top to bottom, which can result only in the downward mass movement of fluid. Though this system differs greatly in every detail from the circulation in animals, the two show strong formal analogies. Both pump fluids through a confined system of vessels to secure a convective flow of solutes and to distribute them.

5

THE KIDNEY

HOMER W. SMITH   January 1953

THE URINE of man is one of the animal matters that have been the most examined by chemists and of which the examination has at the same time furnished the most singular discoveries to chemistry, and the most useful applications to physiology, as well as the art of healing. This liquid, which commonly inspires men only with contempt and disgust, which is generally ranked amongst vile and repulsive matters, has become, in the hands of the chemists, a source of important discoveries and is an object in the history of which we find the most singular disparity between the ideas which are generally formed of it in the world, and the valuable notions which the study of it affords to the physiologist, the physician and the philosopher.

So wrote the chemist Count Antoine Francois de Fourcroy, a disciple of Antoine Lavoisier, in 1804. For all his enthusiasm Fourcroy scarcely knew how significant his remarks really were. The study of the urine had indeed yielded many "singular discoveries," and largely in consequence of Fourcroy's own researches. But the studies up to that time had considered the urine merely as a vehicle for the excretion of waste products; they had not even begun to realize how much this fluid had to tell about the wonderful world of the body.

Three quarters of a century later the great French physiologist Claude Bernard caught the first vision of the remarkable nature of that world. He pointed out that every higher animal lived not in the external environment but in an internal liquid environment of its own—"a kind of hothouse" with a controlled, unchanging atmosphere which made the organism independent of outside conditions. Bernard wrote: "All the vital mechanisms, however varied they may be, have only one object, that of preserving constant the conditions of life in the internal environment."

It has been said that no more preg-nant sentence was ever framed by a physiologist. We know today that the internal environment—the blood and plasma which the heart keeps in constant circulation throughout the body—contains a multitude of compounds the concentrations of which are regulated with remarkable precision, and that in this regulation the kidneys play a crucial role. Where Fourcroy saw the kidneys merely as organs for excretion, the modern physiologist finds them the chief agents for the control of the vitally important body fluids.

The importance of the kidneys in our vital economy can be judged in part from the fact that the two kidneys in man, although representing less than one half of 1 per cent of the body weight, receive about 20 per cent of the blood volume pumped by the heart. Each day more than 1,700 quarts of blood flow through the kidneys. Yet only about one thousandth of this huge flow is converted into urine. To understand this extravagant procedure, the first question we must ask is: How does a kidney manufacture urine from blood?

Malpighi's Corpuscles

All the higher vertebrates normally have two kidneys of about equal size, one attached on either side of the spinal column on the posterior abdominal wall. Each kidney is supplied with blood through a single large artery, from which an elaborate system of smaller arteries and a maze of capillaries distribute the blood to the kidney tissues and filtering units. After being filtered, the blood is collected in a confluent system of veins and returned to the circulation through a single large vein.

William Harvey discovered the circulation of the blood, but it was a young Italian of the 17th century named Marcello Malpighi who first fathomed how the blood conducts its traffic with the tissues. Malpighi was one of the earliest biologists to use the microscope, and he became the founder of microscopic anatomy. Apparently from youth he had been amusing himself with this instrument, and shortly after completing his training in medicine he turned it to the study of the fine structure of animal organs. In 1661, four years after Harvey's death, he discovered the capillaries. Watching the lungs and viscera of a living frog under a microscope, he saw the blood "showered down in tiny streams through the arteries, after the fashion of a flood." He saw also that instead of escaping into empty spaces in the tissues, as had been thought, the blood flowed in definite channels. "Hence it was clear to the senses," he wrote, "that the blood flowed away along tortuous vessels and was not poured into spaces, but was always contained within [vessels], and that its dispersion is due to the multiple winding of the vessels."

When he later came to examine the microscopic structure of the kidneys, Malpighi found some other highly interesting formations. Not long before, a 19-year-old student of physics named Lorenzo Bellini had discovered minute hollow ducts in the cut surface of the kidney. He called them urinary *canaliculi* (little canals), and he observed that they coalesced into larger canals which in turn emptied into the hollow space, or pelvis, of the kidney. Here, then, was the route by which urine was excreted; it was clear that urine collecting in the kidney pelvis would drain by way of the ureter into the bladder. Malpighi now went on to show that Bellini's ducts drained a system composed of thousands of still smaller tubes in the kidneys. He had found that the kidney capillaries were bunched in innumerable little spherical tufts, which he called corpuscles. Malpighi inferred, though he did not prove, that each of his capillary tufts was connected with one of the tiny tubes ("tubules").

His surmise was correct, and he had actually located the place where urine is formed in the kidney, but not until nearly 200 years later was there much further light on the problem. In 1842 William Bowman, a young demonstrator of anatomy at King's College, London, completed the anatomical picture. He showed that each Malpighian corpuscle is in effect formed by the intrusion of a tuft of capillaries into the end of a tubule, wholly in the manner in which the fist would be enveloped if thrust into a large inflated rubber ball. The thin, greatly expanded walls of the tubule form a capsule surrounding the capillary tuft. The fluid in the space within this capsule (corresponding to the space within the indented ball) can drain freely into the tubule. Blood flows into the capillary tuft by one artery and flows out through another which breaks up into a second series of capillaries that is closely intertwined with the tubules [*see drawing on page 47*].

We now know that each of the two kidneys in a human being contains about one million such units, called nephrons. The spherical tuft of each unit, known as the glomerulus (the diminutive of *glomus*, meaning ball), is located in the outer zone or cortex of the kidney. The tubule leading from it twists in a complicated manner around the glomerulus and then plunges in a more or less straight course into the interior of the kidney, where it makes a sharp hairpin turn (loop of Henle) and runs back to the glomerulus again; there it winds about in a second series of twists and finally discharges into a urine-collecting duct. These intricate twistings of the tubule in the kidney cortex apparently serve no purpose other than to give it length; the average tubule in a human kidney is about an inch and a quarter long. If the two million nephrons in the two kidneys were put end to end, they would stretch for nearly 50 miles.

The typical mammalian tubule has a different cell structure in its various sections. The first segment, where it twists around the glomerulus, is composed of thick, irregular cells possessing brushlike filaments on the inner side, forming a so-called brush border. The second segment, which plunges into the kidney to the loop of Henle, has a much smaller diameter and much thinner walls. The third segment, which leads into the collecting ducts, is made of somewhat flatter cells than the first and lacks the brush border. The glomeruli and the twisting segments of the tubules form the cortex, or outer layer, of the kidney. while the thin segments of the loop of Henle, together with the collecting ducts, make up the kidney's medulla, or central part.

The Filter

The picture as Bowman saw it was this: Blood flows into the glomerulus, and somehow urine is formed in the tubule connected with the glomerulus. He proposed a working theory of how this was done. The tubule cells, he said, "secrete" the substances that make up the urine, and the glomeruli secrete the water necessary to wash these substances into the urine. In Bowman's day the word "secretion" implied that the cells carried out this operation by virtue of their "vital activity."

But two years after Bowman's paper was published, the German physiologist Carl Ludwig advanced another idea. He suggested that the glomeruli are simply mechanical filters, beautifully contrived to permit the filtration from the blood of a cell-free and protein-free fluid. This filtrate passes down the tubules in large amounts, and as it does so a considerable part of its water is reabsorbed through the tubule walls into the bloodstream. The residue, greatly reduced in volume and increased in concentration, passes into the kidney cavity as the urine.

Ludwig's hypothesis was ably defended many years later by Arthur Cushny of University College, London,

and it was validated by direct experimental evidence in 1921 by A. N. Richards and his co-workers at the University of Pennsylvania. They collected and analyzed by exquisite micromethods minute quantities of fluid removed from the glomerular capsules of the frog and the salamander. Supplementary if indirect evidence supporting the filtration hypothesis in other species of animals has since been supplied by many other investigators, and it may now be taken as well established that water and all other diffusible substances in the blood (which excludes the plasma proteins, fats and the cellular elements) pass through the glomerular membrane into the capsular space by a simple process of filtration. The energy for this filtration is supplied by the heart and transmitted to the glomeruli by the hydrostatic pressure of the blood.

The Tubules

Yet filtration through the glomeruli does not account for all the substances

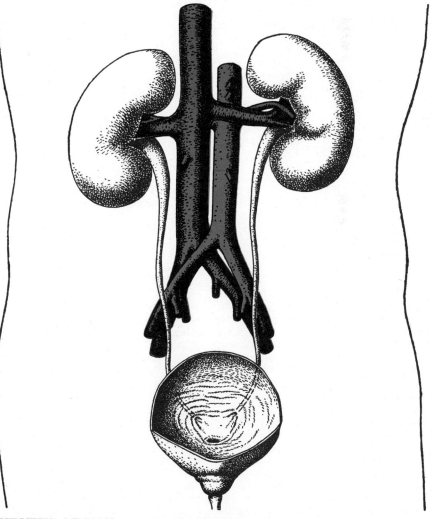

KIDNEYS OF MAN are located behind the other abdominal organs. The vessels that connect them with the circulatory system are shown in red. The large vessel at the right is the aorta; that at the left, the vena cava. Descending from the kidneys are the ureters, which empty into the bladder (*bottom*).

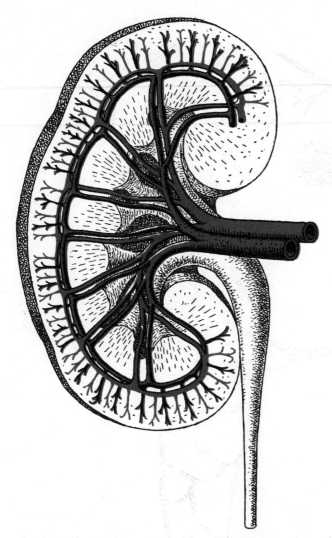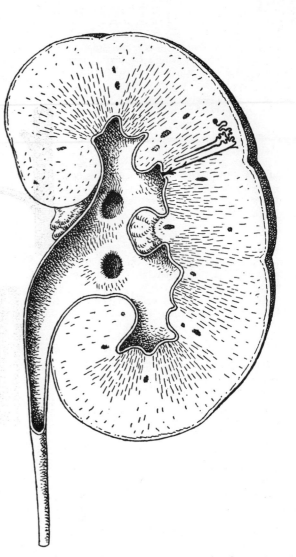

CROSS SECTION of the kidney shows two aspects of its internal structure. Depicted at the left is the circulation of the kidney. At the right the blood vessels have been eliminated to show the renal pelvis and the ureter, into which the urine empties. At the upper right in the right-hand drawing is the outline of the individual unit of kidney function, the nephron. Each kidney of man contains about a million such structures.

found in the urine, nor indeed for the removal of all waste products from the blood. The tubules also play a part in this transfer of substances. The determination of precisely what substances the tubules subtract from and add to the urine became possible when methods were found for measuring the rate of filtration of plasma through the glomeruli. A convenient way to do this is to inject into the blood the starchlike substance inulin, obtained from dahlia tubers and other vegetable sources. All of the inulin injected is filtered from the blood plasma into the urine through the glomeruli, and the tubules do not add to or subtract from it. From the known proportion of this test substance in the plasma it is possible to compute the over-all rate of filtration of the substances filtered by the glomeruli in the two kidneys.

These measurements show that the filtration rate in man averages about 125 cubic centimeters per minute or about 180 quarts per day—a figure which would have seemed scarcely credible to Ludwig or Cushny, the pioneer proponents of the filtration theory. Of the 180 quarts of filtrate, 178 are reabsorbed into the blood through the tubules, and only one to two quarts are excreted as urine. Most of the salt and of many other substances in the filtrate is reabsorbed. The excreted urine contains urea, creatinine and other products of protein metabolism; it also contains foreign substances that enter the body and the excess of sodium, potassium, phosphate, sulfate and other substances beyond the body's needs.

The extravagant nature of this filtration-reabsorption process is emphasized by the fact that about 2.5 pounds of sodium chloride passes from the glomeruli into the tubules each day, but normally only from one sixth to one third of an ounce of this leaves the body in the urine. Similarly nearly one pound of sodium bicarbonate and a third of a pound of glucose are filtered, but only trivial fractions of these amounts are excreted. The tubules likewise reabsorb from the filtrate substantial quantities of potassium, calcium, magnesium, phosphate, sulfate, amino acids, vitamins and other substances valuable to the body. These reabsorptive processes operate to recover valuable constituents from the glomerular filtrate which otherwise would be lost from the body, and it is in part by these processes that the tubules maintain the constancy of the chemical composition of the plasma and body fluids. They synthesize the body's internal environment by operating in reverse, so to speak, on the glomerular filtrate. To this end, they filter and reabsorb our entire fluid internal environment nearly 15 times a day.

It has been noted that the energy for glomerular filtration is supplied by the heart and that the glomeruli play an entirely passive role in this process. But in reabsorption the tubules play an active role. They must remove each reabsorbed substance from a low concentration in the filtrate and transport it to a higher concentration in the blood. This operation requires the expenditure of energy by the tubule cells. The energy must be made available within the cells by the metabolism of suitable fuel stuffs, and then put to work by suitable enzyme systems. These circumstances impose a limit on the rate at which any given substance can be reabsorbed. In many cases the maximal rate of reabsorption can be measured quite accurately by presenting to the tubules more of the substance than they can handle and finding the saturation level. The transport system for each substance is independent of the others; for example, when reabsorption of glucose reaches the saturation level, this does not interfere with reabsorption of another substance, such as phosphate.

Two-way Traffic

As I have already mentioned, we now know that the tubules not only take substances out of the filtrate and transport them back to the blood, but they also work in the opposite direction: they take some of the waste products from the blood and deposit them in the urine. E. K. Marshall, Jr., and his co-workers at the Johns Hopkins University first proved this in 1924 by quantitative excretion studies with the dye phenol red, known to all physicians as PSP. Marshall and others later showed that certain marine fishes, whose kidneys have no glomeruli or arterial blood supply, form urine solely by tubular excretion.

This process of tubular excretion supplements glomerular excretion and thus increases the over-all efficiency of extraction from the blood. The list of substances known to be excreted by the tubules in man is rapidly expanding. It includes hippuric acid and other derivatives of benzoic acid, and many other organic waste products which are difficult to metabolize. As in the case of tubular reabsorption, tubular excretion requires the local expenditure of energy by the tubule cells, and its rate is limited. Unlike tubular reabsorption, it apparently involves only a few enzymatic transport systems, each of which handles a particular type of compound.

Tubular excretion supplies the physiologist with a most valuable subject for study. It presents for examination some highly specific enzyme systems, the exploration of which throws light on fundamental protoplasmic processes. It affords useful quantitative measurements of tubular function in health and disease and of the effects of the endocrine glands, particularly the anterior pituitary, which excretes a hormone with a powerful influence on the activity of the tubules. But most importantly it provides a method of measuring the blood flow in the kidneys of man or experimental animals without disturbing the

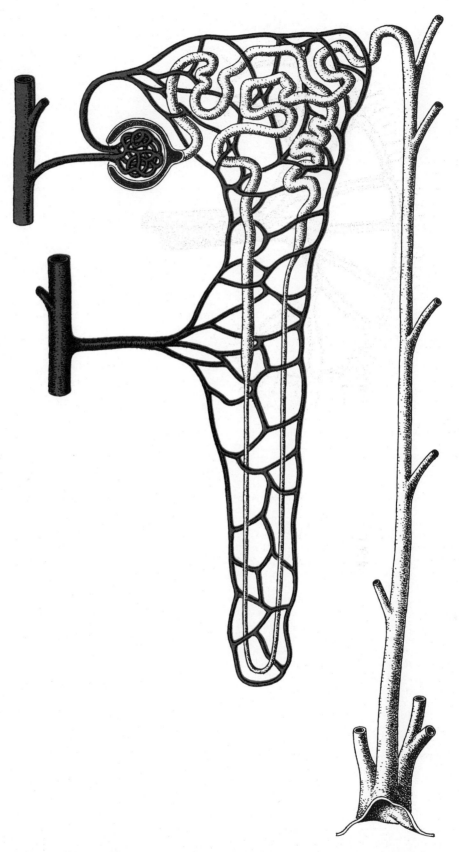

NEPHRON consists of (1) the glomerulus, the ball of capillaries at upper left, (2) the tubule, which twists from the glomerulus to the urine-collecting duct at right, and (3) the bed of capillaries around the tubule. The water of the blood, containing small molecules such as urea and sodium chloride, is filtered out of the glomerulus into the tubule. There virtually all of the water and the other useful substances are reabsorbed.

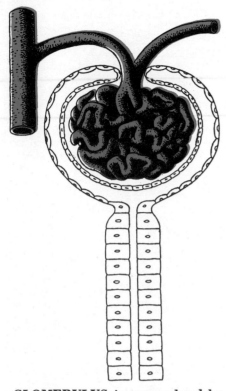

GLOMERULUS is encapsulated by the cells at the end of the tubule.

subject's physiological functions, so that one can observe the changes in blood flow produced by natural hormones, disease and drugs.

The method of measuring the renal blood flow is as simple as that of measuring the filtration rate. Given a test substance which is excreted both by the glomeruli and by the tubules, the substance may be completely or almost completely removed from the blood and excreted into the urine during a single passage of blood through the kidney, provided it is administered in less than saturation amount. If one divides the rate of excretion of the test substance by the quantity contained in each cubic centimeter of plasma, one obtains the total flow per minute through the kidneys. A simple calculation then gives the flow of whole blood. The substance now most widely used for measuring the renal blood flow is p-aminohippuric acid (PAH), a compound which lends itself to easy and accurate chemical determination. When given to a human subject, about 91 per cent of this substance is removed from the blood during a single circulation through the kidneys.

The Kidney's Evolution

We have seen that of all the blood pumped out of the heart (some 8,000 quarts per day) one fifth, or nearly 1,700 quarts, goes to the kidneys for the purpose of making a trifling one to two quarts of urine. One would have thought that nature would have found

a more efficient method of making urine than this! To understand how this circuitous method developed, we must look into the evolutionary history of the kidney.

It is now established that the first chordates, the forerunners of the animals with backbones, arose in the brackish estuaries or fresh-water rivers of the continents. Their ancestors unquestionably came from the sea, and the migration from salt to fresh water required many physiological adjustments, the most important of which concerned the osmosis of water. A salt solution, if separated from fresh water or a more dilute solution by a membrane permeable to water, draws water into itself until the osmotic pressure on both sides of the membrane is the same. Because of the salts contained in their tissues and body fluids, these early chordates tended to absorb water from their freshwater environment by simple osmosis through the permeable gills and oral membranes, which had to be left naked for the purposes of respiration. Consequently provision had to be made to excrete this water from the body. The excretory system which the vertebrates inherited from their marine ancestors consisted of a series of tubules which opened freely into the primitive body cavity. These tubules served to drain the fluid in the cavity to the exterior, and they doubtless reabsorbed some substances from the fluid and added others before it was discharged from the body. In retrospect it seems that it was physiologically beyond the capacity of either the cavity membranes or the primitive tubules to excrete water as such. Hence the vertebrates evolved a device to pump water out of the body. This device consisted simply of a tuft of capillaries, juxtaposed to the mouth of the tubule, through which water could be pumped into the body cavity by the heart. Since all the salts and other valuable substances in the plasma water also passed into the filtrate, the introduction of the filtering device required that the tubules step up their reabsorptive processes to conserve these substances. By the time the true fresh-water fishes evolved, the glomus had become invested within the closed, expanded end of the tubule, forming the typical nephron.

Even as the kidney was being improved as a device for excreting water, the armor in which the early vertebrates had enclosed themselves acquired new flexible articulations, permitting the animal to swim or to crawl upon the bottom. Certain spines in the armor became paddles or fins to promote mobility, and the armor about the head acquired new articulations in the form of jaws so that the animal could eat. All these changes required major reconstruction of the animal's muscles, nerves

and sense organs, as well as of the excretory and reproductive ducts. Thus the evolution of the glomerular kidney as a device for excreting excess water from the body constituted but one in a large number of adaptive changes which arose in response to the requirements for living in fresh water. The entire sequence of such adaptations may be said to comprise the evolution of the vertebrates.

The situation with respect to water balance was scarcely changed in the amphibia that evolved from the air-breathing fresh-water fishes, for these amphibia spent most of their lives in fresh water or very moist areas, just as do frogs, toads and salamanders today. However, when the reptiles, the first truly terrestrial animals, arrived, the balance did change drastically. The reptiles once again acquired waterproof scales (as in lizards and snakes), but now it was not to keep water out but to prevent excessive loss of water by evaporation. For the first time the egg became covered with a waterproof shell—a device to prevent the embryo from drying up during its development. This required that the egg be fertilized in the body of the female before it was enclosed by the shell. Internal fertilization became the rule among the reptiles, birds and mammals.

The reptiles went a step further in the conservation of water: they overhauled the intimate biochemical machinery for the combustion of protein. Instead of degrading the nitrogen of their protein food to the highly soluble compounds urea and ammonia, they degraded it to uric acid, a substance which is almost insoluble in water. Under appropriate conditions uric acid may form supersaturated solutions with a high uric acid content. In this form uric acid is excreted by the tubules directly into the tubular urine. In the renal collecting ducts the uric acid precipitates out of solution, leaving the water free to be reabsorbed, so that the acid is left as a white paste (guano) which is discharged from the cloaca with a minimum of water loss. Birds, like reptiles, degrade their protein nitrogen to uric acid and excrete this waste product by the renal tubules; it is often said that birds are only reptiles that have acquired wings and lost their teeth. In reptiles and birds, which have little water available for urine formation, the glomeruli are degenerate, the capillary tuft being reduced to one or two relatively short channels.

Backward Step

The story of water conservation has been quite different among the fishes. Some of the ancient fishes returned to the sea, and in so doing reversed the osmotic circumstances that had fostered the evolution of the glomerular kidney. The osmotic pressure of sea water is greater than that of the body fluids; consequently water tends to be drawn out of the body by osmosis. The organism is perpetually faced with a deficit of this precious substance. Adaptation to this circumstance among the marine fishes has taken two routes.

One of these is exhibited by the class represented by the sharks, skates, rays and chimaeras. Like the primitive fresh-water fishes, they produce urea, but instead of treating it as a waste product and excreting it as fast as it is formed, as all other vertebrates do, they reabsorb urea through the renal tubules until the concentration in the blood reaches the spectacular figures of 2,000 to 2,500 milligrams per 100 cubic centimeters. This accumulation of urea in the blood raises the osmotic pressure of the latter to a level above that of the sea water, so that the animals absorb water by osmosis through the gills and oral membranes in small but adequate quantities. Here again, however, the young must mature within the body of the mother, or the egg must be covered with an impermeable shell to protect the embryo until it has developed to the point where it can take care of its own water balance. In either case internal fertilization is required. To aid the process of internal fertilization the males of this class of fishes generally have enlarged pelvic fins or specialized organs called claspers. From the presence of such pelvic fins and claspers in the fossil record we can infer that internal fertilization, and therefore the tubular reabsorption of urea, goes back as far as the late Silurian or early Devonian period.

The other great group of marine fishes, the bony fishes, did not return to the sea until the Mesozoic or even later times. Never having acquired the trick of conserving urea to increase the osmotic pressure of the blood, they could obtain water for urine formation only by drinking sea water. The fish kidney cannot, however, make a urine more concentrated than the body fluids. Consequently with every gulp of sea water these fishes are potentially worse off than they were before, because they have taken in more salt than water. They solve their dilemma by excreting the excess salt directly from the blood through the gills, leaving free water available to the body for urine formation. The more urine they excrete, the more sea water they must drink and the more salt they must excrete through the gills. It is vitally necessary, therefore, to keep urine formation to the lowest possible level, and it is not surprising to find that in many marine fishes the glomeruli are degenerate, while in those whose ancestral lines perhaps have had the longest continuous marine history

(*e.g.*, the goose fish, toad fish, sea horse and many deep-sea fishes) the glomeruli are gone, leaving a kidney which functions entirely by tubular excretion.

The Mammalian Kidney

Among all these animals living under conditions where fresh water is at a premium (the reptiles, birds, marine fishes) the development and activity of the glomeruli declined. But among the warm-blooded mammals the adjustment was in the other direction. Elevation of the body temperature above that of the environment entailed a considerable increase in metabolism, which in turn entailed increased respiration and increased circulation of the blood. The speed-up in circulation of the blood was accomplished in part by an increase in mean blood pressure. All this promoted development of the glomerular capillary tuft and a greater amount of filtration. And the more fluid was filtered, the more the body required tubular reabsorption to recover valuable constituents. By the time the mammals were

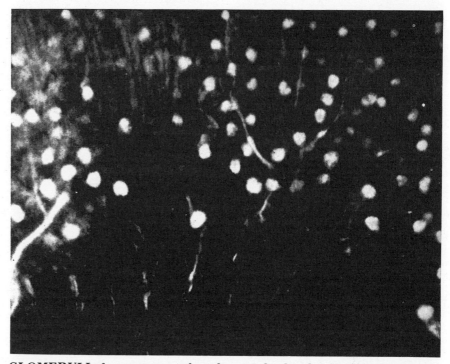

GLOMERULI of a cat appear when they are dyed with a fluorescent material and photomicrographed with ultraviolet light. The method was developed by J. U. Schlegel and J. B. Moses at the University of Rochester.

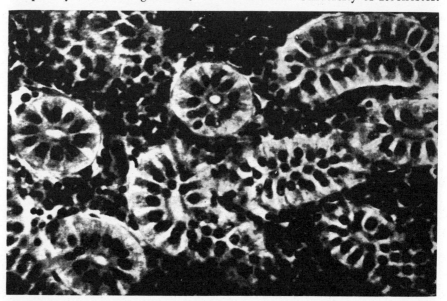

TUBULES of a toad fish are shown in cross section by a conventional photomicrograph. The kidney of this species has no glomeruli; it functions solely by means of tubules (*see evolutionary chart on next page*).

evolved, it was too late to change either the basic pattern of the kidney or the mode of protein metabolism; the mammals had to improve the functioning of the kidney they had inherited from the amphibia and the fresh-water fishes. They simply speeded up its processes—glomerular filtration and tubular reabsorption—and subjected them to more delicate and more precise control.

One of the new things added to the kidney by the mammals is a mechanism for improving the control of water balance. It is a matter of common knowledge that the rate of urine formation is quickly and substantially increased by the ingestion of water, and is substantially reduced by dehydration. This regulation is achieved by changes in water reabsorption in the tubule. Normally the first section of the tubule, which twists around the glomerulus, reabsorbs about seven eighths of the water of the glomerular filtrate, along with sodium chloride and other substances, leaving about one eighth to be reabsorbed optionally by the last segment of the tubule. The latter reabsorption is mediated by a hormone secreted by the pituitary gland: in the absence of the hormone, reabsorption drops to zero and the water which would otherwise have been reabsorbed is excreted in the urine. In the presence of maximally effective concentrations of the hormone, almost all the water is reabsorbed and the urine flow decreases to a minimal level. For this reason the hormone is called the antidiuretic hormone.

In a classic series of studies extending over many years at Cambridge University, E. B. Verney and his colleagues have shown that an increase in the osmotic pressure of the blood causes increased secretion of the antidiuretic hormone into the blood. Thus lack of water makes the rate of hormone secretion

EVOLUTION of the nephron is depicted in the chart at the right. On the tree running from left to right are various vertebrate forms in the order of their development. At the bottom is the time scale in years, eras, periods and mountain-building revolutions. The creatures at the far left lived in salt water; most of their descendants, in fresh water. The latter, however, gave rise to forms which returned to salt water. Here salt water is indicated by closely spaced lines; fresh water, by open lines. The diagrams of the nephrons are highly schematic, with the tubule straightened out for clarity. The earliest nephrons are composed only of a tubule. The latest nephrons of land animals have a complex glomerulus and a tubule with two segments.

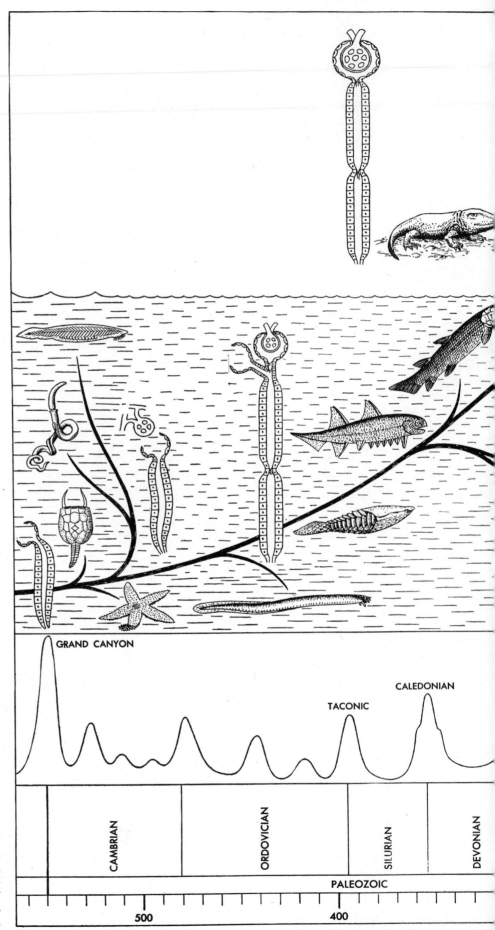

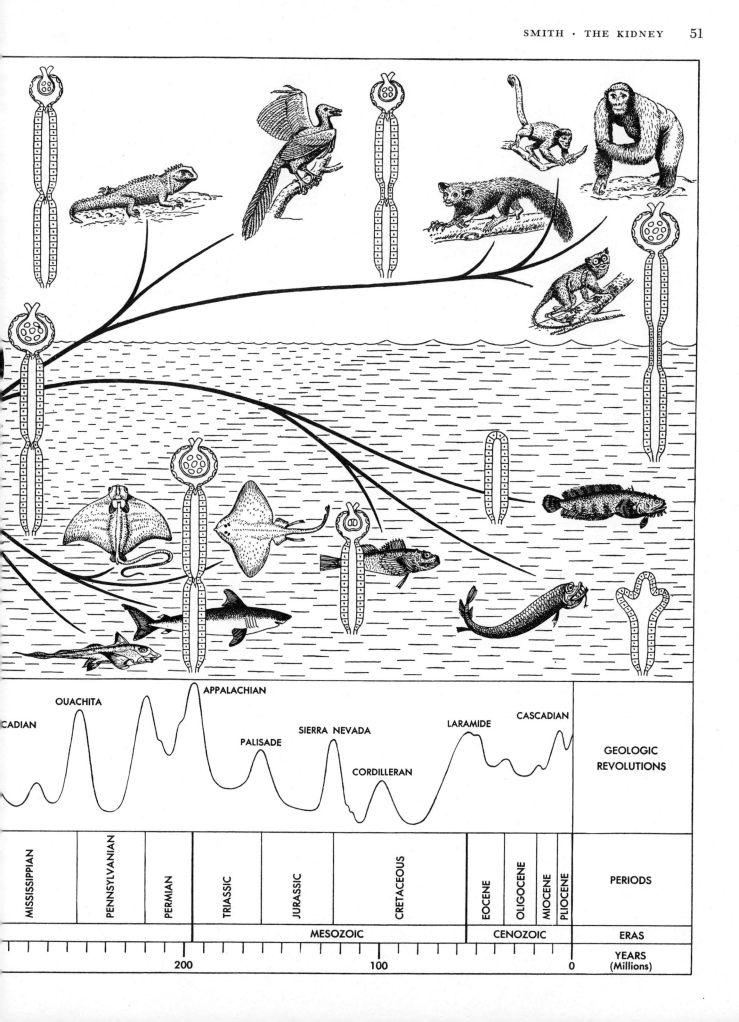

CADIAN

OUACHITA

APPALACHIAN

PALISADE

SIERRA NEVADA

CORDILLERAN

LARAMIDE

CASCADIAN

GEOLOGIC
REVOLUTIONS

MISSISSIPPIAN	PENNSYLVANIAN	PERMIAN	TRIASSIC	JURASSIC	CRETACEOUS	EOCENE	OLIGOCENE	MIOCENE	PLIOCENE	PERIODS
			MESOZOIC			CENOZOIC				ERAS

YEARS
(Millions)

200   100   0

rise, conserving water, while drinking water causes the secretion rate to fall nearly to zero, so that more water is excreted into the urine. So delicate is the mechanism that, aided by the sensation of thirst, and assuming that water is freely available, the body is normally maintained in an almost perfect state of water balance despite wide changes in fluid intake or correspondingly large changes in our external environment.

Alcohol inhibits the secretion of the antidiuretic hormone and therefore increases the urine output. M. G. Eggleton reports that 100 c.c. of water are excreted for every 10 grams of alcohol consumed (within a physiological range). The diuresis associated with beer-drinking is attributable chiefly to the water consumed rather than to the alcohol, since beer contains only 4 per cent alcohol. But in the case of whiskey, which is 40 to 45 per cent alcohol, the alcohol increases the loss of fluid from the body and has a dehydrating effect; undiluted whiskey will not quench one's thirst.

A proper view of the kidney sees that from the beginning of its evolution in the lowly Paleozoic chordates this organ has been more than just a device for excreting waste products: it has always been charged with the regulation of the volume and composition of the body fluids. It fulfills this task by operating in reverse—conserving some constituents by reabsorbing them from the glomerular filtrate, while rejecting others. The kidney acquired its glomerular structure because of the necessity of removing large quantities of water from the body in the Paleozoic fishes. To carry out this operation by a filtration-reabsorption system the kidney tubules had to regulate with great precision the excretion of sodium chloride, water and other substances. Thus the kidney came to be the master chemical laboratory controlling the composition of our internal environment—a laboratory working in reverse by overhauling all the blood many times a day.

The great importance of the kidney is revealed not only by the serious consequences of specific diseases of the organ itself, but also by the many remarkable adjustments in its functioning that accompany various other diseases and physiological disorders which do not directly involve the kidney. It is not surprising that the kidney is one of the most extensively studied organs in the body, having as much interest for the obstetrician, pediatrician, internist, surgeon and geriatrician as for the physiologist—and, as Fourcroy said, for the philosopher.

6

THE HEART

CARL J. WIGGERS   May 1957

The blood bathes the tissues with fluid and preserves their slight alkalinity; it supplies them with food and oxygen; it conveys the building stones for their growth and repair; it distributes heat generated by the cells and equalizes body temperature; it carries hormones that stimulate and coordinate the activities of the various organs; it conveys antibodies and cells that fight infections—and of course it carries drugs administered for therapeutic purposes. No wonder that William Harvey, the discoverer of the circulation, ardently defended the ancient belief that the blood is the seat of the soul.

The blood cannot support life unless it is kept circulating. If the blood flow to the brain is cut off, within three to five seconds the individual loses consciousness; after 15 to 20 seconds the body begins to twitch convulsively; and if the interruption of the circulation lasts more than nine minutes, the mental powers of the brain are irrevocably destroyed. Similarly the muscles of the heart cannot survive total deprivation of blood flow for longer than 30 minutes. These facts emphasize the vital importance of the heart as a pump.

The work done by this pump is out of all proportion to its size. Let us look at some figures. Even while we are asleep the heart pumps about two ounces of blood with each beat, a teacupful with every three beats, nearly five quarts per minute, 75 gallons per hour. In other words, it pumps enough blood to fill an average gasoline tank almost four times every hour just to keep the machinery of the body idling. When the body is moderately active, the heart doubles this output. During strenuous muscular efforts, such as running to catch a train or playing a game of tennis, the cardiac output may go up to 14

barrels per hour. Over the 24 hours of an average day, involving not too vigorous work, it amounts to some 70 barrels, and in a lifetime of 70 years the heart pumps nearly 18 million barrels!

The Design

Let us look at the design of this remarkable organ. The heart is a double pump, composed of two halves. Each side consists of an antechamber, formerly called the auricle but now more commonly called the atrium, and a ventricle. The capacities of these chambers vary considerably during life. In the human heart the average volume of each ventricle is about four ounces, and of each atrium about five ounces. The used blood that has circulated through the body—low in oxygen, high in carbon dioxide, and dark red (not blue) in color—first enters the right half of the heart, principally by two large veins (the superior and inferior venae cavae). The right heart pumps it via the pulmonary artery to the lungs, where the blood discharges some of its carbon dioxide and takes up oxygen. It then travels through the pulmonary veins to the left heart, which pumps the refreshed blood out through the aorta and to all regions of the body [see diagram on next page].

The thick muscular walls of the ventricles are mainly responsible for the pumping action. The wall of the left ventricle is much thicker than that of the right. The two pumps are welded together by an even thicker dividing wall (the septum). Around the right and left ventricles is a common envelope consisting of several layers of spiral and circular muscle [see diagrams on page 56]. This arrangement has a number of mechanical virtues. The blood is not

merely pushed out of the ventricles but is virtually wrung out of them by the squeeze of the spiral muscle bands. Moreover, it is pumped from both ventricles almost simultaneously, which insures the ejection of equal volumes by the two chambers—a necessity if one or the other side of the heart is not to become congested or depleted. The effectiveness of the pumping action is further enhanced by the fact that the septum between the ventricles becomes rigid just before contraction of the muscle bands, so that it serves as a fixed fulcrum at their ends.

The ventricles fill up with blood from the antechambers (atria). Until the beginning of the present century it was thought that this was accomplished primarily by the contractions of the atria, i.e., that the atria also functioned as pumps. This idea was based partly on inferences from anatomical studies and partly on observation of the exposed hearts of frogs. But it is now known that in mammals the atrium serves mainly as a reservoir. The ventricles fill fairly completely by their elastic recoil from contraction before the atria contract. The contraction of the latter merely completes the transfer of the small amount of blood they have left. Indeed, it has been found that the filling of the ventricles is not significantly impaired when disease destroys the ability of the atria to contract.

Factors of Safety

Since most of us believe that every biological mechanism must have some purpose, the question arises: Why complicate the cardiac pump with contractions of the atria if the ventricles alone suffice? The answer is that they provide what engineers call a "factor of safety."

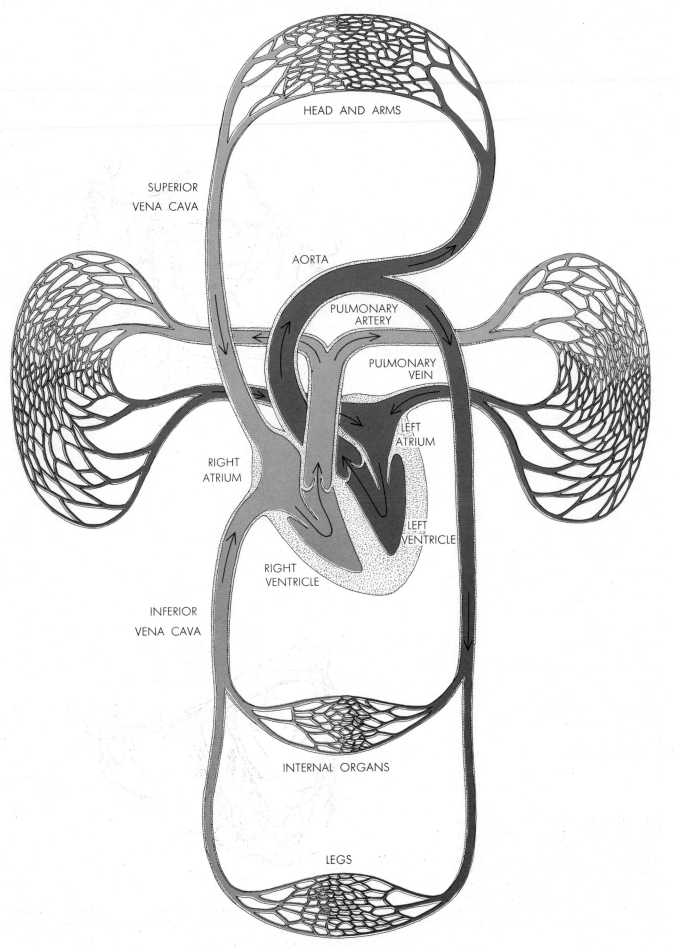

HEAD AND ARMS

SUPERIOR
VENA CAVA

AORTA

PULMONARY
ARTERY

PULMONARY
VEIN

LEFT
ATRIUM

RIGHT
ATRIUM

LEFT
VENTRICLE

RIGHT
VENTRICLE

INFERIOR
VENA CAVA

INTERNAL ORGANS

LEGS

While the atrial contractions make only a minor contribution to filling the ventricles under normal circumstances, they assume an important role when disease narrows the valve openings between the atria and the ventricles. Their pumping action then is needed to drive blood through the narrowed orifices.

The ventricular pumps also have their factors of safety. The left ventricle can continue to function as an efficient pump even when more than half of its muscle mass is dead. Recently the astounding discovery was made that the right ventricle can be dispensed with altogether and blood will still flow through the lungs to the left heart! An efficient circulation can be maintained when the walls of the right ventricle are nearly completely destroyed or when blood is made to by-pass the right heart. Obviously the heart is equipped with large factors of safety to meet the strains of everyday life.

This applies also to the heart valves. Like any efficient pump, the ventricle is furnished with inlet and outlet valves; it opens the inlet and closes the outlet while it is filling, and closes the inlet when it is ready to discharge. The pressure produced by contraction of the heart muscles mechanically closes the inlet valve between the atrium and ventricle: shortly afterward the outlet valve opens to let the ventricle discharge its blood—into the pulmonary artery in the case of the right ventricle and into the aorta in the case of the left. Then as the muscles relax and pressure in the chambers falls, the outlet valves close and shortly thereafter the inlet valves open. The relaxation that allows the ventricles to fill is called diastole; the contraction that expels the blood is called systole.

While it might seem that competent valves are indispensable for the forward movement of blood, they are in fact not absolutely necessary. The laws of hydraulics play some peculiar tricks. As every farmhand knew in the days of hand well-pumps, if the valves of the pump were worn and leaky, one could

ANATOMY OF THE HEART and its relationship to the circulatory system is schematically depicted on the opposite page. The arterial blood is represented in a bright red; venous blood, in a somewhat paler red. The capillaries of the lungs are represented at left and right; the capillaries of the rest of the body, at top and bottom. The term atrium is now used in preference to auricle.

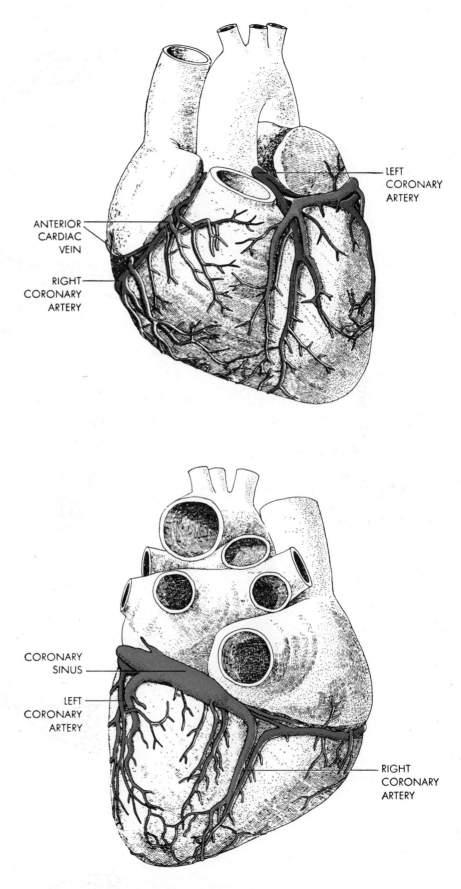

ARTERIES AND VEINS which carry blood to and from the muscles of the heart are shown from the front (*top*) and back (*bottom*). The arteries are bright red; the veins, pale red.

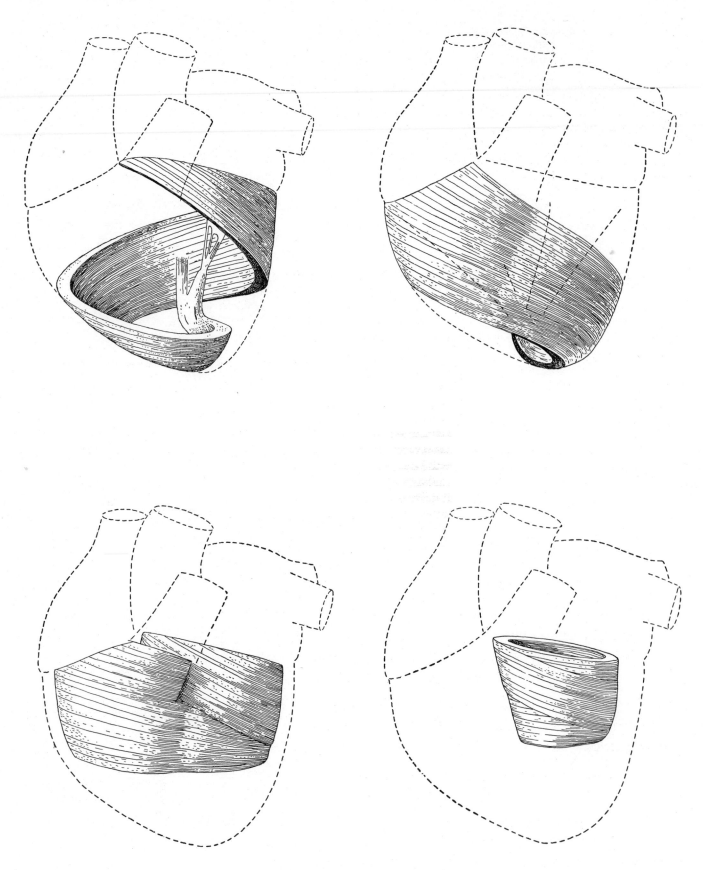

MUSCLE FIBERS of the ventricles are divided into four groups, one of which is shown in each of these four drawings. Two groups of fibers (*two drawings at top*) wind around the outside of both ventricles. Beneath these fibers a third group (*drawing at lower left*) also winds around both ventricles. Beneath these fibers, in turn, a fourth group (*drawing at lower right*) winds only around the left ventricle. The contraction of all these spiral fibers virtually wrings, rather than presses, blood out of the ventricles.

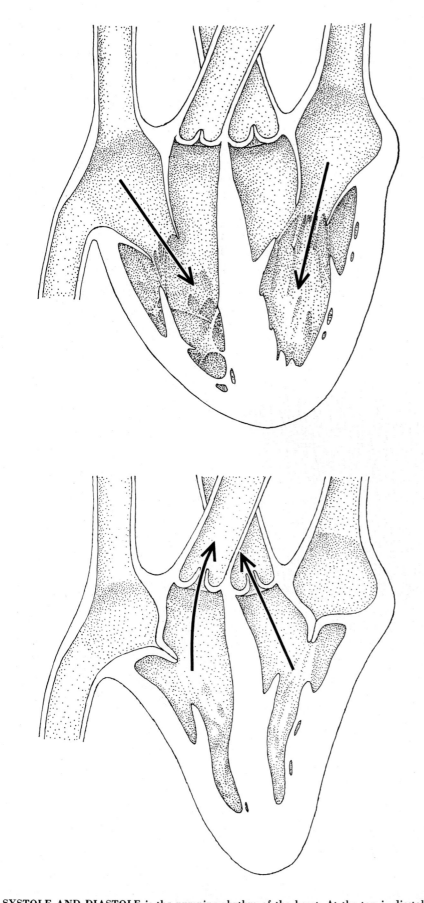

still draw water from the well by pumping harder. Similarly doctors have long been aware that patients can maintain a good circulation despite serious leaks in the heart valves. The factors of safety concerned are partly physical and partly physiological. The physical factor is more vigorous contraction of the heart muscles, aided by a structural arrangement of the deep muscle bands which tends to direct the blood flow forward rather than backward through the leaky valve. The physiological factor of safety is the mechanism known as "Starling's Law." In brief the rule is that, the more a cardiac muscle is stretched, the more vigorously it responds, of course within limits. The result is that the more blood the ventricles contain at the end of diastole, the more they expel. Of course they will fill with an excess of blood when either the inlet or the outlet valves leak. By pumping an extra volume of blood with each beat, the ventricles compensate for the backward loss through the atrial valves. In addition, the sympathetic nerves or hormones carried in the blood may spur the contractile power of the muscles. Under certain circumstances unfavorable influences come into play that depress the contractile force. Fortunately drugs such as digitalis can heighten the contractile force and thus again restore the balance of the circulation even though the valves leak.

Like any sharp closing of a door, the abrupt closings of the heart valves produce sounds, which can be heard at the chest wall. And just as we can gauge the vigor with which a door is slammed by the loudness of the sound, so a physician can assess the forces concerned in the closing of the individual heart valves. When a valve leaks, he hears not only the bang of the valves but also a "murmur" like the sigh of a gust of wind leaking through a broken window pane. The quality and timing of the murmur and its spread over the surface of the chest offer a trained ear considerable additional information. Sometimes a murmur means that the inlet and outlet orifices of the ventricles have been narrowed by calcification of the valves. In that case there is a characteristic sound, just as the water issuing from a hose nozzle makes a hissing sound when the nozzle is closed down.

Blood Supply

In one outstanding respect the heart has no great margin of safety: namely, its oxygen supply. In contrast to many other tissues of the body, which use as

SYSTOLE AND DIASTOLE is the pumping rhythm of the heart. At the top is diastole, in which the ventricles relax and blood flows into them from the atria. The inlet valves of the ventricles are open; the outlet valves are closed. At the bottom is systole, in which the ventricles contract, closing the inlet valves and forcing blood through the outlet valves.

little as one fourth of the oxygen brought to them by the blood, the heart uses 80 per cent. The amount of the blood supply is therefore all-important to the heart, particularly when activity raises its demand for oxygen.

Blood is piped to the heart muscles via two large coronary arteries which curl around the surface of the heart and send branchings to the individual muscle fibers [see diagrams on page 55]. The left coronary artery is extraordinarily short. It divides almost immediately into two branches. A large circumflex branch runs to the left in a groove between the left atrium and ventricle and continues as a large vessel which descends on the rear surface of the left ventricle. It supplies the left atrium, the upper front and whole rear portion of the left ventricle. The other branch circles to the left of the pulmonary artery and then runs downward in a furrow to the apex. It supplies the front wall of the left ventricle and a small part of the rear right ventricle. Close to its origin the left coronary artery gives off several twigs which carry blood to the septum. The right coronary artery, embedded in fat, runs to the right in a groove between the right atrium and ventricle. It carries blood to both of them.

From the surface branches vessels run into the walls of the heart, dividing repeatedly until they form very fine capillary networks around the muscle elements. Eventually three systems of veins return the blood to the right heart to be pumped back to the lungs.

In the normal human heart there is little overlap by the three main arteries. If one of them is suddenly blocked, the area of the heart that it serves cannot obtain a blood supply by any substitute route. The muscles deprived of arterial blood soon cease contracting, die and become replaced by scar tissue. Now while this is the ordinary course of events, particularly in young persons, the amazing discovery was made some 20 years ago that the blocking of a main coronary artery does not always result in death of the muscles it serves. It has since been proved that new blood vessels grow in, from other arteries, if a main branch is progressively narrowed by atherosclerosis over a period of months or years. In other words, if the closing of a coronary artery proceeds slowly, a collateral circulation may develop. This biological process constitutes another factor of safety. Recent experiments on dogs indeed indicate that exercise will accelerate the development of collaterals when a major coronary artery is constricted. If this indication is con-

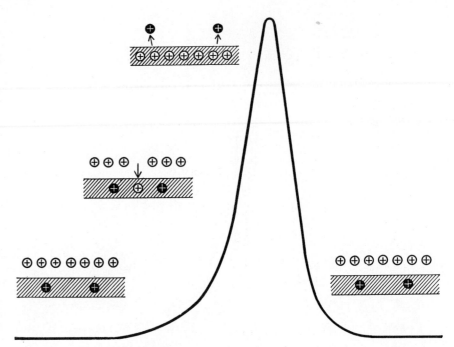

ELECTRICAL IMPULSE IS GENERATED by a cell in the "pacemaker" of the heart, a system of specialized muscle tissue (*see diagram on next page*). At lower left is a cross section of the cell; positively charged potassium ions (*black*) are inside it and a larger number of positively charged sodium ions (*white*) are outside. Because there are more positive charges outside the cell than inside, the inside of the cell is negatively charged with respect to the outside. When the cell is stimulated (*second cross section*), a sodium ion leaks across the membrane. Then many sodium ions rush across the membrane and potassium ions rush out (*third cross section*); this reverses the polarity of the cell and gives rise to an action potential (*peak in curve*). The original situation is then restored (*fourth cross section*)

firmed, it may well be that patients with atherosclerosis will be encouraged to exercise, rather than to adopt a sedentary life.

We have seen that there are many structural and functional factors of safety which enable the heart to respond, not only to the stress and strain of everyday life, but also to unfavorable effects of disease. Their existence has long been recognized; modern research has now thrown some light on the fundamental physiological and chemical processes involved.

The heart's transformation of chemical energy into the mechanical energy of contraction has certain similarities to the conversion of energy in an automobile engine; but there are also essential differences. In both cases a fuel is suddenly exploded by an electric spark. In both the fuel is complex, and the explosion involves a series of chemical reactions. In each case some of the energy is lost as unusable heat. In each the explosions occur in cylinders, but in the heart these cylinders (the heart muscle cells) not only contain the fuel but are able to replenish it themselves from products supplied by the blood. The mechanical efficiency of these cells, *i.e.*, the fraction of total energy that can be converted to mechanical energy, has not been equaled by any man-made

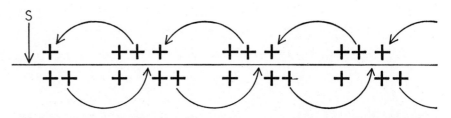

ELECTRICAL IMPULSE IS TRANSMITTED through the pacemaker system not as an electric current but as an electrical chain reaction. When a pacemaker cell is stimulated (S), it discharges and generates a local current which causes the depolarization and discharge of adjacent cells. In effect a wave of positive charge passes through the system (*curved arrows*).

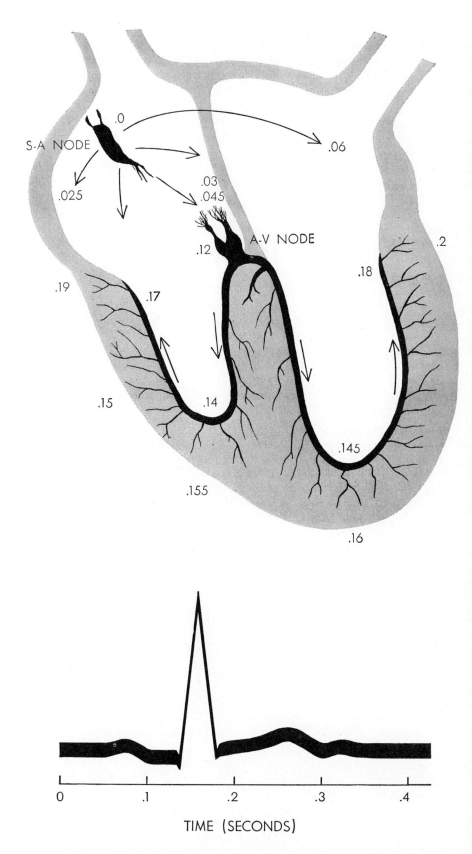

PACEMAKER SYSTEM generates and transmits the impulses which cause the contraction of the heart muscles. The impulse is generated in the sino-atrial (S-A) node and spreads across the atria, causing their contraction and stimulating the atrio ventricular (A-V) node. This in turn stimulates the rest of the system, causing the rest of the heart to contract. The numbers indicate the time (in fractions of a second) it takes an impulse to travel from the S-A node to that point. The electrocardiogram curve at bottom indicates the change in electrical potential that occurs during the spread of one impulse through the system.

machine designed in the pre-atomic age. The mechanism responsible for this efficiency is unique and very complex.

Under the microscope we can see that cardiac muscle consists of long, narrow networks of fibers, with connective tissue and tiny blood vessels filling the spaces between. Each muscle fiber is made up of innumerable fibrils embedded in a matrix. It has been demonstrated that these fibrils are responsible for the contraction of the muscle as a whole. By special and clever techniques the fibrils can be washed free of the matrix, and it has been shown that when brought into contact with the energy-rich substance ATP, the fibrils shorten.

Examinations with the polarizing and electron microscopes and with X-rays have produced a fairly good picture of the ultimate design of these microscopic fibers. Each fibril is composed of many smaller filaments, or "protofibrils," just distinguishable under the highest microscopic magnification. The fibrils of a single muscle fiber may contain a total of some 10 million such filaments. The filaments are the smallest units known to stiffen and shorten. It has been possible to extract the actomyosin of which they are composed and to reconstitute filaments by squirting the extracted protein into a salt solution. These synthetic filaments can be made to contract.

The filaments themselves remain straight during contraction; therefore the kinking or coiling necessary for their contraction must take place at a still lower level—the level of molecules. Here the picture is clouded. We know from X-ray diffraction analyses that the molecules composing myosin filaments are arranged as miniature stretched spiral springs or stretched rubber bands. But just how they effect the filaments' contraction can only be guessed. Regardless of the mechanism, there is no doubt that the stiffness and shortening which are the features of contraction are mediated by changes in the molecular arrangement. This rearrangement requires energy. The consensus is that a tiny electric spark delivered to each individual cell causes the explosion of ATP. Not all the energy released is used for shortening of the actomyosin filaments. Some of it is converted to heat and some is used to initiate a series of complex chemical reactions which replenishes the fuel by reconstituting ATP. The explosion of ATP differs from that of gasoline in that no oxygen is required. But oxygen is indispensable for the rebuilding of ATP.

The millions of cardiac cylinders, as

MUSCLE FIBRILS of the heart, which make up its muscle fibers, are revealed in this electron micrograph made by Bruno Kisch of the American College of Cardiology in New York. The fibrils, which are from the heart of a guinea pig, are the long bands running diagonally across the micrograph. The round bodies between the fibrils are sarcosomes, which are found in large numbers in heart muscle and which appear to supply the enzymes that make possible its tireless contractions. The fibrils themselves are made up of protofibrils, which may barely be seen in the micrograph. The micrograph enlarges these structures some 50,000 diameters.

in an automobile engine, must fire in proper sequence to contract the muscle effectively. When they fire haphazardly, there is a great liberation of energy but no coordinated action. This chaotic condition is called fibrillation—*i.e.*, independent and uncoordinated activity of the individual fibrils.

The Beat

What causes the heart to maintain its rhythmic beat? The ancients, performing sacrificial rites, must have noticed that the heart of an animal continues to beat for some time after it has been removed from the body. That the beat must originate in the heart itself was apparently clear to the Alexandrian anatomist Erasistratus in the third century B.C. But anatomists ignored this evidence for the next 20 centuries because they were convinced that the nerves to the heart must generate the heartbeat. In 1890, however, Henry Newell Martin at the Johns Hopkins University demonstrated that the heart of a mammal could be kept beating though it was completely separated from the nerves, provided it was supplied with blood. And many years before that Ernst Heinrich Weber of Germany had made the eventful discovery that stimu-

lation of the vagus nerve to the heart does not excite it but on the contrary stops the heart. In short, it was established that the beat is indeed generated within the heart, and that the nerves have only a regulating influence.

The nature and location of the heart's "pacemaker" remained enigmatic until comparatively recent times. Within the span of my own memory there was considerable evidence for the view that the pacemaking impulses were generated by nerve cells in the right atrium and transmitted by nerve fibers to the heart-muscle cells. At present the evidence is overwhelming that the impulses are actually generated and distributed by a system of specialized muscle tissue consisting of cells placed end to end. Seventy-two times per minute—more or less—a brief electric spark of low intensity is liberated from a barely visible knot of tissue in the rear wall of the right atrium, called the sino-atrial or S-A node. The electric impulse spreads over the sheet of tissue comprising the two atria and, in so doing, excites a succession of muscle fibers which together produce the contraction of the atria. The impulse also reaches another small knot of specialized muscle known as the atrioventricular or A-V node, situated between the atria and ventricles. Here

the impulse is delayed for about seven hundredths of a second, apparently to allow the atria to complete their contractions; then from the A-V node the impulse travels rapidly throughout the ventricles by way of a branching transmission system, reaching every muscle fiber of the two ventricles within six hundredths of a second. Thus the tiny spark produces fairly simultaneous explosions in all the cells, and the two ventricles contract in a concerted manner.

If the heart originates its own impulses, of what use are the two sets of nerves that anatomists have traced to the heart? A brief answer would be that they act like spurs and reins on a horse which has an intrinsic tendency to set its own pace. The vagus nerves continually check the innate tempo of the S-A node; the sympathetic nerves accelerate it during excitement and exercise.

Normally, as I have said, the S-A node generates the spark, but here, too, nature has provided a factor of safety. When the S-A node is depressed or destroyed by disease, the A-V node becomes the generator of impulses. It is not as effective a generator (its maximum rate is only 40 or 50 impulses per minute, and its output excites the atria and ventricles simultaneously), but it suffices to keep the heart going. Patients

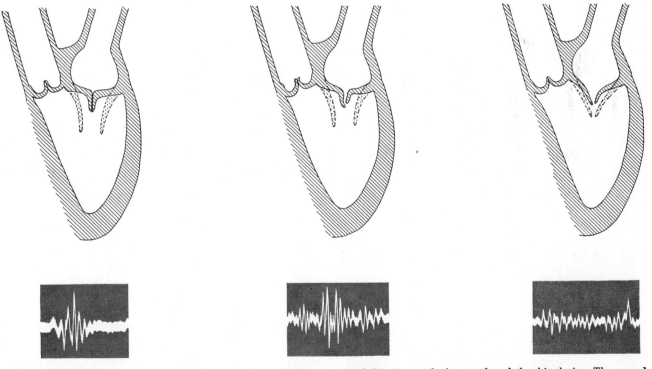

HEART SOUNDS indicate the normal and abnormal functioning of the heart valves. The three drawings at the top show the left side of the heart in cross section. The aortic, or outlet, valve is at upper left in each drawing; the mitral, or inlet, valve is at upper right. The first drawing shows the normal closing of the mitral valve; the trace below it records the sound made by this closing. The second drawing shows the partial closing of a leaky mitral valve; the trace below it records the murmur of blood continuing to flow through the valve. The third drawing shows the partial closing of a mitral valve with stiff leaves; the trace below it records a fainter murmur.

have survived up to 20 years with the A-V pacemaker substituting for a damaged S-A node.

There are still lower pacemakers which can maintain a slow heartbeat when the higher ones fail. When all the pacemakers are so weakened that, like an old battery, they are barely able to emit impulses, an anesthetic administered during an operation may stop the heart. In that case the beat can often be restored by rhythmic electric shocks —a system now incorporated in an apparatus for revival of the heart.

The Spark

What sort of mechanism exists in nodal tissues that is able to emit electric sparks with clockwork regularity 104,-000 times a day? We must look first to the blood. The fact that an excised heart does not long continue to beat unless supplied artificially with blood shows that the blood must supply something essential for preservation of its beat. In the 19th century physiologists began to experiment on isolated hearts, first of frogs and turtles and later of rabbits and cats, to determine what constituents of blood could be spared without halting the rhythmic heart contractions. They found that the serum (the blood fluid without cells) could maintain the beat of a mammalian heart, provided the serum was charged with oxygen under pressure. What constituents of the serum, besides the oxygen, were necessary? Attention first focused on the proteins, on the theory that the beating heart required them for nourishment. The heart was, in fact, found to be capable of maintaining its beat fairly well on a "diet" of blood proteins or even egg white or oxygenated milk whey. But the nourishment idea received a blow when it was discovered that the heartbeat could be maintained on a solution of gum arabic! It was then suggested that serum proteins act by virtue of their viscosity. This is an example of how experimenters are sometimes led astray in trying to uncover nature's secrets.

An eventful discovery in 1882 by the English physiologist Sydney Ringer changed the direction of thinking. He showed that a solution containing salts of sodium, potassium and calcium and a little alkali, in the concentrations found in the blood, would sustain the beat of a frog's heart. It was but a step to show that Ringer's solution, when oxygenated, also keeps the mammalian heart beating for a short time. Later it was found that the addition of a biological fuel—glucose

or, better yet, lactic acid—would extend the heart's performance.

Summing up the evidence, it was known at the beginning of the present century that the beat of the mammalian heart, and obviously also the generation of the spark, depends primarily on a balanced proportion of sodium, potassium and calcium plus a supply of oxygen and an energy-yielding substance such as glucose.

During the present century the scientific minds have sought to learn how these inorganic elements are involved in the initiation and spread of impulses. In order to understand the intricate mechanisms we must recall what most of us learned in high school: *viz.*, the theory that, when a salt is dissolved in water, the elements are dissociated and become ions charged with positive and negative electricity.

The delicate enclosing membrane of all cells is differentially permeable: that is, ordinarily (at rest) it allows potassium ions to enter the cell but excludes sodium ions. We may say that the potassium ions have admission tickets, while the sodium ions do not. Since sodium ions predominate in the body fluids, the positively charged potassium ions within a cell are greatly outnumbered by positively charged sodium ions around the outside of the cell; the net result is that the outside is more positive than the inside, and the interior can therefore be regarded as negative with respect to the exterior. The potential difference is about one tenth of a volt. Each cell thus becomes a small charged battery. Now in the case of cells of the S-A node, the membrane leaks slightly, allowing some sodium ions to sneak in. This slowly but steadily reduces the potential difference between the inside and the outside of the membrane. When the difference has diminished by a critical amount (usually about six hundredths of a volt), the tiny pores of the membrane abruptly open. A crowd of sodium ions then rushes in, while some of the imprisoned potassium ions escape to the exterior. As a result the relative charges on the two sides of the membrane are momentarily reversed, the inside being positive with respect to the outside. The action potential thus created is the release of the electric spark.

As soon as activity is over, the membrane repolarizes, *i.e.*, reconstitutes a charged battery. How this is accomplished is little understood, beyond the fact that oxidation of glucose or its equivalent is required. The mechanism

is pictured as a kind of metabolic pump which ejects the sodium ions that have gained illegal admittance, allows potassium ions to re-enter and closes the pores again. Then the cells are ready to be discharged again.

A little reflection should make it evident that the frequency with which such cells discharge depends on at least two things: (1) the rate at which sodium ions leak into the cell, and (2) the degree to which the potential across the membrane must be reduced in order to discharge it completely. The rate of sodium entry is known to be increased by warming and decreased by cooling, which accounts in part for the more rapid firing of the pacemakers in a patient who has a fever. The magnitude of the potential difference required to discharge the cell depends on the characteristics of the membrane. In this the concentration of calcium ions plays a basic role. Calcium favors stability of the membrane: if its concentration falls below a certain critical value, the rate of discharge by the cells increases; if calcium ions are too abundant, the rate is slowed. The rate of discharge is also affected by other factors. The vagus nerves tend to reduce it, the sympathetic nerves to increase it. The blood's content of oxygen and carbon dioxide, its degree of alkalinity, hormones and drugs—these and other influences can change the stability of the membranes and thus alter the rate of cell discharge.

Transmission of Impulses

The spark from a pacemaker is transmitted to the myriads of cylinders constituting the ventricular pumps by way of the special conducting system. When we say that the impulse travels over this system, this does not mean that electricity flows, as over wires to automobile cylinders. The electric impulse spreads by a kind of chain reaction involving the successive firing of the special transmitting cells. When pacemaker cells discharge, they generate a highly localized current which in turn causes the depolarization and discharge of an adjacent group of cells, and thus the impulse is relayed to the muscle cells concerned with contraction. An advantage of this mechanism is that the strength of the very minute current reaching the contracting fibers is not reduced.

Such a mode of transmission is not unknown in the inanimate world. There is a classic experiment in chemistry which illustrates an analogous process. An iron wire is coated with a microscopic film of iron oxide and suspended in a cylinder of strong nitric acid. Protected by this coating, the iron does not dissolve. But if the coat is breached (by a scratch or by an electric current) at a spot at one end of the immersed wire, a brown bubble immediately forms at this spot and a succession of brown bubbles then traverses the whole length of the wire. An electrical recorder connected to a number of points along the wire shows that a succession of local electric currents is generated down the wire as it bubbles. At each spot the current breaks the iron oxide film and the ensuing chemical reaction generates a new action potential. The contact between bare iron and nitric acid is only momentary, because the break in the film is quickly repaired.

Summarizing, the passage of electric impulses over the conduction system of the heart represents a series of local bio-electric currents, relaying the impulses step by step over special tissue to the contracting cells. On arrival at these cells the electric charges trigger the breakdown of ATP and so release the chemical energy needed for contraction.

Diagnosis

Considering the complexity of the cardiac machinery, it is remarkable that the heartbeat does not go wrong more often. Like a repairman for an automobile or a television set, a physician sometimes has to make an extensive hunt for the source of the disorder. It seems appropriate to close this article with a list of the points at which the machinery is apt to break down.

1. The main (S-A) pacemaker, or in rare instances all the pacemakers, may fail.

2. There may be too many pacemakers. The secondary pacemakers occasionally spring into action and work at cross purposes with the normal one, producing too rapid, too slow, or ill-timed beats.

3. The system for conduction of the pacemaker impulses may break down, leading to "heart block."

4. The heart muscle cylinders may respond with little power because of poor fuel, lack of enough oxygen for building fuel, fatigue or lack of adequate vitamins, hormones or other substances in the blood.

5. Some of the cylinders may be put out of commission by blockage of a coronary artery.

6. The heart valves may leak, and in its gallant effort to compensate, the heart may be overworked to failure.

7

THE MICROCIRCULATION OF THE BLOOD

BENJAMIN W. ZWEIFACH

January 1959

When we think of the circulatory system, the words that first occur to us are heart, artery and vein. We tend to forget the microscopic vessels in which the blood flows from the arteries to the veins. Yet it is the microcirculation which serves the primary purpose of the circulatory system: to convey to the cells of the body the substances needed for their metabolism and regulation, to carry away their products—in short, to maintain the environment in which the cells can exist and perform their interrelated tasks. From this point of view the heart and the larger blood vessels are merely secondary plumbing to convey blood to the microcirculation.

To be sure, the entire circulatory system is centered on the heart. The two chambers of the right side of the heart pump blood to the lungs, where it is oxygenated and returned to the chambers of the left side of the heart. Thence the blood is pumped into the aorta, which branches like a tree into smaller and smaller arteries. The smallest twigs of the arterial system are the arterioles, which are too small to be seen with the unaided eye. It is here that the micro-circulation begins. The arterioles in turn branch into the capillaries, which are still smaller. From the capillaries the blood flows into the microscopic tributaries of the venous system: the venules. Then it departs from the microcirculation and is returned by the tree of the venous system to the chambers in the right side of the heart.

The vessels of the microcirculation permeate every tissue of the body; they are never more than .005 inch from any cell. The capillaries themselves are about .0007 inch in diameter. To give the reader an idea of what this dimension means, it would take one cubic centimeter of blood (about 14 drops) from five to seven hours to pass through a capillary. Yet so large is the number of capillaries in the human body that the heart can pump all the blood in the body (about 5,000 cubic centimeters in an adult) through them in a few minutes. The total length of the capillaries in the body is almost 60,000 miles. Taken together, the capillaries comprise the body's largest organ; their total bulk is more than twice that of the liver.

If all the capillaries were open at one time, they would contain all of the blood in the body. Obviously this does not happen under normal circumstances, whereby hangs the principal theme of this article. How is it that the flow of blood through the capillaries can be regulated so as to meet the varying needs of all the tissues, and yet not interfere with the efficiency of the circulatory system as a whole?

It was William Harvey, physician to Charles I of England, who first demonstrated that the blood flows continuously from the arterial system to the venous. In 1661, 33 years after Harvey had published his famous work *De Motu Cordis* (*Concerning the Motion of the Heart*), the Italian anatomist Marcello Malpighi

CAPILLARY from a cat's leg muscle is shown in cross section by this electron micrograph, which enlarges the structure some 20,000 diameters. The band running around the picture is the wall of the capillary. The large, dark object in the center is a single red blood cell. The micrograph was made by George D. Pappas and M. H. Ross of Columbia University.

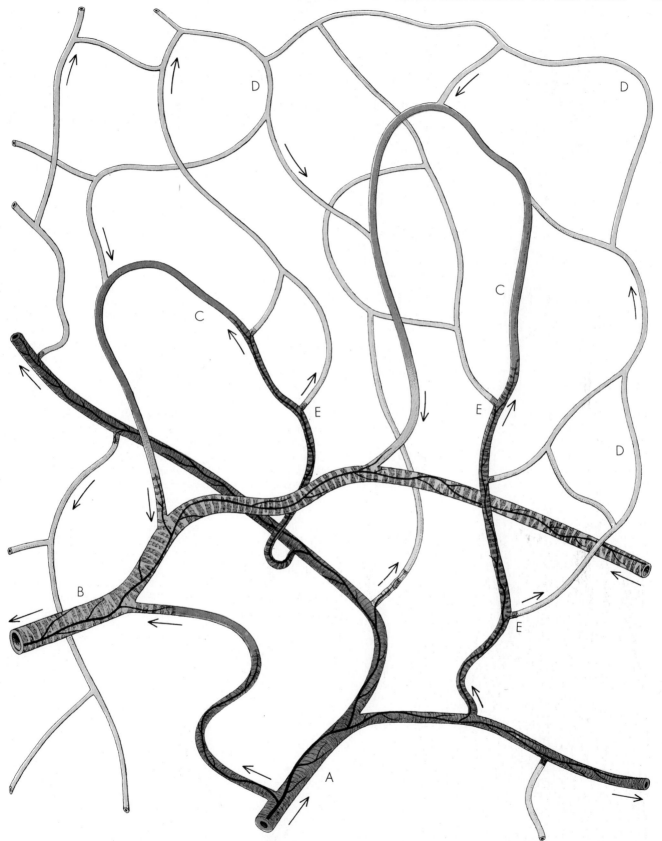

TYPICAL CAPILLARY BED is depicted in this drawing. The blood flows into the bed through an arteriole (A) and out of it through a venule (B). Between the arteriole and the venule the blood passes through thoroughfare channels (C). From these channels it passes into the capillaries proper (D), which then return it to the channels. The arteriole and venule are wrapped with muscle cells; in the thoroughfare channels the muscle cells thin out. The capillaries proper have no muscle cells at all. The flow of blood from a thoroughfare channel into a capillary is regulated by a ring of muscle called a precapillary sphincter (E). The black lines on the surface of the arteriole, venule and thoroughfare channels are nerve fibers leading to muscle cells. At lower left, between the arteriole and venule, is a channel which in many tissues shunts blood directly from the arterial system to the venous when necessary.

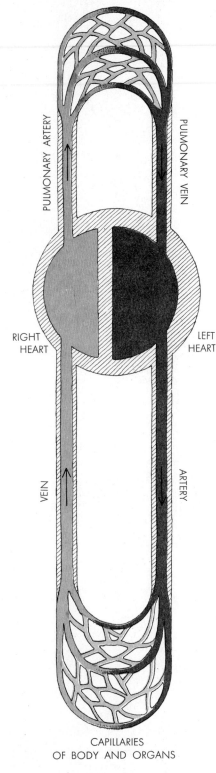

PULMONARY CAPILLARIES

PULMONARY ARTERY

PULMONARY VEIN

RIGHT
HEART

LEFT
HEART

VEIN

ARTERY

CAPILLARIES
OF BODY AND ORGANS

CIRCULATORY SYSTEM is schematically outlined. The blood is pumped by the right heart through the pulmonary artery into the capillaries of the lungs. It returns from the lungs through the pulmonary vein to the left heart, which pumps it through the arteries to the capillaries of the internal organs and of the rest of the body. It finally returns to the right heart through the veins.

first observed through his crude microscope the fine conduits which link the two systems. These vessels were named capillaries after the Latin word *capillus*, meaning hair. Since Malpighi's time the capillaries have been intensively examined by a host of microscopists. Their work has established that not all the vessels in the network lying between the arterioles and the venules are the same. Indeed, we must regard the network as a system of interrelated parts. Hence it is preferable to think not of capillaries, but of a functional unit called the capillary bed.

The capillary bed, unlike muscle or liver or kidney, cannot be removed from an experimental animal and studied as an intact unit outside the body of the animal. By their very nature the capillaries are interwoven with other tissues. It is possible, however, to examine the capillary bed in a living animal. For example, one can open the abdomen of an anesthetized rat and carefully expose a thin sheet of mesentery: the tissue that attaches the intestine to the wall of the abdominal cavity. In this transparent sheet the capillary bed is displayed in almost diagrammatic form.

The tube of a capillary is made of a single layer of flat cells resembling irregular stones fitted together in a smooth pavement. The wall of the tube is so thin that even when it is viewed edge-on at a magnification of 1,000 diameters it is visible only as a line. When the wall is magnified in the electron microscope, it may be seen that the wall is less than .0001 inch thick. This so-called endothelium not only forms the walls of the capillaries but also lines the larger blood vessels and the heart, so that all the blood in the body is contained in a single envelope.

In a large blood vessel the tube of endothelium is sheathed in fibrous tissue interwoven with muscle. The fibrous tissue imparts to the vessel a certain amount of elasticity. The muscle is of the "smooth" type, characterized by its ability to contract slowly and sustain its contraction. The muscle cells are long and tapered at both ends; they coil around the vessel. In the tiny arterioles, in fact, a single muscle cell may wrap around the vessel two or three times. When the muscle contracts, the bore of the vessel narrows; when the muscle relaxes, the bore widens.

The muscular sheath of the larger blood vessels does not continue into the capillary bed. Yet as early as the latter part of the 19th-century experimental

physiologists reported that the smallest blood vessels could change their diameter. Moreover, when the flow of blood through the capillary bed of a living animal is observed under the microscope, the pattern of flow constantly changes. At one moment blood flows through one part of the network; a few minutes later that part is shut off and blood flows through another part. In some capillaries the flow even reverses. Throughout this ebb and flow, however, blood passes steadily through certain thoroughfares of the capillary bed.

If the capillaries have no muscles, how is the flow controlled? Some investigators suggested that although the endothelium of the capillaries was not true muscle, it could nonetheless contract. Indeed, it was demonstrated that in many lower animals blood vessels consisting only of endothelium contract and relax in a regular rhythm. However, contractile movements of this kind have not been observed in mammals.

Another explanation was advanced by Charles Rouget, a French histologist. He had discovered peculiar star-shaped cells, each of which was wrapped around a capillary, and he assumed that they were primitive muscle cells which opened and closed the capillaries. Many investigators agreed with him, among them the Danish physiologist August Krogh, who in 1920 won a Nobel prize for his work on the capillary system. It was not possible, however, to prove or disprove the contractile function of the Rouget cells by simple observation.

There the matter rested until methods were developed for performing microsurgical operations on single cells [see "Microsurgery," by M. J. Kopac; SCIENTIFIC AMERICAN, October, 1950]. Now it was possible to probe the cell with extremely fine needles, pipettes and electrodes. Microsurgery established that in mammals neither the capillary endothelium nor the Rouget cells could control the circulation by contraction. The endothelium did not contract when it was stimulated by a microneedle, or by the application with a micropipette of substances that cause larger blood vessels to contract. When one of the star-shaped Rouget cells was stimulated, it became thicker but did not occlude the capillary. When the same stimulus was applied to the recognizable muscle cell of an arteriole, on the other hand, the cell contracted and the arteriole was narrowed.

The microsurgical experiments established an even more significant fact: not

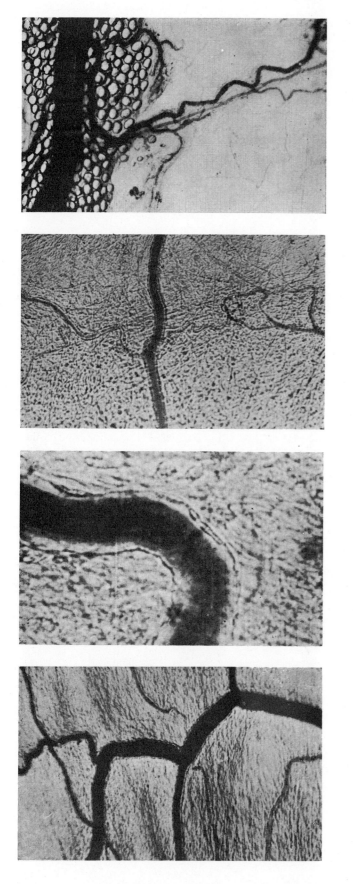

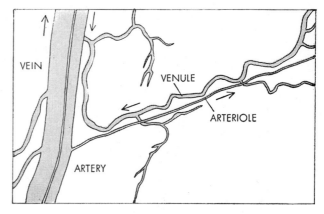

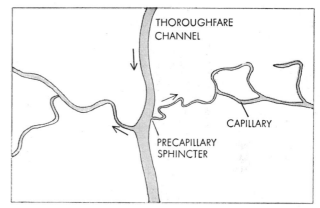

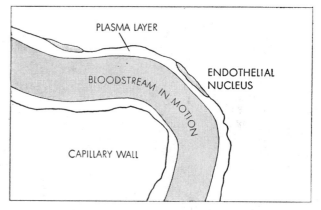

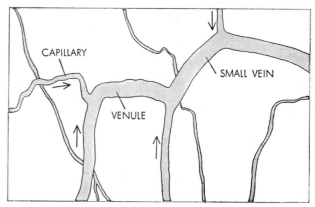

MESENTERY of a rat is photographed at various magnifications to show the characteristic structures of the microcirculation. The drawings at right label the structures. The magnification of the photomicrograph at top is 100 diameters; of the photomicrograph second from top, 200 diameters; of the third photomicrograph, 1,000 diameters; of the photomicrograph at bottom, 200 diameters.

all the vessels in the capillary bed entirely lack muscle. For example, if epinephrine, which causes larger blood vessels to contract, is injected into the capillary bed with a micropipette, some of the vessels in the bed become narrower. Even when no stimulating substances are added, the same vessels open and close with the ebb and flow of blood in the capillary bed. It is these vessels, moreover, through which the blood flows steadily from the arterial to the venous system.

So the arterial system, with its muscular vessels, does not end at the capillary bed. The blood is continuously under muscular control as it flows into the venous system. To be sure, the muscle cells along the thoroughfare are sparsely distributed. As the arterial tree branches into the tissues the muscular sheath of the endothelium becomes thinner and thinner until in the smallest arterioles it is only one cell thick. In the thoroughfare channel of the capillary bed the muscle cells are spaced so far apart that the channel is almost indistinguishable from the true capillaries. The major portion of the capillary network arises as abrupt side branches of the thoroughfare channels, and at the point where each of the branches leaves a thoroughfare channel there is a prominent muscle structure: the muscle cells form a ring around the entrance to the capillary. It is this ring, or precapillary sphincter, which acts as a floodgate to control the

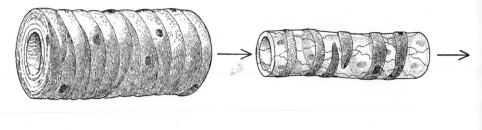

ARTERY   ARTERIOLE

WALLS OF BLOOD VESSELS of various kinds reflect their various functions. The wall of an artery consists of a single layer of endothelial cells sheathed in several layers of muscle cells interwoven with fibrous tissue. The wall of an arteriole consists of a single layer of

flow of blood into the capillary network from the thoroughfare channel.

The muscular specialization of the circulatory system is illuminated by its embryonic development. In the early embryo the circulatory system is a network of endothelial tubes through which the primitive blood cells flow in an erratic fashion. The tubes are at first just large enough to pass the blood cells in single file. Attached to the outer wall of the tubes are numerous star-shaped cells which have wandered in from the surrounding tissue.

As the development of the embryo proceeds, those tubes through which the blood flows most rapidly are transformed into heavy-walled arteries and veins. In the process the star-shaped cells evolve through several stages into typical muscle cells. The outer reaches of the adult

circulatory system possess a graded series of muscle-cell types, which are a direct representation of this developmental process. Thus the star-shaped cells of the capillary bed—the Rouget cells—are primitive muscle elements which have no contractile function.

From this point of view the capillary bed can be considered the immature part of the circulatory system. Like embryonic tissue, it has the capacity for growth, which it exhibits in response to injury. It also ages to some extent, and ultimately becomes less capable of dealing with the diversified demands of the tissue cells.

When we put these various facts together, we see the capillary bed not as a simple web of vessels between the arterial and venous systems, but as a

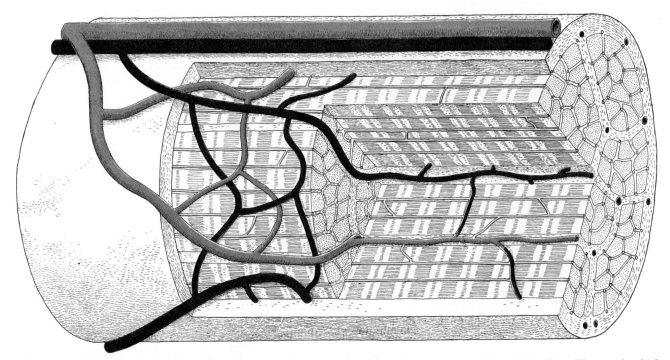

MUSCLE FIBER is richly supplied with capillaries. Lying atop this dissected muscle fiber are two blood vessels, the smaller of which is an artery and the larger a vein. Most of the capillaries run parallel to the fibrils which make up the fiber. The vessels which cut across two or more capillaries are thoroughfare channels. The system is shown in cross section at the right end of the drawing.

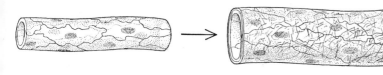

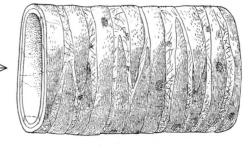

CAPILLARY   **VENULE**   **VEIN**

endothelial cells sheathed in a single layer of muscle cells. The wall of a capillary consists only of a single layer of endothelial cells. The wall of a venule consists of endothelial cells sheathed in fibrous tissue. The wall of a vein consists of endothelial cells sheathed in fibrous tissue and a thin layer of muscle cells. Thus a layer of endothelial cells lines the entire circulatory system.

physiological unit with two specialized components. One component is the thoroughfare channel, into which blood flows from the arteriole. The other is the true capillaries, which form a secondary network connected to the thoroughfare channel. The precapillary sphincters along the channel open and close periodically, irrigating first one part of the capillary network, then another part. When the sphincters are closed, the blood is restricted to the thoroughfare channel in its movement toward the venous system.

The structure of the physiological unit varies from one tissue to another in accordance with the characteristic needs of the tissues. For example, striated muscle, which unlike the smooth muscle of the blood vessels and other organs contracts rapidly and is under voluntary control, requires over 10 times more blood when it is active than when it is at rest. To meet this wide range of needs each thoroughfare channel in striated muscle gives rise to as many as 20 or 30 true capillaries. Glandular tissues, on the other hand, require only a steady trickle of blood, and each of their thoroughfare channels may give rise to as few as one or two capillaries. In the skin, which shields the body from its outer environment, there are special shunts through which blood can pass directly from the arteries to the veins with minimum loss of heat. Still other tissues require specialized capillary beds. The capillary beds of all the tissues, however, have the same basic feature: a central channel whose muscle cells control the flow of blood into the true capillaries.

But what controls the muscle cells? To answer this question we must draw a distinction between the control of the larger blood vessels and the control of the microcirculation. The muscle cells of the arteries and veins are made to contract and relax by two agencies: (1) the nervous system and (2) chemical "messengers" in the blood. These influences not only cause the vessels to constrict and dilate but also keep the muscle cells in a state of partial contraction. This muscle "tone" maintains the elasticity of the vessels, which assists the heart in maintaining the blood pressure. The operation of the system as a whole is supervised by special regulatory centers in the brain, working in collaboration with sensory monitoring stations strategically located in important vessels.

In the capillary bed, on the other hand, the role of the nervous system is much less significant. Most of the muscle cells in the capillary bed have no direct nerve connections at all. A further circumstance sets the response of the microscopic vessels apart from that of the larger vessels. Whereas the muscle cells of the large vessels are isolated from the surrounding tissues in the thick walls of the vessels, the muscle cells of the arterioles and the thoroughfare channels are immersed in the environment of the very tissues which they supply with blood. This feature introduces another chemical regulatory mechanism: the continuous presence of substances liberated locally by the tissue cells. As a consequence the contraction and relaxation of muscle cells in the microcirculation are under the joint control of messenger substances in the blood and specific chemical products of tissue metabolism.

The chemical substances that influence the function of the blood-vessel muscle cells comprise a subtly orchestrated system which is still imperfectly understood. Among the more important messengers are those released into the bloodstream by the cortex of the adrenal gland. These corticosteroids are essential to all cells in the body, notably maintaining the cells' internal balance of water and salts. (They have also been used with spectacular results in the treatment of degenerative diseases such as arthritis.) When the corticosteroids are deficient or absent, the muscles of the blood vessels lose their tone and the circulation collapses.

Another substance of profound importance to the circulatory system is epinephrine, which is secreted by the core of the adrenal gland (as distinct from its cortex). Epinephrine is one of two principal members of a family of substances called amines; the other principal member is norepinephrine, which is released both by the adrenal gland and by the endings of nerves in the muscles. All the amines cause the contraction of the muscle cells of the blood vessels, with the exception of certain vessels such as the coronary arteries of the heart. Also liberated at the nerve endings is acetylcholine, the effect of which is directly opposite that of the amines: it causes muscle cells to relax.

Many workers have suggested that it is norepinephrine and acetylcholine which control the flow of blood through the small vessels. Our own work at the New York University–Bellevue Medical Center leads us to conclude that such an explanation is too simple. The mechanism could not by itself account for the behavior of the small vessels.

In our view the function of the muscle cells of the small vessels is regulated not only by substances that directly cause them to contract and relax, but also by other substances that simply modify the capacity of the cells to react to stimuli and do work. It is known that a wide variety of substances extracted from tissues cause the small vessels to dilate. We postulate that when the metabolism of tissue cells is accelerated, the cells produce substances of this sort. When such substances accumulate in the vicinity of a precapillary sphincter, they depress the capacity of its muscle cells to respond to stimuli. As a result the sphincter relaxes, and blood flows from the thoroughfare channel into the capillary which nourishes the tissue.

The reaction limits itself, because the blood flow increases to the point where it is sufficient to meet the nutritional requirements of the tissue cells. This leads to a gradual disappearance of the substances liberated by accelerated metabolism, and to a gradual lessening of the inhibition of the precapillary sphincter. As the muscle cells regain their tone, the sphincter shuts off the capillary.

The muscle cells of the arterioles and the capillary bed are extraordinarily sensitive to chemical stimuli, so sensitive that they respond to as little as a hundredth of the amount of substance required to constrict or dilate a large blood vessel. This sensitivity is dramatically demonstrated by microsurgical experiments on the capillary bed of a living rat. As little as .000000001 gram (.001

microgram) of epinephrine, injected into the capillary bed by means of a micropipette, is sufficient to close its capillary sphincters completely. Such substances reduce the flow of blood through the capillary bed by an orderly sequence of events: first the precapillary sphincters are narrowed, then the thoroughfare channels, then the arterioles, and finally the venules. Substances that cause the blood vessels to dilate, such as acetylcholine, set in motion a similar sequence: first the precapillary sphincters are opened, then the thoroughfare channels, and so on. The sensitivity of the arterioles and the capillary bed to such stimuli contributes to their independent behavior. An amount of substance sufficient to cause dramatic changes in the micro-

circulation simply has no effect on the larger vessels.

The tone of the muscle cells of the microcirculation may well be maintained by norepinephrine continuously discharged from the nerve endings, and by the level of epinephrine circulating in the blood. Our work indicates that the tone is also influenced by the local release of sulfhydryl compounds, which are key substances in the regulation of the oxidations conducted by cells. Now epinephrine and norepinephrine lose their activity when they are oxidized. Thus the actual level of these substances in the vicinity of muscle cells is not only dependent on their formation but also on their removal or destruction. Sulfhydryl compounds have been found to reduce the rate at which epinephrine and norepinephrine are oxidized. In this way the local release of such compounds could regulate the tone of smooth muscle.

Recently it has been suggested that a role in the local control of the microcirculation is played by the so-called mast cells, large numbers of which adjoin the small blood vessels. Various investigators have shown that the mast cells release at least three substances that strongly affect blood vessels: histamine, serotonin and heparin. It has been proposed that these substances, working alone or in certain combinations, are local regulatory factors.

It must be borne in mind that, even though the control of the microcirculation is largely independent of the rest of the circulatory system, the small blood vessels depend upon the nervous controls of the larger blood vessels for the shifting of blood from one organ to another as it is needed. Obviously the nervous controls of the larger vessels and the chemical controls of the microcirculation must be linked in some fashion. Under normal circumstances tissues that are inactive are perfused with a minimal amount of blood to allow the flow to be diverted to the tissues that need it most. During shock and acute infections, on the other hand, the demands of the tissues may be so great that the circulatory system cannot meet them, and the circulation collapses. In such conditions the effect of substances released locally to relax the muscle cells of the capillary bed has superseded the efforts of the nervous system to restrict the blood flow by the release of substances such as norepinephrine. It is ironic that this primitive response, in striving to insure the survival of individual cells, frequently overtaxes the circulation and brings about the death of the organism.

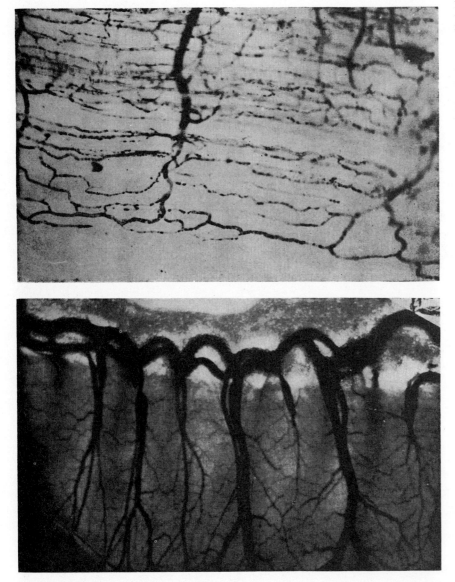

CAPILLARY BEDS OF TWO TISSUES in a living rat are enlarged 200 diameters in these photomicrographs. At top is the capillary bed of a striated muscle; the capillaries run parallel to underlying muscle fibers. At bottom are capillaries in the surface of intestine.

8

HOW SAP MOVES IN TREES

MARTIN H. ZIMMERMANN   March 1963

The topmost leaves of a tree reach for the sunlight dozens and sometimes even hundreds of feet in the air above the deepest-probing rootlets in the ground. From root to leaf an ascending stream of water maintains the living tissues of a tree in the aquatic environment in which plant life originated and in which more primitive plants still flourish. From leaf to root a descending stream of water bears carbon, fixed by photosynthesis in sugar compounds, to build the supporting structure of the tree. Aquatic plants that live entirely submerged do not need a water-conducting system or a self-supporting structure. Practically every part of the plant is capable of photosynthesis, and the cells absorb water and minerals directly from their surroundings. The specialization of tissue to leaf, stem and root came with the conquest of the land. As the competition for sunlight carried the leaves farther and farther from the source of water and minerals in the ground, no evolutionary development was more decisive than that of the two-way water transportation system that appears in its fullest elaboration in the tree.

The mechanism of the ascending stream—which in a tree such as the Douglas fir can lift a prodigious quantity of water more than 200 feet above the ground—was the first to attract the curiosity of investigators. The problem of "the ascent of sap" can now be regarded as solved in its essentials, although many questions remain. The water moves upward, driven by more or less straightforward physical forces, through open conduits formed by cells that have died. What drives the return stream from the leaves is not nearly so well understood, even though the movement is largely downward. Significantly, the cells that convey this stream have not lost their living cytoplasm; physiology as well as physics appears to be involved. For some of the insights into this process that have been gained in recent years, investigators are indebted to the aphid, an insect that feeds on the products of photosynthesis in the stream.

In most trees the water-conducting tissue, called the xylem, and the tissue that transports the products of photosynthesis, called the phloem, appear as distinct systems. Both derive from the cambium, the thin layer of actively growing cells between the bark and the wood. The life cycle of the xylem moves inward into the wood to form the water-conducting conduits and the heartwood of the tree, while the life cycle of the phloem moves outward to form its channels in the innermost layers of the bark. In another large class of plants, which includes the stately palm as well as the grasses and lilies, the two systems can be distinguished only under the microscope, because the strands of xylem and phloem are joined in conducting bundles distributed throughout the cross section of the stem. Movements in the two systems are mostly in opposite directions, but in some cases movement is in the same direction: growing shoots and fruits have to be supplied by both systems simultaneously.

It is a remarkable fact that the xylem cells fulfill their vital function only after their death. By the consequent loss of their cytoplasm and the cytoplasmic membrane that segregates one from the other during growth they form continuous conduits. This, however, is only one of the steps in the transformation that prepares them for their function. Living cells possess a certain rigidity due to the turgor of the fluid contents enclosed in the cell membrane; this turgor is of course lost with the disappearance of the membrane. Before the xylem cells die their cell walls are greatly strengthened with cellulose fibrils and encrusted with lignin. The resulting structure, comparable to reinforced concrete, prevents collapse of the cells when water is pulled through them. Thus the xylem serves a dual function: it not only conducts water but also provides the plant with structural rigidity. It is as though the plumbing system of a house were used as its structural framework as well.

The xylem of the conifers (needle-bearing trees, such as the pine and spruce) shows these trees to be somewhat less advanced on the evolutionary scale than the deciduous, or broadleaf, trees. It is composed of spindle-shaped tracheid cells that overlap one another along their thinner end portions. Water connection between one cell and the cell next in line is provided by small holes called "bordered pits," which allow the passage of water but trap bubbles of air [see illustration on page 74]. The wood of the more highly developed broadleaf trees shows the result of another evolutionary breakthrough—a quite literal one. Butted end to end, the water-conducting cells have partly or completely lost their transverse end walls and so form capillaries up to several meters in length. At the same time special fibrous elements have evolved to give the tissue extra mechanical strength.

In certain trees in the early spring, before the leaves are out, water can be found moving in the xylem under positive pressure. If the xylem of the sugar maple is punctured, for example, the sap will flow from the wood. The maple sirup and the sugar that can be boiled from the sap show that it is rich in reserve products of photosynthesis that

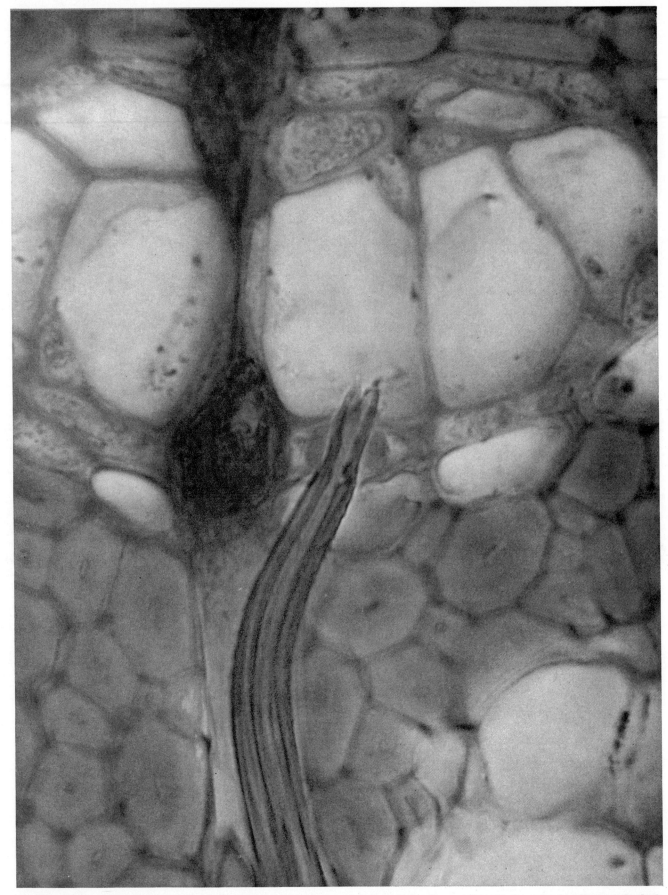

MICROSCOPIC APHID STYLETS (*vertical orange streak at the bottom of this photomicrograph*) penetrate a single cell in the phloem, or sugar-conducting tissue, of a linden tree. High internal pressures force the sap through the stylet bundle into the body of the feeding insect. The direction of normal sap flow is perpendicular to the plane of this section. Magnification is 2,250 diameters.

have been stored during the previous growing season, ready to support the growth of shoots in the spring. The presence of these materials in the sap of the maple and other trees contributes to the positive pressure—a high osmotic pressure with respect to the water in the ground, which forces the water inward and upward through the trunk.

During most of the vegetation period, however, water is pulled up into the trees and the pressure in the xylem is lower than that of the atmosphere. Under the right circumstances, when the xylem is cut, one can even hear the hissing sound of air being drawn into the injured vessels. But this is not the only indication that water is pulled upward in the tree. One of the most beautiful pieces of evidence was developed by the German botanist Bruno Huber in 1935. Huber heated the sap with a small electric element inserted into the xylem and measured the time it took for the ascending wave of warm sap to pass a thermocouple placed a few inches higher on the stem. He found that in the morning water begins to move in the twigs earlier than it does in the stem. In the afternoon, as photosynthetic activity in the leaves begins to lessen, sap movement falls off first in the twigs and only later in the stem. Hence the "motor"

of sap ascent must be in the crown of the tree.

This motor is powered, of course, by sunlight. When the leaves are engaged in photosynthesis, they liberate water vapor to the air by transpiration. In fact, they transpire more than 90 per cent of the water that is delivered to their tissues through the xylem. By this apparently wasteful process sufficient quantities of the dilute soil minerals are carried upward in the water.

What puzzled early investigators is the fact that most trees are more than 33 feet tall; a vacuum pump cannot pull a column of water beyond this height. Some trees reach 10 times higher. How could they pull water into their crowns without the breakage of the water column that causes a vacuum pump to fail? The answer is that water in the xylem of trees is pulled up directly and not by vacuum. Through the cell membranes of the tiny stomata, or pores, on the under surface of the leaves, the water is transpired a molecule at a time; the molecules that escape into the air are replaced by molecules pulled up from below by surface-tension forces. The water columns are continuous, all the way from the rootlets to the submicroscopic capillaries in the leaves.

They do not, therefore, depend on the pressure of the atmosphere for support but are held up by cohesive forces within the water itself and adhesion between the water and the cell walls.

The underlying principle here was first demonstrated experimentally in 1893 by the Austrian botanist Josef Böhm. By evaporating water from a closed system of tubing connected to a bowl of mercury he was able to lift the mercury in a column to heights of more than 100 centimeters, considerably above the height of 76 centimeters to which it can be pulled by a vacuum [see illustration on page 76]. The Irish botanist H. H. Dixon and his collaborator J. Joly repeated this experiment, harnessing the transpiration of a pine twig to lift a column of mercury, and got the same result. These two workers gave the cohesion theory of sap ascent its formal statement in 1895.

The power of the transpiration pump is dramatically demonstrated in an experiment first performed in 1897 by Josef Friedrich at the Forestry Research Institute in Mariabrunn, Austria. Using a sensitive instrument designed to measure the cross-section growth of a tree, he found that the upper portion shrinks in the morning when photosynthesis begins, showing that the loss of water by

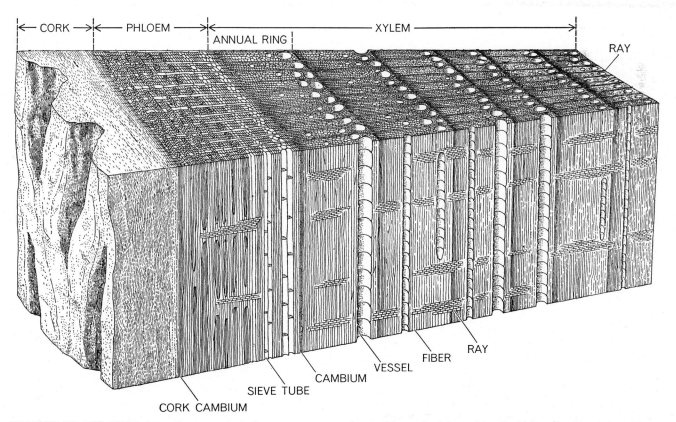

SECTION OF ASH STEM shows the two principal transport systems of a typical deciduous tree. Water ascends from the roots through tubes of dead cells in the xylem. Photosynthetic products descend from the leaves through the living cells of the phloem. Both xylem and phloem cells are produced in the cambium. Only six of this tree's 30 or more annual xylem growth rings are shown.

transportation runs ahead of delivery from below.

Acceptance of the cohesion theory called for more precise determination of the tensile strength of water and the stress to which the water columns are subjected in the xylem conduits. Theoretical calculations, from heats of evaporation and surface tension, indicate a tensile strength for water of several thousand atmospheres. Experimental values, however, are somewhat lower, ranging from 25 to 300 atmospheres. In one illustrative experiment the British investigator H. M. Budgett wrung two polished steel plates together with a film of water between them. To pull the plates apart required tensions of up to 60 kilograms per square centimeter, or 60 times the atmospheric pressure of one kilogram per square centimeter.

As for the stress to which the water columns are subjected in the tree, these can be measured and calculated only indirectly, because any tampering with the integrity of the xylem conduits necessarily breaks the continuity of the water column. The formation of even the tiniest bubble in the water may break the column. It is apparent, however, that a portion of the tension in the column must correspond to the static stress needed to hold the water at a given height. At one atmosphere for every 33 feet above the barometric height of 33 feet, this would come to nine atmospheres in a 330-foot redwood. To this must be added a dynamic component, corresponding to the force necessary to cause the water to flow in the xylem conduits.

The dynamic tension may approach zero during a rainy night, when transpiration has practically ceased, and will reach a maximum value during a hot, dry summer's day, when the velocity of flow is estimated to run as high as 200 feet per hour in some trees. A theoretical value for the maximum dynamic tension can be computed from this estimated maximum velocity and from the measured diameter of the xylem capillaries. Since the theory assumes an ideal smooth-walled capillary of unlimited length, a correction must be made for the resistance set up by irregularities in the actual xylem conduits. Experimentally one can approximate a measurement of conditions in the living tree by forcing water through a freshly cut log of known dimensions and plotting the volume of flow against the pressure required to attain it. In a tree such as the oak, which has rather long xylem conduits, the experimental resistance proves to be twice the theoretical resistance,

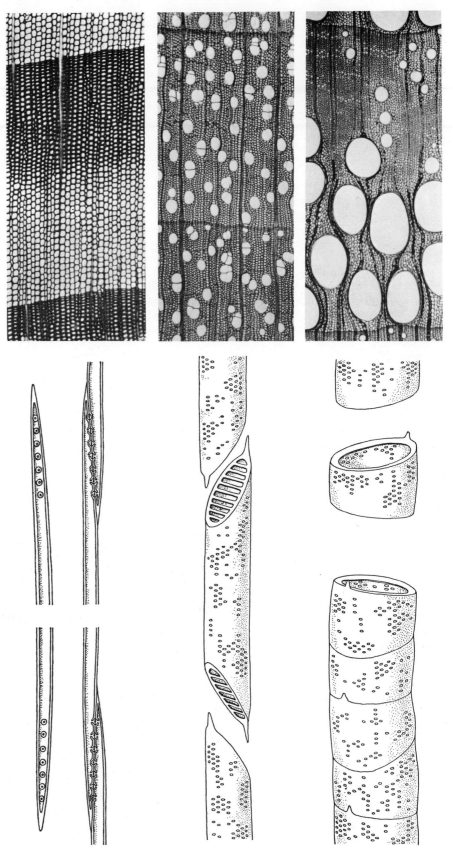

THREE STAGES in the evolution of xylem conduits are represented by the photomicrographic cross sections at top and by the corresponding longitudinal drawings below. In pine (left), a primitive conifer, spindle-shaped cells called tracheids conduct water through small "bordered pits" in their lateral walls. In birch (center), an intermediate, diffuse-porous wood, conduction takes place through partially dissolved end walls. In oak (right), an advanced, large-porous wood, end walls are absent and water passes through a series of squat vessel segments arranged into long, rigid tubes. None of the above cells are alive.

whereas in the birch or maple, with somewhat shorter conduits, it is three times higher. The tortuous tracheid conduits of the conifers, surprisingly enough, offer less resistance—only 1.5 times the theoretical—and the stem of the grapevine, with its very long xylem conduits, shows a resistance even closer to the theoretical value. At its greatest the dynamic component of tension turns out to be smaller than the static. For the tallest trees, therefore, a tensile strength of water of no more than 20 atmospheres is ample to allow smooth operation of the transpiration pump as described by the cohesion theory. This is well within the lowest value found by experiment.

One might still wonder how a mechanism so delicate can function reliably in the high, wind-tossed branches of a tree. The answer undoubtedly lies in the minute subdivision of the chambered structure of the wood. If a column is broken by the formation of a gas bubble, the resulting break remains confined to

that column. The Norwegian physiologist P. F. Scholander, now at the University of California at La Jolla, has pointed out a more serious hazard to trees in cold climates. This is the freezing of the water in the xylem; freezing inevitably causes bubbles to form because air is practically insoluble in ice. Scholander has shown that trees do freeze and that bubbles do form in all vessels and tracheids. A re-examination of the anatomy of trees prompted by this observation has shown that the cold-climate species are variously accommodated to survive.

In the conifers, for example, the bubbles in the thawing ice are trapped within individual tracheids. As the weather warms, the gas in the bubbles undoubtedly redissolves in the water. In the birches, maples and grapevines the positive pressures that make the sap flow in the spring also force the gas back into solution in the xylem water. The oaks, ashes, elms and other trees with vessels of large diameter simply lose the

past season's water-conducting system by rupture of the water columns; they replace them by forming a new growth ring in the early spring before the leaves come out. During the summer the transpiration stream flows almost entirely in this new growth ring. It is apparent that winter freezing of the xylem plays an important role in the geographical distribution of trees.

Sooner or later, of course, the tracheids and vessels irreversibly lose their conductive capacity and go out of function. Air embolism is the first step in the gradual formation of the heartwood, which ends with the deposition of pigmented excretion substances in the adjacent cells. The center of the trunk is the "dump" of the metabolic processes of the tree.

As these observations suggest, all the essential parts of a tree are renewed at the beginning of each growing season. Buds open, new shoots appear, leaves

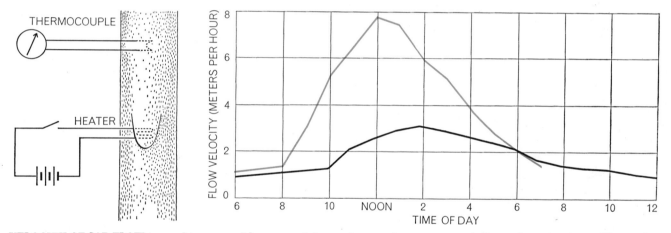

VELOCITY OF SAP FLOW in wood is measured by means of the apparatus at left. A small heating element inserted into the xylem heats the ascending sap for a few seconds. A thermocouple farther up the stem records the passing wave of heat. The time interval between these two events indicates the sap's velocity. The graph at right shows that in the morning sap begins to flow first in the twigs (*colored curve*) and later in the trunk (*black curve*). In the evening sap flow diminishes in the twigs sooner than it does in the trunk.

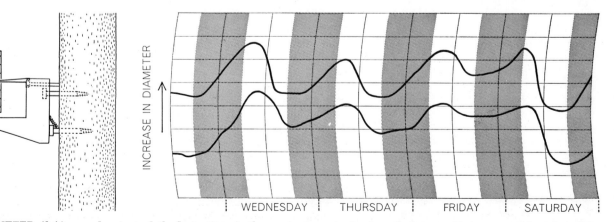

DENDROMETER (*left*) records minute daily fluctuations in the diameter growth of a tree trunk. Simultaneous measurements made at two different elevations (*right*) indicate that morning shrinkage of the upper trunk slightly precedes that of the lower. Early-morning transpiration from the leaves pulls water out of the xylem of the upper trunk before it can be replenished from the roots. As transpiration lessens later in the day, expansion of the upper trunk again precedes that of the lower. The shaded strips signify nighttime.

unfold and the cambium generates new xylem and phloem over the entire surface of the tree, from the twigs down the stem into the roots. These internal changes occur in the evergreens just as they do in the trees that drop their leaves. Even in a 2,000-year-old redwood most of the vital functions are carried out in tissue that is only a few weeks to a few years old. The rhythmic renewal of conductive tissue displayed in the growth rings of Temperate Zone trees makes a record not only of the age of the tree but also of conditions prevailing from season to season.

Growth rings appear in the bark as well. Unlike the wood, however, the bark keeps an impermanent, short-term record. The growth of new phloem tissue taking place inside the cylinder of bark around the tree constantly disturbs and breaks up the outer tissues. Three different regions can be discerned in the cross section of the bark. The most recently produced phloem layer, lying immediately outside the cambium, is the conducting phloem. Outside are the older rings of phloem that have lost their capacity for long-distance transport and serve for a while as storage places for the products of photosynthesis. Farther toward the outer surface of the bark one can see the cork cambium, a layer of rapidly dividing cells that produces the dermal tissue of the stem. All the tissues outside the cork cambium are dead, and in this region the growth rings are disturbed and indistinct.

As in the case of the xylem of various species, the conducting elements in the phloem can be classified into primitive and more advanced types. Conifers show individual sieve cells, similar in shape to the tracheids of their xylem. These cells make intimate contact in "sieve areas" but do not open mechanically into one another. In the hardwoods the sieve cells are lined up in a continuous series, forming sieve tubes comparable to the conduits in their xylem. The abutting end walls of these cells form sieve plates, in the pores of which, as electron micrographs have recently shown, the cell wall has disappeared. In all species the phloem cells retain their cytoplasm, evidently in modified form. The most striking change in the maturation of a sieve element is the loss of the cell nucleus.

The mechanism of phloem transport is, if anything, less accessible to direct study than that of the xylem. It is evident that the fluid in this system, bearing a rich concentrate of the products of photosynthesis, moves under positive

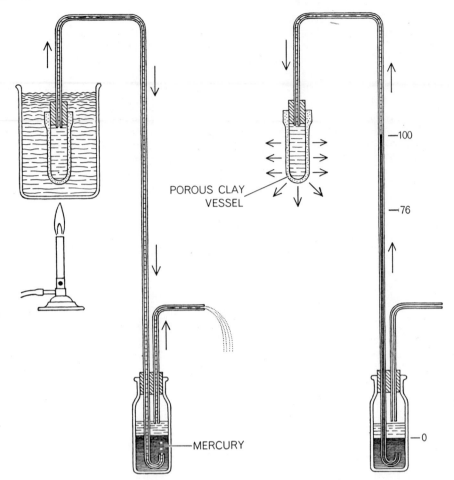

POROUS CLAY VESSEL

—100

—76

MERCURY

—0

TENSILE STRENGTH OF WATER was first demonstrated by the Austrian botanist Josef Böhm in 1893 with an apparatus similar to the one shown here. A porous clay vessel is immersed in a beaker of boiling water (*left*), forcing a continuous stream of water through a glass capillary into a bottle containing a layer of mercury and out through an exhaust tube. When the beaker is removed (*right*), evaporation through the walls of the vessel causes the water to flow up the capillary, pulling the mercury behind it to a height of more than 100 centimeters. The appearance of an air bubble anywhere in the system will cause the mercury column to fall back to a normal barometric height of 76 centimeters. Böhm's device was a simplified mechanical analogue of the xylem transport system of a tree.

pressure. But the pressure and the velocity of flow have been calculated from indirect evidence only. In many species of trees the sieve tubes will produce intense exudations on being punctured. In other species the exudation is sucked into the closely adjacent xylem system by the negative pressure that drives the transpiration stream. The phloem of all species is highly reactive to injury; the sieve tubes quickly interrupt their continuity and shut off the flow of fluid. Certain components of the cytoplasm may instantly plug the sieve plates when pressure is released, and callose, a sugar of high molecular weight, closes the sieve pores with a more permanent seal. It is callose formation that brings phloem transport to a halt in the fall after the leaves have dropped. The same process will plug up the entire phloem in a branch that has been cut from a tree.

Parasites have found a rich source of food in the phloem of trees and lesser plants. Some plants, like the mistletoe, live on the phloem of others and some have lost their ability to carry on photosynthesis. Porcupines, beavers and bark beetles eat the whole bark. Of greatest interest to the plant physiologist are the aphids that are specialized as feeders on the sieve tubes. Small green species live on leaves, and larger brownish or gray ones on the phloem of branches. Their mouth parts consist of a bundle of stylets with which they penetrate the bark to the phloem, where they tap a single sieve cell or tube [see illustration on page 72]. The food they obtain in this way is so ample they exude a surplus in the form of "honeydew," which is collected in turn by ants and honeybees.

Entomologists long ago noticed that the stylet bundles may continue to exude fluid from the plant after the aphid has been severed from its mouth parts. Recently Tom E. Mittler, now at the University of California at Berkeley, developed this observation into a technique for the study of phloem transport. The aphid is immobilized with a gentle stream of carbon dioxide and its body is then cut away with a sharp knife, leaving the stylets in place. If the operation is successful, exudation continues, often for days, and the exudate can be collected with a micropipette. As the success of the technique indicates, it works with so little injury to the plant that it does not provoke the defense mechanisms of the phloem. The purest samples of the phloem stream can be obtained in this way and even some indication of rates of flow.

Analysis of the exudates from 250 species of trees shows that a group of compound sugars is carried in the phloem stream in concentrations of 10 to 30 per cent by volume. In addition to sugars the cargo of the phloem stream is made up of sugar alcohols, amino acids, phosphorus compounds and inorganic ions. Some of these substances, particularly the nitrogen- and phosphorus-containing compounds, increase in concentration during the fall. These materials are salvaged from the leaves, which are soon to be lost.

The question of what forces bring about the long-distance transport of materials in the phloem is still far from settled. It is evident that simple diffusion—the universal tendency of solutes to come to equal concentration everywhere in a solution—is inadequate. Yet there is still argument over whether or not phloem transport actually takes place as the mass flow of a solution. The strongest evidence for mass flow is the exudation from phloem tubes, particularly as observed with the help of aphid stylets. The rate of exudation from the stylets is remarkable, in certain cases exceeding five cubic millimeters per hour. Such a rate of flow requires refilling of the sieve element three to 10 times per second. From this and other evidence it is estimated that translocation velocities in the phloem are of the order of 100 centimeters per hour. Some mass-flow process is necessary to explain movement of the fluid at this speed.

Assuming that phloem transport does

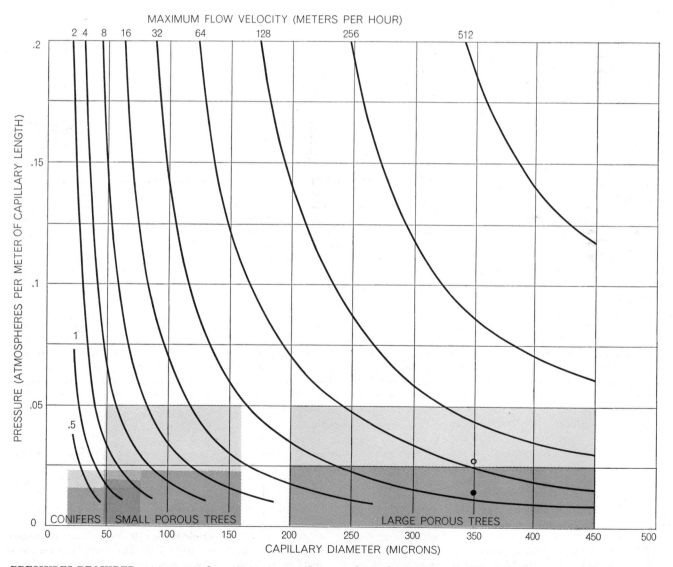

PRESSURES REQUIRED to overcome the resistance to sap flow in xylem capillaries are calculated with the aid of this graph. The dark-colored strips below represent the range of pressures calculated for the three principal types of xylem conduits. The light-colored strips above each of these are the corresponding pressures obtained experimentally. The difference is caused by the fact that most woods do not contain ideal capillaries. A calculated value for a single oak is indicated by the black dot, an experimental value for the same tree by the open circle. In spite of their small diameters, conifer tracheids are surprisingly efficient water conductors.

proceed by mass flow, the next step is to explain what drives it. The German forest botanist Ernst Münch proposed that differences in osmotic pressure are responsible for the activation and maintenance of the flow. The crucial points in the pressure gradient are located at opposite ends of the tree, in the leaves and in the roots. High osmotic pressure, created by the high concentration of the products of photosynthesis in the leaf, draws water from the capillary end vessels of the xylem into the capillaries of the phloem. Low osmotic pressure in the phloem of the roots, caused by the withdrawal of the products of photosynthesis on the way down, forces the water to flow from the phloem into the negative-pressure system of the xylem or out into the surrounding soil. In effect, it is the initial concentration gradient in the leaves, maintained by photosynthesis, that creates the pressure gradient and causes the mass flow of the phloem solution.

There is a good deal more to the system, however, than this sketchy statement of the hypothesis suggests. To make such a system function, the side-wall membranes of the sieve tubes must be differentially permeable to the passage of the molecules. At the same time the passage from one sieve element to the next must not be hindered by a semipermeable membrane. Electron micrographs of the fine structure of the sieve plates are playing a decisive role in the elucidation of this problem. The main difficulty is presented by the sensitivity of the sieve tubes to injury.

More recently it has been suggested that the electric potentials across the sieve plates (or, in the conifers, sieve areas) may supply the driving force. Osmotic pressure gradients would then merely supply the trigger for electro-osmosis.

The mechanism of phloem transport is as unresolved in grasses and lilies as it is in trees. Because of their size trees supply ideal subjects for experimental studies of this phenomenon. They offer clear lengths of 30 feet or more of uniform conducting tissue, with all the products of photosynthesis concentrated at one end along with the peak of the osmotic pressure thereby generated. For simultaneous, or nearly simultaneous, measurement of pressures and flows at many points on the tree, the aphid and its stylets do not provide a manageable technique. It is therefore necessary to resort to the cruder method of making incisions in the bark. The larger the tree, however, the less disturbing is the effect of these traumas. In a tree with a diameter of eight inches or more one may

APHID (*Longistigma caryae*) feeds on the underside of a linden branch. The stylet sheath, from which the stylets are projected into the bark, is clearly visible in this photograph. Surplus sugar is released in the form of a "honeydew" droplet about once every half hour.

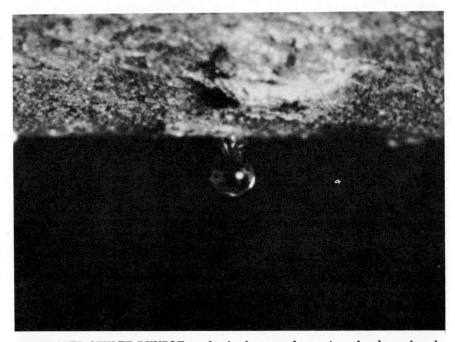

AMPUTATED STYLET BUNDLE exudes for hours, and sometimes for days, after the aphid has been cut away. The high rate of exudation from amputated aphid stylets supports the mass-flow theory of phloem transport. Stylet exudate is the purest phloem sap obtainable.

make as many as 100 useful incisions, if they are properly placed. The concentration gradients of the materials transported in the phloem can thus be measured throughout a great length of tissue before and after such experimental treatments as defoliation, interruption of phloem transport and locally applied temperature.

In the Harvard Forest at Petersham, Mass., we have carried on work of this kind over several growing seasons. We invariably find that the concentration of photosynthetic products decreases down the length of the tree during the summer, when the leaves are exporting this material and growth is taking place in the stem and roots. This gradient disappears soon after defoliation, be it natural leaf fall in autumn or artificial defoliation done at any time in the summer. Analysis of the fluid tapped at intervals down the trunk soon after defoliation shows that the sugars are being converted and removed from the sieve tubes. The measured decline in concentration of the sugars along the trunk provides an index of the velocity of phloem transport. In agreement with findings made by other techniques, this shows values of 50 to 100 centimeters per hour.

It is clear that the two-way water transportation system of trees presents questions that can be approached only in the living organism as a whole. Some of the most interesting questions lie in the cross transfer of substances from the phloem to the xylem, by which trees and other plants distribute vital substances throughout their entire bodies.

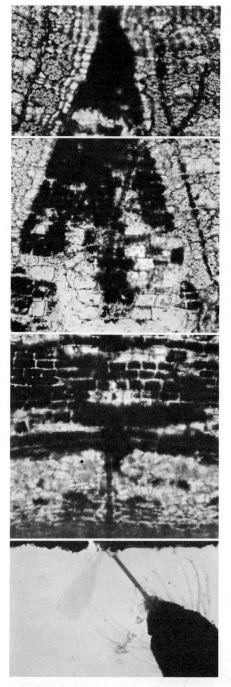

STYLET PATH extends from sheath (*bottom*) up through several layers of outer bark to an individual sieve element in the conducting phloem (*see enlarged section on page 72*). The four transverse sections that make up this composite micrograph were made by Gerda Aerni of Harvard University.

PORES OF SIEVE PLATE in this photomicrograph have been partially plugged with callose following an artificial reduction in pressure at one end of the sieve tube. Ordinarily callose formation interrupts the flow of sap through the phloem of deciduous trees in late fall after the leaves have dropped. Magnification is 1,350 diameters.

Part III

METABOLIC
REGULATION

III

Metabolic Regulation

INTRODUCTION

The systems of exchange and transport described in Section II are adjusted to the needs of the organisms that possess them, and each system has a rate of functioning which, given a particular set of conditions, is optimum for its owner. Since that optimum rate will depend upon a variety of conditions—such as the availability of oxygen or foodstuffs, the outside temperature, or the level of internal metabolic activity—the organism must exert a fine control over the rates of these processes. Most of the rates are controlled by means of a negative feedback system; that is, the rate of a process is controlled in such a way that its output or result is held constant. This necessitates a system with a sensing device that registers a certain constant fraction of the output and returns it as a negative signal to the input, thus reducing the rate. In the feedback circuit of the familiar home thermostat, a thermometer is the sensor to record temperature. If the thermometer registers too high a value, the thermostat actuates a relay to turn off the furnace; this step represents the negative feedback. In this way the temperature hovers about the value that was selected by the householder when he adjusted the thermostat. The value is called the "set point" of the control system. If a window is left open and it is cold outside, the furnace will have to remain on most of the time to hold the set point; conversely, a warm day will keep the house temperature above the set point, and the control system will be inoperative.

From this description we may draw several points of comparison with biological control systems. The optimum, or set point, which a given organism reaches in the course of evolution allows a number of biological systems to narrow their properties in order to function at high efficiency under these specified conditions. Thus the body temperature of mammals and the osmotic pressure of their body fluids are held at values to which the properties of enzyme molecules and of other internal chemical systems conform nicely. This circumstance suggests an economy in arrangement whereby overall control protects and thus permits the specialization of subsidiary elements. Though the control may often be quite tight, some organisms can vary the set point from time to time. Finally, most organisms would be in a bad way if they had to depend, like the house in our example, upon a control system that can respond to changes in only one direction. Almost all of the internal regulations performed by living organisms can compensate for positive and negative variations; these regulations would be more properly analogous to the control system of a house in which the thermometer was connected both to a furnace and to an air-conditioning system, so that responses to heating as well as to cooling would be possible. In this way an organism can maintain the constancy of a certain internal variable in the face of any change in the external environment. Physiologists have been aware of this principle of control for a long time. It was first recognized by the Frenchman Claude Bernard, and later by the American Walter Cannon, who termed this sort of regulation *homeostasis.*

Just as a stranger to the world of thermostats would choose to learn about them by studying the behavior of houses under extremes of temperature, biologists have learned about physiological regulation by analyzing animals whose control systems are under unique stresses from the environment. There are two approaches to this kind of analysis. The first is the experimental approach, in which one may work with a control system that normally contends with a certain range of external variation and may then subject it to a much wider variation than it is accustomed to. Frequently, this method can yield information about the nature of the signals from the thermometer to the relay, or about the identity of the thermometer itself. Secondly, one may examine certain control systems that are designed to operate at unusual extremes of external variation. Examples of such control systems might include the temperature regulation machinery of an arctic animal, the water-balance control of a camel, the salt-regulation mechanism of a whale that eats shrimps, or the respiratory system of a seal. This is the comparative approach: by studying one of these special cases, one seeks to comprehend both the evolutionary die that molded the device and the general purposes of the device itself.

The first selection, "The Human Thermostat" by T. H. Benzinger, is an example of the experimental approach successfully applied to an old problem. The internal temperature of a man, or of any adult bird or mammal, is normally controlled within a very narrow range. The outputs of this control system are easy to detect. When it gets too cold, mammals shiver, thus warming themselves by accelerating their metabolic rate of heat production. When it becomes too warm, they sweat, cooling themselves by evaporation, and dilate their peripheral blood vessels to dissipate their internal heat more efficiently. Until recently, the identity of the sensing device that initiates these reflexes at the appropriate time was uncertain. Benzinger describes the careful temperature measurements that revealed its location in the hypothalamus of the brain. The device for temperature regulation, moreover, is a useful illustration of the general features of those biological control sequences that operate within narrow limits.

The special problems that develop in animals existing in an environment unusual for their general kind are illustrated by the diving mammals discussed by P. F. Scholander in "The Master Switch of Life." Like other animals subjected to sudden oxygen deprivation, they perform a variety of responses which together conserve the available oxygen and restrict it to those tissues needing it the most. The countermeasures include a variety of adjustments: slowing of the heart rate (*bradycardia*), peripheral constriction of blood vessels, and reductions in metabolic rate. These adjustments are triggered by a variety of stimuli. In man as well as in seals, they can result from merely submerging the face. Oddly enough, the same kinds of responses occur in fish that have been brought from their aquatic home into air—a change as distressing to their oxygen supply system as total immersion is to a man's.

A different account of managing life under harsh circumstances is related by Knut and Bodil Schmidt-Nielsen in "The Desert Rat." For this animal, the problem is not the conservation of oxygen but of water, a commodity difficult enough to obtain in any terrestrial environment but especially scarce in the great deserts of the world, where standing water is nonexistent, temperature is high, and humidity is low. Despite these handicaps, the kangaroo rat is able to maintain the osmotic pressure of its blood, as well as its total body-water content, at approximately the same levels as those of its relatives that live in less in less challenging surroundings. One of the most important countermeasures is behavioral: the kangaroo rats stay in burrows during the hot part of the day. This avoids the problem resulting from the unfortunate but necessary compromise between the systems of temperature and fluid regulation, which makes it impossible for mammals to indulge in evaporative cooling by panting or sweating without losing precious water. Also, the kangaroo rat's kidney can excrete a urine having twice as high a salt content as sea water. Thus conservation measures overcome the scarcity of water, and the kangaroo rat can balance its fluid budget with the water derived from metabolic oxidation.

Though it is not discussed in "The Desert Rat," Schmidt-Nielsen and others have inspected similar problems of water conservation in animals occupying dry environments. An almost infinite variety of solutions exist. The camel, for example, simply endures the punishment that the kangaroo rat avoids by adroit management. Because the camel's kidney is not exceptional in any way, the rate of water loss through that organ is significant; moreover, contrary to popular legend, these animals cannot store water. They do avoid evaporative loss, but again more by tolerance than by special mechanism: camels will not begin to sweat or to pant until their body temperature rises to a level that would be fatal to a man. Finally, camels tolerate substantial losses in body water, ranging up to 30% of the body weight—a loss which they replace in one dramatic drinking bout at the next oasis.

Other vertebrate animals have difficulty with their water balance not because they live in deserts but because they live in the sea, an environment which, because the high salt concentration tends to withdraw water osmotically from the animals' bodies, is appropriately described as physiologically dry. There is clearly no water shortage, because the water can be drunk; but the excess salts must be discharged somehow, and the kidneys of most marine vertebrate inhabitants of the ocean are not able to excrete a urine in which the salts are more concentrated than in sea water. In recent years, it has been shown that most of these animals have special salt-secreting organs outside the kidney that enable them to eliminate the excess salt. In some of the birds of the open ocean, such as petrels and albatrosses, the salt-secreting glands are located on top of the bill near its base. Marine turtles discharge the excess salt from a set of glands opening near the corner of the eye. In most fish, the gills themselves serve the purpose of extrarenal salt secretion.

In the fourth article, "Hormones and Genes" by Eric H. Davidson, the problem of metabolic regulation is approached from different aspect. Many if not most of the control systems regulating physiological homeostasis rely upon hormones. For example, the retention of water in the kidney to maintain a constant osmotic pressure in the blood is promoted by a hormone, vasopressin, from the pituitary gland; blood glucose is regulated by insulin. Surely these hormones, and the control processes they mediate, must work at the molecular level in influencing the activity of their target cells. A variety of recent efforts have been made to relate these regulatory phenomena to the molecular level. Although not many conclusions have yet been reached, the advances reported in this article suggest that before long we will know something of the way in which hormones influence the activities of the genome.

9

THE HUMAN THERMOSTAT

T. H. BENZINGER   January 1961

Fever is usually the first symptom to arouse concern in illness. The rise in body temperature is not great in the absolute sense. On the contrary, the attention it attracts is a measure of the constancy with which the body temperature is normally maintained. Compared to the daily and seasonal variation in the temperature of "cold-blooded" animals, whose internal temperature depends upon that of the environment, a fever represents a tiny variation in temperature. Yet it is many times greater than the normal variation in the regulated temperature of the healthy body. In spite of large differences in environmental temperature—from the arctic tundra and windswept highlands to fiery deserts and steaming jungles, from season to season and from day to night—the body temperature departs little from the norm of 37 degrees centigrade (98.6 degrees Fahrenheit). Life in the cells continues undisturbed, although the metabolic processes are irrevocably linked to temperature by the laws of thermodynamics and the kinetics of chemical reactions.

Heat is a by-product of these processes. With the body at rest, the heat of basal metabolism easily supplies the necessary interior warmth when external conditions are comfortably cool. Only under extreme conditions does the system fail; as when, in a hot environment, physical effort fans the flame of the metabolic furnace beyond control by the regulatory system; or when, in a very cold environment, the loss of heat by radiation, conduction and convection overbalances the metabolic production of heat and reduces body temperature to a fatal degree. Man of course shares this vital capacity with other mammals and with birds. Favored in consequence with nervous systems maintained at op-

timal working temperatures under all environmental conditions, the "warm-blooded" animals have become masters of the living world on our planet.

The question of how the body keeps its temperature constant within such narrow limits has engaged the efforts of an astonishing number of investigators. It is only recently, however, that one of the two parts of the regulating mechanism—defense against overheating—has been clarified. Max Rubner of Berlin had recognized in 1900 that sweating and the dilation of the peripheral blood vessels constitute the effector mechanisms for the dissipation of excess heat from the body. E. Aronsohn and J. Sachs, two medical students at the University of Berlin, came upon the center of control in the brain as long ago as 1884, when they damaged in animals an area "adjacent to the corpus striatum toward the midline." A few months earlier Charles Richet of Paris had also produced excessive body temperature by puncturing the forebrain. It now seems certain that in both cases the investigators damaged the hypothalamus, an area at the base of the brain stem just above the crossing of the optic nerves.

But how does the body sense and measure its temperature and bring the control center into action? The investigation of this question was confused for a long time by the conspicuous part that the temperature-sensitive nerve endings in the skin play in the feeling of warmth and cold. In recent years, however, a new approach to the problem has been made possible by the development of a new principle of measurement called gradient calorimetry and of the instrumentation to go with it. Experiments employing this instrumentation have now located the sensory end-organ at which the body "takes" its own tempera-

ture when it becomes too warm. The discovery is an unusual one at this late date in the history of physiology. The body's "thermostat" must now be included in the short list of major sensory organs adapted to the primary reception and measurement of physical or chemical quantities. Moreover, it now becomes possible to measure the characteristic responses of the thermostat and perhaps to produce or to suppress those responses artificially. Such investigation will lead to a better understanding not only of the aberration of fever but also of the precise regulation of internal temperature that is so important to the vital function of the body, particularly to the function of the delicate nervous system. With the thermostat identified, it has also become possible to explain

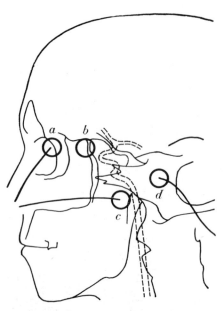

TEMPERATURE AT THERMOSTAT in the brain is measured by thermocouples

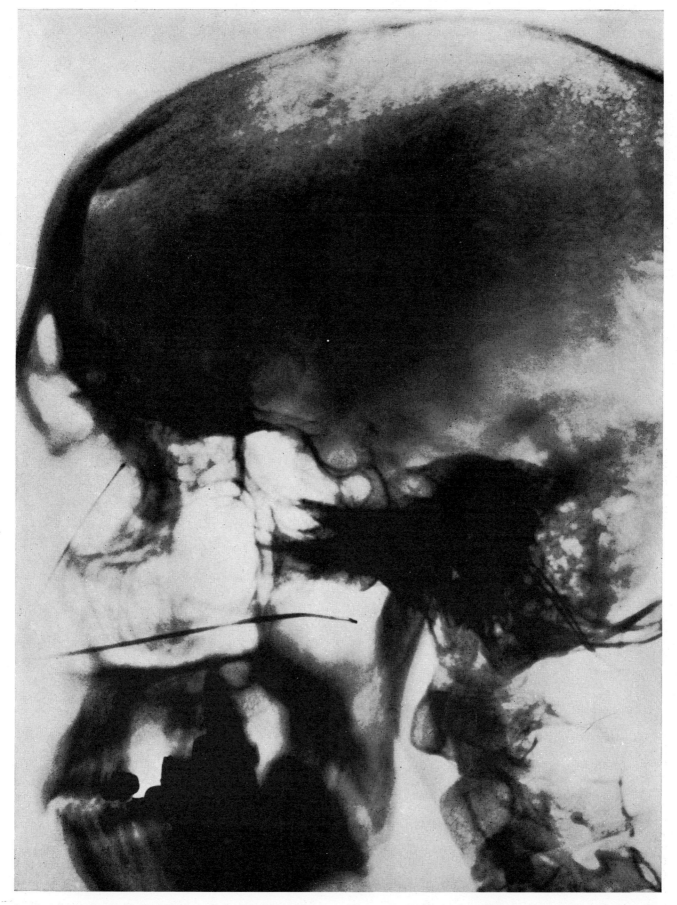

placed at forward wall of ethmoid sinus ("*a*" *in diagram at left*), deep in rear wall of nasopharyngeal cavity (*c*) and at eardrum (*d*). Thermostat itself is in hypothalamus behind sphenoid sinus (*b*), at which temperature has also been measured by thermocouple.

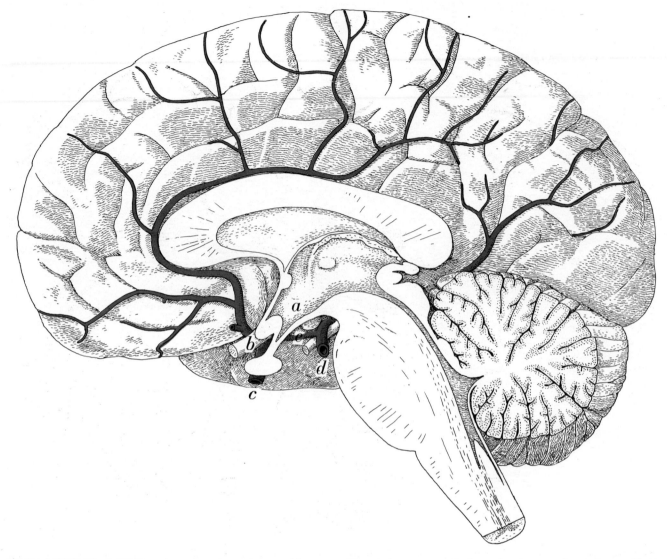

HEAT CONTROL CENTER is located in forward part of hypo-thalamus (*a*), shown in cross section (*left*) and from below (*right*).

Hypothalamus, centrally located under great hemispheres of brain, rests on the Circle of Willis (*e*), an arterial ring through which

the effects of such mundane factors as a hot meal, a cold drink, a hot bath and a cold shower.

When one encounters a physical or chemical quantity in technology or in a living organism that is maintained at a constant level against disturbances from outside, one looks for a "servo-mechanism." Pressure, rate of flow, chemical composition or temperature are automatically controlled by such mechanisms in the realm of engineering. The servomechanisms of the body control the same kinds of variable. In man-made devices the chain of control begins with a "sensory" instrument, perhaps a thermometer, which measures the variable in question. The measurement is relayed to a "controller" which compares it with a set point to which the variable is to be held. Whenever the need arises,

the controller sends instruction to an "effector" mechanism, perhaps to the heating system, which brings the temperature into accord with the set point in the controller [see "Feedback," by Arnold Tustin; SCIENTIFIC AMERICAN, September, 1952].

The corresponding elements in biological servomechanisms and the nervous and chemical pathways that interconnect the sites of stimulus and response constitute systems of far greater complexity. They are nonetheless put together in a similar way. A servomechanism in the human body may be considered to be clarified when the sensory organ, the controller and the effector mechanism are known, and when a reproducible, inseparable and quantitative relation has been established between the magnitudes of the stimuli and of the responses they induce. The net

effect of the response must be the restoration of "homeostatic" equilibrium; that is, the variable in question must return to the optimum, stable level essential to the life of the cells.

In the control circuit that prevents overheating of the body, classical physiology had identified the effector and the controller mechanism, but not the sensory organ. In a hot environment, as Rubner showed, the effector is the dilation of the blood vessels in the skin, which increases the transport of heat from the interior of the body to the surface; and sweating, which increases the rate of total heat loss from the surface to the environment as energy is absorbed in evaporation. In a cold environment the corresponding mechanism is increased metabolic heat-production.

The location of the controller in the hypothalamus by Aronsohn and Sachs

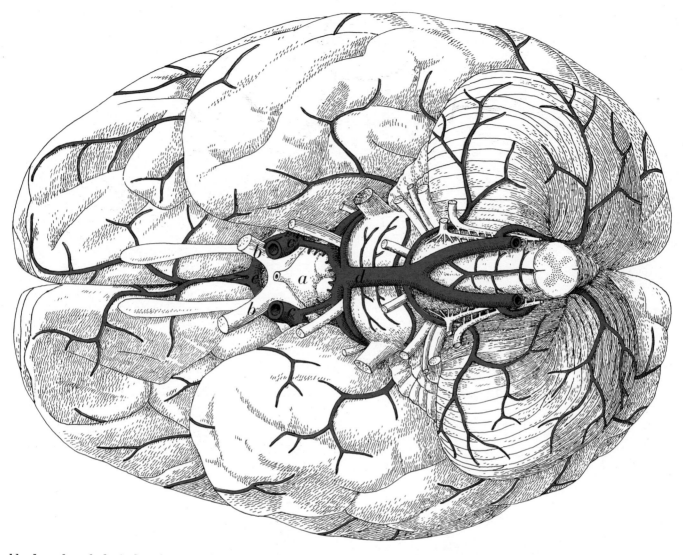

blood supply to the brain flows from carotid arteries (*c and c'*) and basilar artery (*d*). Hypothalamus, optic nerves (*b and b'*) and ret- **ina derive from same tissue matrix. Bulb of pituitary gland attached to hypothalamus appears at left, but is cut away at right.**

was confirmed by later investigators. Some of them applied the stimulus of temperature directly to the site. In 1904 Richard Hans Kahn of the German University in Prague found that heating the head arteries of a dog lowered its body temperature. In 1912 Henry Gray Barbour, a young American physician working in Vienna, carrying out an experiment designed by the pharmacologist H. Meyer, applied warm and cold probes to the general area of the hypothalamus. He observed the expected thermoregulatory responses. In 1938 Horace W. Magoun, now at the University of California at Los Angeles, discovered that this function is mediated by a circumscribed area in the forward part of the hypothalamus. Bengt Andersson of the Royal Swedish Veterinary Institute in 1956 delineated the organ with unprecedented precision in goats. In 1950 Curt

von Euler of the Nobel Neurophysiological Institute in Stockholm even succeeded in recording, in parallel with temperature changes, slow electrical "action potentials" from this area of the hypothalamus of cats.

But the body is also equipped with an elaborate system of millions of tiny sensitive nerve endings, distributed throughout the skin, which produce conscious sensations of warmth. The scientific literature tended to support the view that the skin and not the hypothalamus furnishes the primary temperature measurements to the control center for sweating and the dilation of the arteries. Some investigators held that both systems were involved; a rise in the temperature of the "heat center" in the hypothalamus supposedly made it more responsive to incoming impulses from the temperature-sensing organs of the skin. The

question, in this view, was one of determining the relative importance of the two sites. It was also possible, as some believed, that the body possessed a third area sensitive to temperature or heat flow, and that neither the skin nor the brain was involved.

It was not easy to design a conclusive experiment. In experimental animals one might destroy the nervous pathways from the thermoreceptors in the skin to the heat center. But the results of such an experiment would not exclude the possibility that the temperature of these centers played a role in heat regulation under normal conditions. It would still be necessary to carry out the reverse experiment and destroy the heat-sensitive part of the hypothalamus. Since this structure is intimately involved with the temperature-control center itself, it seems impossible to secure the final evidence by surgical procedures alone.

To observe the operation of sensory receptors in the skin and in the brain independently of each other in the intact organism presented comparable difficulties. No one, apparently, had succeeded in keeping one of the two sites at a constant temperature while observing the effects brought about by a temperature change in the other. This approach called for techniques to measure temperature in the human body at the two sites of presumed temperature-reception—the skin and the hypothalamus—and some way to record, rapidly and continuously, the effector responses of vasodilation and sweating.

The gradient calorimeter has satisfied the second of these two requirements. This rapidly responding and continuously recording successor to the classical calorimeter makes it possible to record for the first time the total output of the effector mechanisms. It measures separately the heat that is carried from the body by radiation and convection and the heat that is dissipated by the evaporation of sweat. From working models made and tested by Charlotte Kitzinger at the Naval Medical Research Institute in Bethesda, Md., the first full-scale human gradient calorimeter was construct-

ed under the direction of Richard G. Huebscher at the laboratory of the American Society of Heating and Ventilating Engineers in Cleveland. Similar units have now been constructed at other laboratories. The gradient calorimeter now operated at Bethesda is a chamber large enough to hold a man stretched out at full length [*see illustration below*] The subject is suspended in an open-weave sling, out of contact with the floor or walls of the chamber, and is free to go through the motions of prescribed exercise when the experiment calls for such exertion.

The new and essential feature of gradient calorimetry is the "gradient layer," a thin foil of material with a uniform resistance to heat flow which lines the entire inner surface of the chamber. Some thousands of thermoelectric junctions interlace the foil in a regular pattern and measure the local difference in temperature (and hence the local heat flow) at as many points across the foil. The junctions are wired in series; their readings are thus recorded in a single potential at the terminals of the circuit. That potential measures the total energetic output from the subject's skin, independent of his position with respect to the surfaces of the gradient layer lining the chamber. The rate of blood flow through the skin

can be derived by computing this measurement against the temperature of the outgoing blood (measured internally) and the temperature of the returning blood (measured on the skin), since the observed transfer of heat per unit time at any given difference between internal and external temperature can be effected by only one calculable rate of blood flow. The energy dissipated by evaporation from the subject's skin is also measured by gradient layers which line heat-exchange meters at the inlet and outlet of the air circuit of the calorimeter. Measurements taken for control make it possible to maintain the same temperature and humidity in the air at these two points, so that the air neither gains nor loses energy as it passes through the system. The unbalanced output from the additional gradient layers thus precisely measures the heat loss by evaporation and hence the sweat-gland activity. Heat loss through the lungs is measured separately and subtracted from the total.

With the help of the gradient calorimeter our group at Bethesda set out to establish the correlations obtaining, on the one hand, between the performance of the effector mechanisms and the temperature of the skin and, on the other hand, between the performance of the effector mechanisms and the internal temperature of the body. In these first experiments it was assumed that rectal temperature provided an adequate index of internal temperature as measured at the internal temperature-sensing organ, wherever that might be located. But no correlation could be found, in either resting or "working" subjects, between rectal temperature and the observed rates of sweating. Measurement of skin temperature against the same heat-dissipation variable yielded equally meaningless plots. For a time it seemed that all the effort that had gone into the design of the gradient calorimeter had been wasted. The results made sense only in terms of the classical notion that the thermostat in the interior of the body and the temperature-sensing nerve endings in the skin have indissolubly interlaced effects upon the vasodilation and sweating responses.

Then we found a way to measure the internal temperature of the body at a site near the center of temperature regulation in the brain. We introduced a thermocouple through the outer ear canal and held it against the eardrum membrane under slight pressure. The eardrum is near the hypothalamus and shares a common blood supply with it from the internal carotid artery. At the

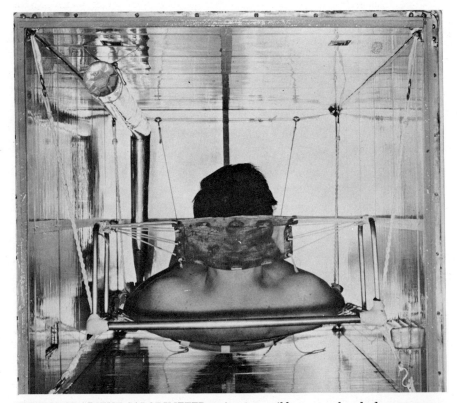

HUMAN GRADIENT CALORIMETER makes it possible to correlate body temperature with dissipation of heat by radiation and convection from skin and by evaporation of sweat. Lining of chamber is interlaced with thermoelectric junctions which measure heat loss from skin; loss by sweating is measured by temperature and humidity control system of calorimeter.

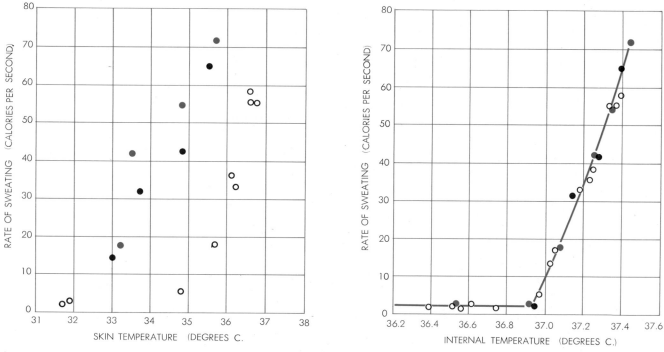

HEAT DISSIPATION by sweating plotted against skin temperature yields senseless graph (*left*) when natural correlation between skin and internal temperature is broken by internal heating through exercise. When the same measurements of sweating are plotted instead against internal head temperature (*right*), an inseparable and always reproducible relation appears between the stimulus of temperature and the response of the sweat glands, whether the subject is sweating or not.

very first attempt we observed temperature changes associated with the eating of ice or the drinking of hot fluids, and we soon found we could detect variations caused by immersion of the limbs in warm water. Parallel rectal measurement did not show these variations at all. To make sure that the entire region of the head supplied by the carotid arteries can be expected to show the same temperature variations as the eardrum, we tried other sites. With the help of local anesthesia H. W. Taylor, a surgeon at the Naval Hospital in Bethesda, placed thermocouples in our heads: at the main trunk of the internal carotid in the rear of the nasopharyngeal cavity, in the nasal cavity below the forebrain and at the forward wall of the sphenoid sinus only one inch away from the hypothalamus [*see illustration on pages 86 and 87*]. Continuous measurements of temperature at these points showed large discrepancies with internal temperature as measured at the rectum.

The discrepancies appeared before and after the subject exerted himself by physical exercise, after internal cooling by the eating of ice, after warming the arms or legs in warm water and cooling them in cold water and after immersing the whole body in warm or cold water. These were precisely the situations in which earlier experimenters had found the same absence of correlation between

rectal temperature and the heat-dissipating responses of vasodilation and sweating. It was clear that the temperature at the rectum could under no circumstances be trusted as reflecting the temperature at the internal temperature-sensing organ. The hypothalamus was plainly the place to look for correlation between changes of internal temperature and the responses that regulate it. Since the eardrum is by all odds the most accessible of the four sites thus measured in the head, it was adopted in the experiments that followed. Readings could be taken here with an error of .01 degree centigrade against a standard of temperature maintained with an error of .002 degree C.

The subjects now spent time in the calorimeter on many different days at different environmental temperatures ranging from almost intolerably cold for the nude body at rest to almost intolerably hot for the subject undergoing exertion. Between these two extremes, measurements were made for all the intermediate levels at five-degree temperature intervals and with the subject at rest and at work. Under each set of circumstances the instruments kept a continuous record as the state of homeostasis was reached and maintained for one hour. This arduous series of experiments, extended over two months, made it possi-

ble to plot for the first time the heat-dissipating responses of vasodilation and sweating independently against skin temperature and against internal head temperature. The volunteer subject for this series, Lawrence R. Neff, was observed with a cool skin and cool interior (resting in a cold environment), with a cool skin and warm interior (working in a cold environment), with a hot skin and a relatively cool interior (resting in a hot environment) and with a hot skin and warm interior (working in a hot environment).

The records showed the familiar disordered relationship between the heat-dissipating responses and skin temperature. But the plot of the responses against eardrum temperature showed an almost perfect undisturbed relation. Whatever the temperature of the skin, one certain specific rate of sweating and no other invariably showed up in association with a given internal temperature measured at the eardrum. A reproducible, inseparable and quantitative relation between the stimulus of temperature and one of the heat-dissipating responses had at last been observed. It exhibited a sharply defined breakoff at 36.9 degrees C. (98.4 degrees F.). This was no doubt the set point of the human thermostat in this subject at the time of the experiment. The response proved to be so forceful that a mere .01-degree-C.

rise in temperature was sufficient to increase the dissipation of heat through sweating by one calorie per second and to raise the blood flow through the skin by 15 milliliters per minute.

The success of this series of experiments in distinguishing between the variations in skin and brain temperature was confirmed in many other experiments that subjected the body to quite different sets of extremes. In one of them the skin and the interior of the body were warmed as the subject accommodated himself in the calorimeter to an environmental temperature of 45 degrees C. (113 degrees F.). With homeostasis attained, the subject gulped down large measured helpings of sherbet three times at suitable intervals. On each occasion, as the melting ice withdrew heat from the internal organs and the circulating blood, the brain temperature declined. No less impressively, the skin temperature was observed to rise. The curve

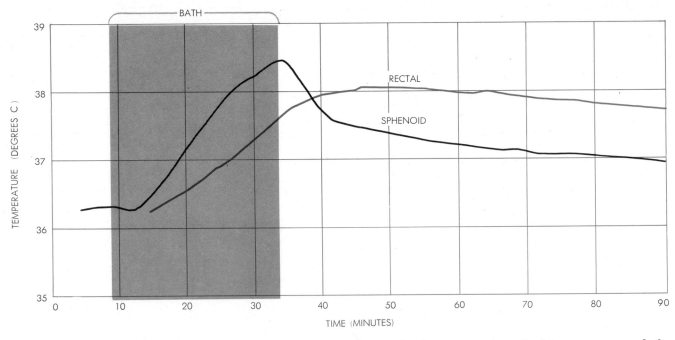

RECTAL TEMPERATURE is shown in this graph to have an uncertain relationship to the hypothalamic temperature measured at the forward wall of the sphenoid sinus. The sphenoid temperature rises sharply as an experimental subject enters a warm bath and falls off sharply when the subject leaves the bath. The rectal temperature reaches a peak only after the subject has left the bath.

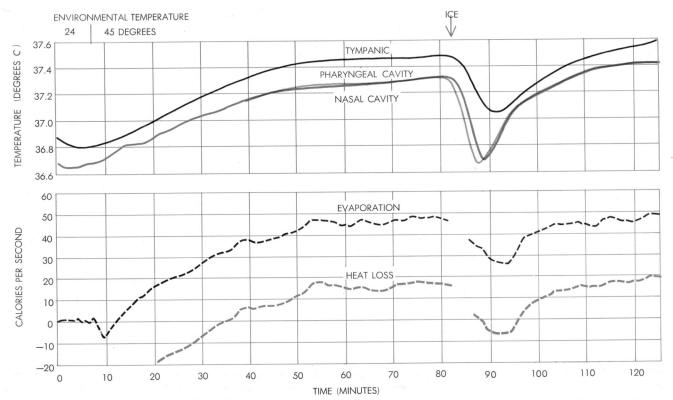

MEASUREMENTS OF HEAD TEMPERATURE at three points near the hypothalamus—at eardrum (*tympanic*), at ethmoid sinus (*nasal cavity*) and in rear wall of pharyngeal cavity—show close correspondence with one another and with the heat dissipation by evaporation of sweat and heat loss by vasodilation. Location of these points is shown in the illustration on pages 86 and 87.

drawn by sweat-gland activity now showed unequivocally which temperature-sensing system controls the heat-dissipating responses: the rate of sweating fell off and rose in perfect parallel with the decline and rise in internal head temperature. It was the consequent drying of the skin that caused the skin to be heated by radiation and conduction in the hot environment. But the sensory reception of heat in the skin brought no response from the heat-dissipating mechanism.

These observations accord well with the familiar constancy of the body temperature. It is difficult to see how it could be maintained within the same narrow range, year in and year out, if the heat-controlling responses were not always triggered at the same set point. As these experiments show, moreover, the responses always closely match the

magnitude of the stimulus. Such precise regulation of temperature could not be achieved by measurement of skin temperature. As in all feedback systems, the quantity which is controlled must itself be measured. An architect who wants to control the temperature of a house does not distribute thousands of thermometers over the outside walls. One thermostat in the living room suffices. It responds not only to warming and chilling from out-of-doors but to overheating from within. The thermostat in the hypothalamus similarly monitors the internal temperature of the body from the inside and thereby maintains its constancy.

This is not to say that the warm-sensitive nerve endings of the skin have no function in the regulation of body temperature. They are the sensory organs for another system which operates via the centers of consciousness in the cortex, bypassing the unconscious control

center in the hypothalamus. To sensations of heat or cold reported by the skin the body reacts by using the muscles as effector organs. Under the stimulus of discomfort from the extremes of both heat and cold, man seeks a cooler or a warmer environment or takes the measures necessary to make his environment comfortably cool or warm. But for all the mastery of external circumstances that follows from this linkage in the body's temperature-sensing equipment, the skin thermoreceptors cannot regulate internal temperature with any degree of precision. They can contribute directly to the regulation of skin temperature alone. The automatic system of hypothalamic temperature regulation takes over from there and achieves the final adjustment with almost unbelievable sensitivity and precision.

In the regulation of internal temperature, therefore, the hypothalamus can

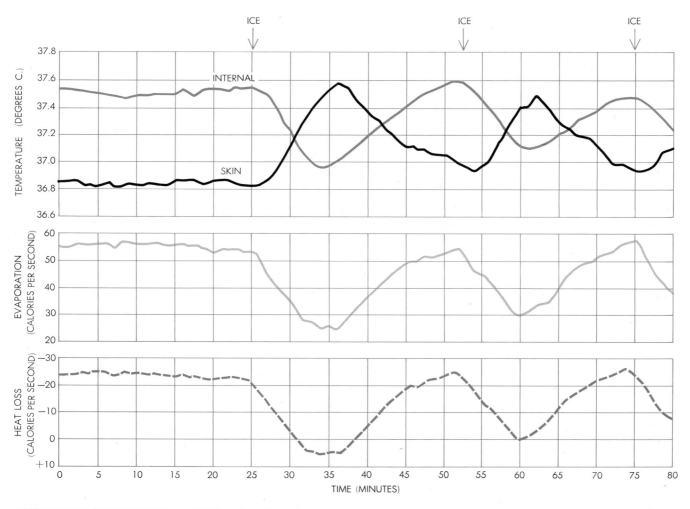

CONCLUSIVE EXPERIMENT establishing hypothalamic temperature as key to control of body temperature is charted. At left in top chart the subject's hypothalamic (internal) temperature is stabilized at the "normal" 37 degrees C. in an environment of 45 degrees C. (113 degrees F.). Upon ingestion of ice at half-hour intervals, internal temperature drops sharply and skin temperature ascends. In middle chart rate of sweating falls and rises in close correlation with hypothalamic temperature; shutdown of sweating accounts for rise in skin temperature. In bottom chart heat loss from skin shows same correlation with hypothalamic temperature.

no longer be regarded simply as a controller which converts incoming sensory stimuli into outgoing impulses to the effector system. It is itself the site of a receptor end-organ, an "eye" for temperature comparable to the retina—the receptor organ for light. This analogy between the temperature eye and the optical eye has, in fact, a sound anatomical basis. Both are derived from the same matrix: the bottom of the third ventricle of the brain. These are two parts of the brain that have a proved sensory receptor function. In the course of evolution the optical eye moved outward to connect with a dioptric apparatus partly derived from the skin, and thereby gained a view of the external world. The temperature eye in the hypothalamus is located in the interior of the head, where it properly belongs. It measures as well as regulates the temperature of the blood which bathes its cells and the rest of the brain, the vital function of which requires a closely maintained optimal temperature.

The feedback system that dissipates heat and thus keeps the body from overheating under normal conditions has thus been elucidated. The same cannot be said, however, of the regulatory system that steps up metabolic heat production and keeps the body temperature from falling below the optimum level. It appears that the two systems operate quite differently and that in the metabolic warming-up of the body the temperature-eye performs its task by inhibition of sensory impulses originating elsewhere.

On the other hand, the sure location of the thermostat in experiments on the "warm side" now makes it possible to renew the study of many interesting questions. Temperature measurements at sites that reliably reflect hypothalamic temperature should replace rectal observations of temperature in all these studies and even in some clinical situations. How bacterial toxins produce fever and how drugs act to reduce it can now be redefined in terms of shifts in the set point of the thermostat and may be made the subject of quantitative investigation. The same direct attack may also be made upon individual or group tolerances and the adaptability of human temperature regulation. These are important objectives in connection with hypothermia (the reduction of the body temperature to low levels) in surgical operations, and with the conquest of new spaces for the life of the human species.

10

THE MASTER SWITCH OF LIFE

P. F. SCHOLANDER   December 1963

In the higher animals breathing and the beating of the heart seem synonymous with life. They implement the central process of animal metabolism: the respiratory gas exchange that brings oxygen to the tissues and removes carbon dioxide. Few events are more dangerous to life than an interruption of breathing or circulation that interferes with this exchange. It is not that all the tissues of an animal need to be continuously supplied with fresh oxygen; most parts of the human body display a considerable tolerance for asphyxia. The tissues of an arm or a leg can be isolated by a tight tourniquet for more than an hour without damage; the kidney can survive without circulation for a similar period and a corneal transplant for many hours. The heart and the brain, however, are exquisitely sensitive to asphyxia. Suffocation or heart failure kills a human being within a few minutes, and the brain suffers irreversible damage if its circulation ceases for more than five minutes.

One might expect that the body would respond with heroic measures to the threat of asphyxia. It does indeed. The defense is a striking circulatory adaptation: a gross redistribution of the blood supply to concentrate the available oxygen in the tissues that need it most. The identification of this defense mechanism has resulted from studies, extending over a number of years, of animals that are specialized to go for an unusual length of time without breathing: the diving mammals and diving birds. Only recently has it become clear that this "master switch" of life is the generalized response of vertebrate animals to the threat of asphyxia from any one of a number of quite different circumstances.

A cat or a dog or a rabbit—or a human being—dies by drowning in a few minutes. A duck, however, can endure submersion for 10 to 20 minutes, a seal for 20 minutes or more and some species of whales for an hour or even two hours. How do they do it? The simplest explanation would be that diving animals have a capacity for oxygen storage that is sufficient for them to remain on normal aerobic, or oxygen-consuming, metabolism throughout their dives. As long ago as the turn of the century the physiologists Charles R. Richet and Christian Bohr realized that this could not be the full story. Many diving species do have a large blood volume and a good supply of oxygen-binding pigments: hemoglobin in the blood and myoglobin in the muscles. Their lungs, however, are not unusually large. Their total store of oxygen is seldom even twice that of comparable nondiving animals and could not, it was clear, account for their much greater ability to remain submerged.

At the University of Oslo during the 1930's I undertook a series of experiments to find out just what goes on when an animal dives. For this purpose it was necessary to bring diving animals into the laboratory, where they could be connected to the proper instruments for recording in detail the physiological events that take place before, during and after submergence. Over the years my colleagues—Laurence Irving in particular—and I have worked with many mammals and birds. We have found seals to be ideal experimental animals: they tame easily and submit readily to a number of diving exercises. At first we confined them to a board that could be lowered and raised in a bathtub full of water. Lately my colleague Robert W. Elsner at the Scripps Institution of Oceanography has trained seals to "dive" voluntarily, keeping their noses under water for as long as seven minutes.

Our first experiments at Oslo confirmed the earlier discovery, by Richet and others, of diving bradycardia, or slowing of the heart action. When the nose of a seal submerges, the animal's heartbeat usually falls to a tenth or so of the normal rate. This happens quickly, indicating that it occurs by reflex action before it can be triggered by any metabolic change. The initiation of bradycardia is affected by psychological factors. It can be induced by many stimuli other than diving, such as a sharp handclap or a threatening movement on the part of the investigator when the seal is completely out of the water. Conversely, bradycardia sometimes fails to develop in a submerged seal if the animal knows it is free to raise its head and breathe whenever it likes. In long dives, however, the slowing down is always pronounced. It is significant that the impulse is so strong it ordinarily continues for the duration of the dive, even when the animal works hard—a situation that would normally cause a rise in the heart rate.

Bradycardia occurs in every diving animal that has been studied. It has been reported in such diverse species as the seal, porpoise, hippopotamus, dugong, beaver, duck, penguin, auk, crocodile and turtle. The same thing happens in fishes when they are taken out of the water. And when such nondivers as cats, dogs and men submerge, bradycardia develops too, although it is often less pronounced than in the specialized divers.

When the heart of a seal beats only five or six times a minute, what happens to the blood pressure? We found that the central blood pressure—in the main artery of a hind flipper, for instance—stays at a normal level. The shape of the pressure trace, however, reveals that

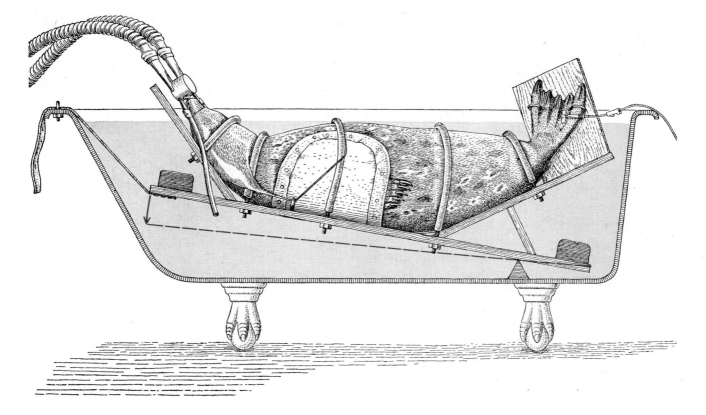

SEAL DIVES IN LABORATORY by being ducked in a bathtub full of water. The animal is strapped loosely to a weighted board. Its head is covered by a mask connected to a device for recording respiration. When the board is tilted down (*broken line*), the mask fills with water and the seal's nose is submerged. An artery in a hind flipper is shown cannulated for removal of blood samples.

VOLUNTARY DIVING eliminates any possibility that restraint affects the seal's responses. This harbor seal (*Phoca vitulina*) is being trained to keep its nose under water until the experimenter lowers his warning finger and instead displays the reward, a fish.

0 MINUTES

2 MINUTES

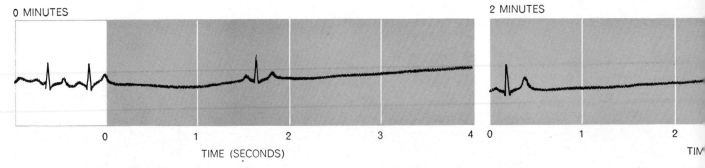

TIME (SECONDS)

TIM

DIVING BRADYCARDIA, the slowing of the heart rate that occurs in vertebrates when they submerge, is quite apparent in this electrocardiogram of a diving seal. Three segments of the record are shown, made at the beginning of, during and at the end of an

whereas the pressure rise with each beat is normal, the subsequent drop in pressure is gradual and prolonged. This indicates that, although the systolic phase of the heartbeat is almost normal, the diastolic phase, during which the blood is forced through the aorta, encounters resistance: the peripheral blood vessels are constricted. Measurements in a small toe artery in the seal's flipper show that the pressure there drops when the dive begins, falling rapidly to the much lower level maintained in the veins. In other words, we found that the circulation in the flippers shuts down to practically nothing during a dive [*see bottom illustration on opposite page*].

For another clue to circulation we measured the level of lactic acid in the muscles and blood of a diving seal. Lactic acid is the end product of the anaerobic metabolic process from which muscles derive energy in the absence of oxygen. The concentration of this metabolite in muscle tissue rises sharply during a dive but the concentration in the blood does not; then, when the seal begins to breathe again, lactic acid floods into the bloodstream. The same sequence of

events has been found to occur in most other animals, showing that the muscle circulation remains closed down as long as the dive continues. Similarly, oxygen disappears from muscle tissue a few minutes after a seal submerges, whereas the arterial blood still contains plenty of oxygen—enough to keep the myoglobin saturated if the muscles are being supplied with blood [*see illustrations on page 100*]. Other experiments revealed that in the seal both the mesenteric and the renal arteries, supplying the intestines and kidneys respectively, close down during diving. All these findings made it apparent that a major portion of the peripheral circulation shuts off promptly on submergence. This was evidently the reason the heart slows down.

At this point our results tied in nicely with some conclusions reached by Irving, who was then at Swarthmore College. His efforts had been stimulated by pioneering studies of circulatory control conducted in the 1920's by Detlev W. Bronk, then at Swarthmore, and the late Robert Gesell of the University of

Michigan. Bronk and Gesell had discovered in 1927 that in a dog rendered asphyxic by an excess of carbon dioxide and a lack of oxygen the muscle circulation slowed down as the blood pressure remained normal and the brain circulation increased. Irving noted in 1934 that this phenomenon might explain a diving animal's resistance to asphyxia, and he proceeded to measure blood flow in a variety of animals by introducing heated wire probes into various tissues and recording the rate at which their heat was dissipated. His data indicated that during a dive the flow in muscle tissue is reduced but the brain blood flow remains constant or even increases. He decided that the essence of the defense against asphyxia in animals would prove to be some mechanism for the selective redistribution of the circulation, with preferential delivery of the decreasing oxygen store to those organs that can least endure anoxia: the brain and the heart.

When the blood flow closes down in most tissues during a dive, what happens to energy metabolism? This is best studied during a quiet dive, with a seal or

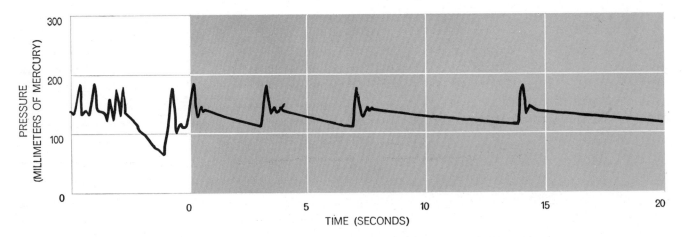

CENTRAL BLOOD PRESSURE stays at about a normal level during a seal's dive (*color*); the rate of increase in pressure during a contraction is also normal. The slow pressure drop between contractions, however, suggests constriction of peripheral blood vessels.

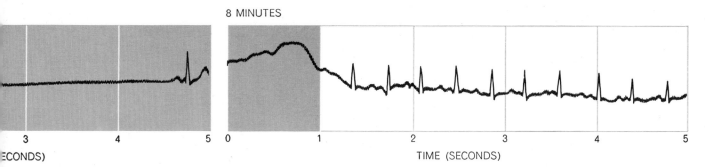

8 MINUTES

3   4   5   0   1   2   3   4   5
ECONDS)   TIME (SECONDS)

eight-minute dive, the duration of which is shown in color. The
heart slows down at the start of the dive. The rate remains as low
as seven or eight per minute during the dive and then returns to
a normal 80 or so per minute as soon as the seal breathes again.

duck trained to remain inactive while
under water. The oxygen stores are large
enough to provide only a quarter of the
energy expended in a predive resting
period of the same length. The next
question was: Do anaerobic processes,
including lactic acid production, substitute fully for the lack of oxygen?
Muscle on anaerobic metabolism incurs
an "oxygen debt" that must be paid off
when oxygen becomes available. The excess oxygen intake on recovery from a
dive is a measure of that debt. If an
animal consumed energy at the same
rate during a dive as before it, this excess intake would be enough to equal
a normal oxygen-consumption rate during the dive. We found that it was characteristic in quiet dives, however, for
the seal or duck to exhibit an oxygen
debt much smaller than this. In the case
of the sloth, a tree-living animal that
is curiously tolerant of submersion, there
was no apparent oxygen debt [see illustration on page 101]. The implication
was that metabolism must slow down.

We could not settle this definitely by
studying the oxygen debt alone; it was
conceivable that the debt was being
paid off so slowly it eluded us. Temperature measurements, however, confirmed the impression of decreased metabolism. We often noticed that after
long dives (20 minutes or so) the seal
would be shivering during the recovery
period. We found that the animal lost
body temperature at a rapid rate while
submerged. Now, this could not be because of increased heat loss, since there
was no substantial change in the thermal
contact between the seal and the water;
only the nostrils were submerged for
the dive. Moreover, the reduction of circulation meant that heat conductivity
was lessened, not increased. The loss in
body temperature therefore meant an
actual decrease in heat production—a
slowing down of metabolism. Apparently the lack of blood in the tissues simply
jams the normal metabolic processes by
mass action; the flame of metabolism
is damped and burns lower. It is quite
logical that submergence should bring
about a progressive reduction in energy
metabolism, considering that the suspen-
sion of breathing ultimately terminates
in death, or zero metabolic activity.

In most dives under natural conditions, of course, this general metabolic
slowing down is masked. The animal is
actively gathering food, and its muscles
probably expend energy at several times
the resting rate for the total animal.
After a few minutes the muscles have
used up the private store of oxygen in
their myoglobin, and then they depend
on anaerobic processes resulting in lactic
acid formation. After such dives there
is a substantial oxygen debt reflecting
the amount of exercise; it is therefore
impossible to detect the subtle lowering
of metabolism that must still occur in
the nonactive tissues deprived of circulation.

It has been fascinating, and of particular interest from the point of view
of evolution, to discover the very same
asphyxial defense in fishes taken out
of the water—diving in reverse, as it
were. The response is found in a variety
of fishes, including many that would
never leave the water under normal conditions. It is most striking in the aquatic
versions of diving mammals and diving
birds: the fishes that routinely make excursions out of the water, such as the fly-

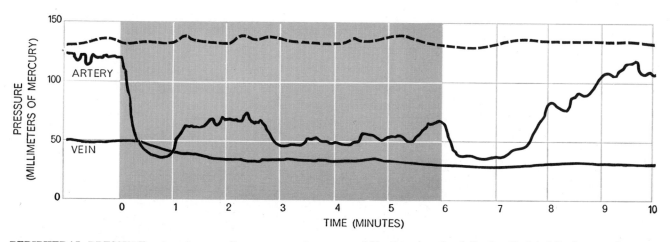

PERIPHERAL PRESSURE, taken in a small toe artery, drops
appreciably during a seal's dive (colored area). From near the central blood-pressure level (broken line) it falls almost to the venous
level, which indicates a closing down of circulation in the flipper.

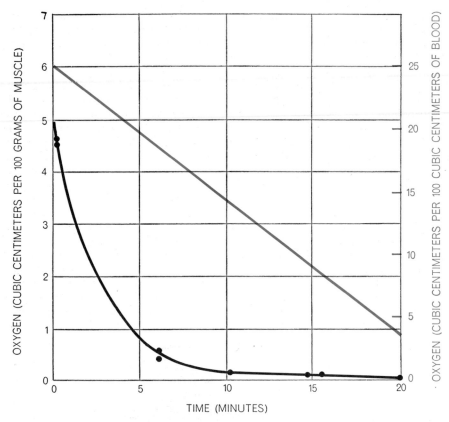

OXYGEN concentration is traced in the muscle (*black curve*) and arterial blood (*colored curve*) of a harbor seal during a dive. The sharp drop in muscle oxygen while the blood is still more than half-saturated suggests that there is no appreciable blood flow in muscle.

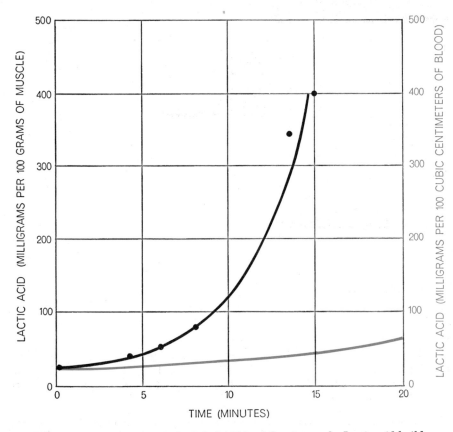

LACTIC ACID concentration confirms lack of blood flow in muscle. Lactic acid builds up in the muscle as the oxygen there is used up, but little enters the blood. The blood lactic acid level rises sharply only after the muscle circulation is restored when breathing resumes.

ing fish. It would be interesting to obtain an electrocardiogram of a flying fish taking off on a natural flight, but this would call for a rather tricky technique. When the leap is simulated, however, by lifting a flying fish out of the water, a profound bradycardia develops immediately.

Another fish that survives on land for some time is the grunion, an amazing little member of the herring family that frequents the coast of California. These fish spawn only on a few nights with maximum tides during the spring. They ride up the beach on a long wave at high tide. As the water recedes the female digs into the sand tail first and deposits her eggs; the male curves around her and fertilizes them. When they have finished, the fish ride out to sea again on another high wave. The spawning procedure can last five or 10 minutes or even longer and is accompanied by much thrashing about; in spite of this activity there is a profound bradycardia during the entire period. Walter F. Garey and Edda D. Bradstreet of the Scripps Institution have studied the lactic acid sequence in grunions caught on the beach and kept overnight in a laboratory tank. The fish are placed in a dish and prodded to keep them wriggling; blood and muscle are sampled during this period and after return to the water. Garey and Miss Bradstreet found that during the anaerobic period lactic acid increases rapidly in the muscles; practically none appears in the circulation until the fish is back in the water. Then, as the peripheral circulation opens up again, lactic acid is flushed out of the muscles and suddenly appears in the blood [see upper illustration on page page 103].

Whereas fishes such as the grunion dive in reverse, the mudskipper (*Periophthalmus*) performs a double reverse. It spends most of its time out of the water in mangrove swamps at the edge of tropical seas, perching on a mangrove root and slithering, if it is frightened, into a burrow in the mud. These mudholes are frequently devoid of oxygen. By dint of heroic and slippery investigations in northern Australian mangrove swamps, Garey has determined that the heart of a mudskipper in its mudhole develops a pronounced bradycardia. It would seem, then, that the creature has turned evolution around: it is more at home as an air-breathing animal than as a proper fish!

In view of the strikingly similar responses to asphyxia in so many quite different vertebrate animals, it would be strange if human beings did not conform to the common scheme. Indeed, a number of recent studies of human divers, of birth anoxia in babies and of several pathological conditions have turned up exactly the same pattern.

My associates and I obtained valuable information by examining the native pearl divers of northern Australia, who are trained from boyhood to make deep dives. (We found, incidentally, that these experts seldom stay down for longer than a minute; many individual divers can remain submerged for twice as long, but this is evidently too strenuous as a regular practice.) A diver develops bradycardia within 20 to 30 seconds whether he remains quiet or swims about. The arterial blood pressure is normal or even elevated; just as in the seal, the diastolic pressure drop is slowed down, apparently by constriction of the peripheral blood vessels. As we expect-

ed, there is little or no rise in the lactic acid level in the blood during the dive, but there is an acute rise in the recovery period. In all these respects human divers respond like other vertebrates. In one respect, however, human beings may be unique: Pathological arrhythmias, or irregularities of the heartbeat, are alarmingly common in man after only half a minute's dive and such arrhythmias have so far not been observed in animals.

In our laboratory at the Scripps Institution, Elsner has been able to demonstrate ischemia, or lack of blood flow, in the muscles of an extremity simply by having a volunteer submerge his face in a basin of water. An electrocardiograph measures the heart rate, and the flow of blood into the calf is measured by plethysmography. In this technique a cuff placed around the thigh

is inflated just enough to occlude the return of blood through the veins while leaving the arteries open to supply blood to the lower part of the leg. As the calf fills with blood its circumference is measured and traced by a recording device. As soon as the subject immerses his face his heart slows down. At the same time there is a sharp decrease in the extent to which the calf expands when the venous return is obstructed; the constriction of the small arteries diminishes and may virtually stop blood flow into the calf. As soon as the subject lifts his face out of the water and breathes, the arterioles open up again and the calf expands [see illustrations on page 104]. If a subject is merely told to hold his breath without submerging his face, all these effects are less pronounced. As in the case of the seal that is free to breathe at will, psychological

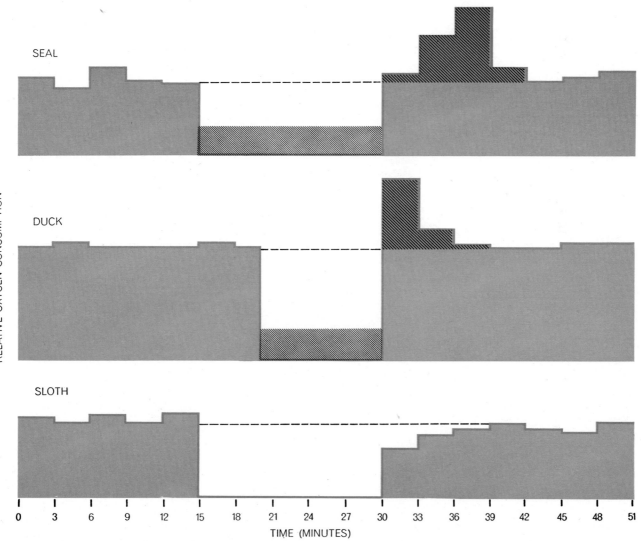

METABOLIC SLOWING DOWN during a dive is demonstrated in three animals by the record of oxygen consumption in successive three-minute periods. In the seal and duck the amount of excess oxygen intake after the dive (*hatching on color*) represents the oxygen debt incurred by anaerobic metabolism during the dive. This debt (*hatching on white*) is clearly not enough to have sustained an energy expenditure at a normal rate (*broken lines*) during the dive. The sloth seems to incur no oxygen debt while diving.

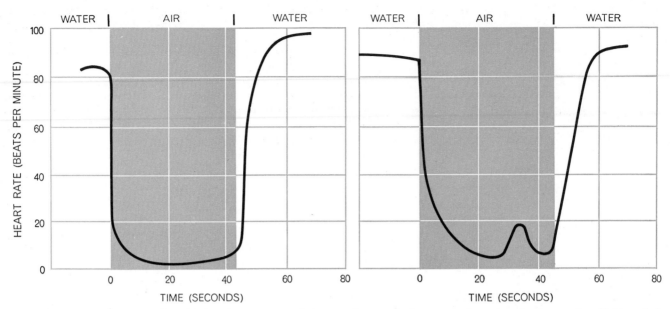

FISH OUT OF WATER develops bradycardia just as a diving animal does when it submerges. These two graphs show the sharp decrease in heart rate that occurs in the grunion (*left*) and the flying fish (*right*) when they are temporarily taken out of water.

factors seem to influence the physiological response to asphyxia.

Physicians have been aware that bradycardia sometimes occurs in babies before, during and immediately after birth and that this can be a sign of asphyxia induced by obstruction or final interruption of the placental blood flow. This concept has been strengthened by lactic acid measurements in newborn infants by Stanley James of the Columbia University College of Physicians and Surgeons. Judging by his data, a normal birth is always followed by a sharp rise in the blood lactic acid. This rise is sharper and higher in babies that have survived a difficult delivery and show clini-

cal symptoms of birth distress; in other words, the longer the period of anoxia, the greater the lactic acid build-up [*see lower illustration on page 103*]. Newborn animals in general have a short period of increased resistance to asphyxia. The sequence of events in babies suggests that selective ischemia is an important asphyxial defense even in newborn infants.

Various pathological conditions that decrease cardiac output, such as arrhythmias and coronary occlusions, are sometimes followed by such apparently unrelated complications as damage to the kidneys or even gangrenous sores in the intestine. Donald D. Van Slyke

and his collaborators at the Rockefeller Institute for Medical Research found in 1944 that severe shock in dogs resulted in decreased kidney function and tissue damage—and that the same symptoms appeared if they simply clamped the renal artery of a healthy dog. Pointing out the analogy to the peripheral vasoconstriction we had reported in diving animals, Van Slyke concluded that under stress the blood supply to the brain is maintained, if necessary, at the cost of restriction of circulation to other areas: the organism, as he said recently, is reduced to "a heart-lung-brain preparation."

More recently Eliot Corday and his

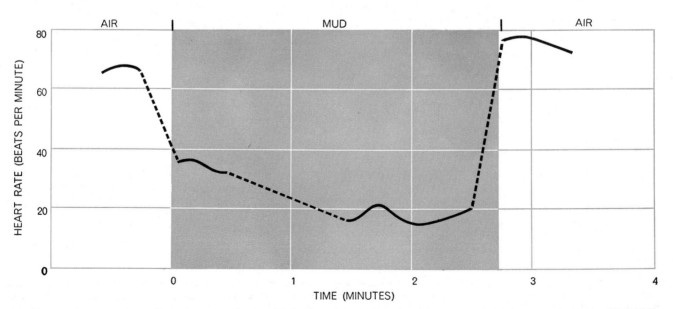

MUDSKIPPER is a curious fish that has become acclimated to breathing air. While it is out of water, its heart rate is normal; when it enters its mud-filled burrow, it develops bradycardia. Broken lines join the various segments of this fragmentary record.

colleagues at the University of California at Los Angeles have found that the same events account for certain gangrenous lesions of the intestine. They impaired the circulation of dogs in various ways, inflicting cardiac arrhythmias by electrical stimulation or decreasing the blood pressure by bleeding the animals. With modern blood-flow-metering techniques and blood-pressure measurements they were able to demonstrate a widespread vasoconstriction that tends to sustain the blood pressure near a normal level but leaves the kidney, the gastrointestinal tract, the muscles and the skin with greatly reduced circulation. These workers again recognized the sequence as a mechanism for maintaining an adequate blood supply to the most sensitive organs.

A quite different physiological event that seems to depend on the same circulatory switch as the prime control is hibernation. In all the relatively few species of mammals and birds that hibernate the body temperature is lowered in the presence of an unfavorable thermal environment. In most animals hibernation is seasonal but in others the temperature drops in a daily cycle. The dormant state is characterized, in any case, by a body temperature only a degree or so warmer than the surroundings; along with this there is a correspondingly low metabolic rate, perhaps a tenth or less of the resting rate in the waking condition. The heart rate is very low—only a few beats per minute—but the central blood pressure remains quite high in relation to this bradycardia. Again the pressure trace shows the slow diastolic emptying of the arteries that suggests a peripheral vasoconstriction. There is good evidence that hibernation is a controlled state; when a decrease in the ambient temperature brings a threat of freezing, the animal increases its heat production and usually emerges from hibernation.

The transition periods during which the animal enters or emerges from hibernation are of particular interest. When a ground squirrel or woodchuck goes into hibernation, the heart rate slows down before the body temperature starts to drop, indicating that the drop in metabolic rate is caused—as in asphyxial defense—by a primary vasoconstriction. Arousal from hibernation is easier to study because it can be precipitated at will by disturbing the animal. This triggers an immediate acceleration of the heartbeat to as much as 100 times the hibernating rate. There follows an intense shivering of the front part of the body, which warms up much more quickly than the rest of the body does as measured by the rectal temperature. Midway through arousal the blood flow in the forelegs of the squirrel is sometimes 10 times greater than in the hind legs. The uneven distribution of metabolic and circulatory activity is apparently accomplished by a dilation of the blood vessels that begins in the forward parts. When the vessels in the rest of the animal finally dilate, the over-all metabolic rate sometimes rises as high as when the animal exercises. The entire sequence is consistent with the idea that the onset and termination of hibernation are triggered in the first instance by

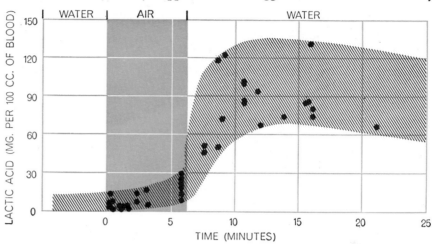

MUSCLE ISCHEMIA, or lack of blood, in grunions results in a lactic acid build-up in muscle while the fish is out of the water. As seen here, the lactic acid does not rise much in the blood until the muscle circulation is restored when the fish re-enters the water.

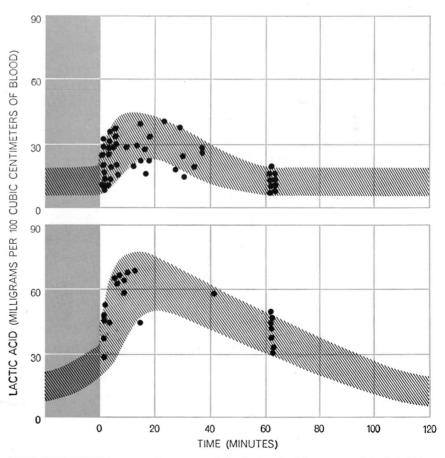

SIMILAR ISCHEMIA apparently protects a baby during the delivery period (color). When breathing begins, the muscle circulation opens up and lactic acid floods the blood. The lactic acid build-up is smaller in a normal delivery (top) than in a long, difficult one (bottom).

vasomotor impulses controlling the size of the small blood vessels. The circulation then throttles metabolism in the tissues to a rate compatible with the blood flow. Going into hibernation seems to call for the same primary vasoconstriction that operates in asphyxial defense.

Any mechanism that operates in many kinds of animals across a wide range of circumstances must be of fundamental physiological significance. In our current work at the Scripps Institution we are trying to learn more about the details of blood flow in animals by im-

planting ultrasonic measuring devices on arteries and veins. We hope to discover just how the autonomic nervous system responds to environmental changes and the threat of anoxia and what sequence of events actually throws the circulatory switch.

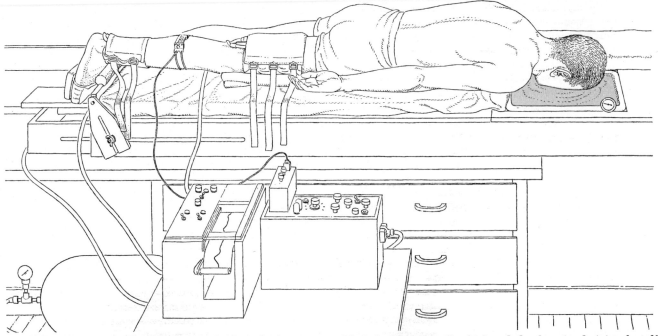

HUMAN DIVING is investigated in the laboratory by having a volunteer immerse his face in a basin of water. In this case the circulation in the lower leg is being measured by plethysmography.

The inflatable cuff on the thigh occludes the veins draining the calf but leaves the arteries open. By measuring the circumference of the calf one can determine the blood flow into the lower leg.

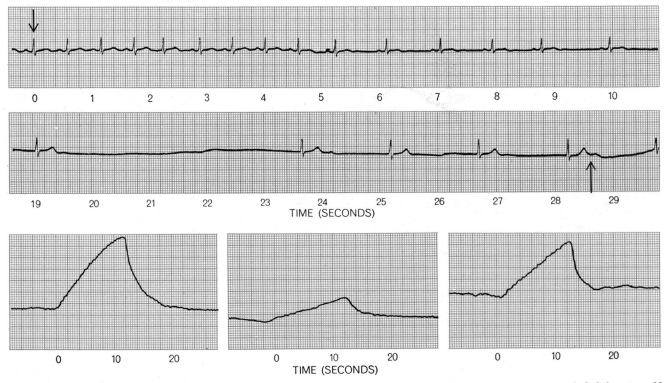

FACE IMMERSION results in bradycardia. The electrocardiogram (*two top strips*) records an extreme case (*arrows mark start and end of dive*). Plethysmographic records (*bottom*) show changes in calf circumference when venous return is occluded for some 12 seconds before (*left*), during (*center*) and after (*right*) face immersion. Blood flow into calf is clearly much reduced during dive.

11

THE DESERT RAT

KNUT AND BODIL SCHMIDT-NIELSEN   July 1953

THERE IS a common impression that no higher animal can live long without drinking water. Certainly this is true of man and many other mammals; we need water at frequent intervals, and in a very hot, dry desert a man without water cannot last more than a day or so. An animal such as the camel can survive somewhat longer, but sooner or later it too must drink to refill its supply.

Yet we know that the waterless desert is not uninhabited. Even in desert areas with no visible drinking water within scores of miles, one will often find a fairly rich animal life. How do these animals get the water they must have to live? The body of a desert mammal has about the same water content (65 per cent of body weight) as that of a drinking animal, and it generally has no more tolerance to desiccation of the tissues, sometimes less. For many desert animals the answer is simple: they get their water in their food. These animals live on juicy plants, one of the most important of which is cactus. The pack rat, for example, eats large quantities of cactus pulp, which is about 90 per cent water. Thus it is easy to account for the survival of animals in areas where cacti and other water-storing plants are available.

There are, however, animals which can live in areas altogether barren of juicy vegetation. An outstanding example is a certain general type of desert rodent which is found in all the major desert areas of the world—in Africa, in Asia, in Australia and in the southwestern U. S. Although the rats of this type seem to have evolved independently in the several areas and are not related to one another, all of them are similar in appearance and habits and all seem to be able to live with a minimum of water. How they do so has long been a puzzle to biologists.

During the past few years we have investigated intensively a rodent of this type—the so-called kangaroo rat that lives in deserts of the U. S. Southwest. In a field laboratory in Arizona and in biological laboratories at Swarthmore College, Stanford University and the University of Cincinnati we have studied the kangaroo rat's habits and physiology, and we now have a good understanding of how this rodent is able to get along on a diet so dry that other animals would soon die of thirst.

THE LITTLE kangaroo rat is not actually related to the kangaroo, though it looks a great deal like one. It hops along on long hind legs, and it has a long, strong tail which it uses for support and steering. It lives in a burrow in the ground by day and comes out for food only at night. The animal thrives in the driest regions, even in the bare sand dunes of Death Valley. Water to drink, even dew, is rarely available in its natural habitat. The kangaroo rat apparently has only a short range of movement—not more than a few hundred yards—and

BANNERTAILED KANGAROO RAT (*Dipodomys spectabilis*) from the Santa Rita Range of Arizona is photographed in the laboratory. It is called the kangaroo rat because of its powerful hind legs and long tail.

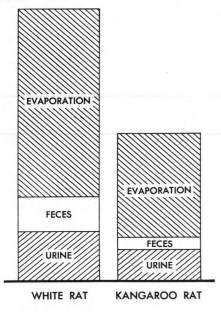

WATER LOSSES of the kangaroo rat (*right*) and the white rat (*left*) are compared at zero humidity.

therefore does not leave its dry area to find juicy plants. Stomach analyses have shown that it seldom or never consumes any succulent vegetation. Its food consists of seeds and other dry plant material. In the laboratory it will live indefinitely without water and with no other food but dry barley seeds.

The first question to be answered was whether the kangaroo rat stores water in its body to carry it over long dry periods. It was found that the animal's water content was always about the same (some 65 per cent of body weight), in the rainy season or the dry season, or after it had been kept on dry food in the laboratory for several weeks. During eight weeks on nothing but dry barley in the laboratory some of the animals increased their body weight, and their water percentage was as high as at the beginning of the experiment; they had actually increased the total amount of water in their bodies. Furthermore, kangaroo rats which were

given watermelon as well as barley to eat showed no higher water percentage in their bodies than animals maintained on a dry diet. They must therefore have eliminated the excess of water at the same rate as it was taken in. Altogether the experiments made clear that the kangaroo rat does not store water or live through dry periods at the expense of its body water.

There is only one way the kangaroo rat can get any substantial amount of water on its dry diet. That is by oxidizing its food. The oxidation of hydrogen or a substance containing hydrogen always forms water. Obviously the amount of water an animal forms in its metabolism will depend upon the hydrogen content of its food. The amount is simply a matter of chemistry and is the same in all animals. Oxidation of a gram of carbohydrate (starch or sugar) yields .6 of a gram of water; of a gram of fat, 1.1 grams of water; of a gram of protein, .3 of a gram of water. Protein produces relatively little water because a considerable part of its hydrogen is not oxidized but is excreted with nitrogen as urea.

Now the experimental diet of dry barley on which the kangaroo rats lived yields 54 grams of water for each 100 grams of barley (dry weight) consumed. If there is any moisture in the air, the barley will also contain a little absorbed water—about 13 grams per 100 grams when the relative humidity is 50 per cent at 75 degrees Fahrenheit. The kangaroo rat consumes 100 grams of barley in a period of about five weeks. Thus during that period its total intake of water is between 54 and 67 grams, depending on atmospheric conditions.

THIS is an astonishingly small amount of water for an animal of its size to subsist on. It can maintain its water balance only if its water losses are correspondingly small. The next step, therefore, was to find out how the animal manages to keep its water loss so low, if indeed it does. We proceeded to measure its losses of water through the three

routes by which an animal eliminates water; in the urine, in the feces and by evaporation from the skin and the respiratory passages.

There is an animal, the African lungfish, which can get along for long periods without excreting any urine at all. When the stream or pond in which it is living dries up, it burrows into the mud and stays there until the next rain. The eminent authority on the kidney, Homer W. Smith [see "The Kidney," by Homer W. Smith, on page 37 in this book], has found that during this time the urea content of the fish's blood may rise to the extravagant level of 4 per cent. Can the kangaroo rat similarly accumulate waste products and avoid urinating during a dry spell? We investigated and found that it could not: the urea and salt content of its blood did not rise when it was on a dry diet or fall when it had a moist diet. And it continued on its dry diet to excrete urea as usual.

However, we learned that it could get rid of its waste products with a very small output of water. The kangaroo rat has an amazingly efficient kidney. The concentration of urea in its urine can be as high as 24 per cent, whereas in man the maximum is about 6 per cent. Thus the kangaroo rat needs only about one fourth as much water to eliminate a given amount of urea as a man would. Its excretion of salts is similarly efficient. The animal can excrete urine about twice as salty as sea water!

The reason that a human being cannot tolerate drinking sea water is that the body is dehydrated in the process of getting rid of the salts. The saltiness of the kangaroo rat's urine suggested that this animal might be able to drink sea water and get a net water gain from it. Of course it was not easy to induce the kangaroo rat, which normally does not drink, to imbibe sea water. But we were able to make it do so by feeding it a high protein diet (soy beans) which formed very large amounts of urea and forced the rat to drink to avoid dehydration. The animal's kidney proved able to excrete both the excess of urea and the salts in the sea water. Drinking sea water actually enabled the animal to maintain its water balance. So far as is known, no other mammal can drink sea water with impunity.

From the known efficiency of the kangaroo rat's kidney we calculated that the animal uses 13 grams of water to excrete the waste products formed from 100 grams of dry barley. Measurement of the water content of its feces, which are exceptionally dry, showed that it loses about three grams of water by this route in metabolizing 100 grams of dry barley. There remained, then, the question of how much water the kangaroo rat loses by evaporation.

Very little escapes through its skin. Rodents have no sweat glands except on the toe pads, and the kangaroo rat has

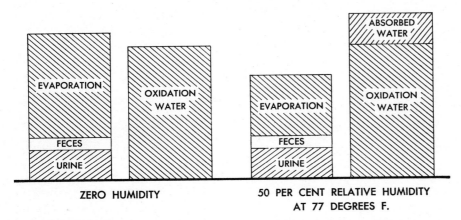

WATER BALANCE of the kangaroo rat is shown under two different conditions of humidity. The first and third bars indicate water loss; the second and fourth, water gain. The term "absorbed water" refers to water in food.

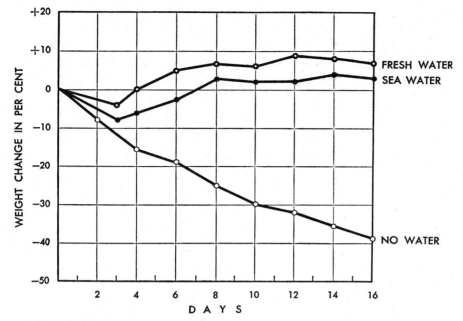

WEIGHT CHANGES of kangaroo rats fed either fresh water, sea water or no water showed that they could drink sea water without ill effects. Although kangaroo rats normally drink no water, they were induced to do so by a protein-rich diet that required water for the excretion of urea.

fewer sweat glands than most rats. All mammals lose a little water from the skin even where there are no sweat glands, and there is reason to believe that the kangaroo rat suffers less loss by this route than other mammals. It does, however, lose a considerable amount of water by evaporation from its respiratory tract. In the extremely dry desert air this loss could be serious. At zero humidity the loss by evaporation from the skin and respiratory tract would amount to some 44 grams of water during the five weeks in which the rat metabolizes 100 grams of barley. At a relative humidity of 50 per cent and a temperature of 75 degrees the loss by evaporation would be about 25 grams.

WE CAN NOW add up the balance sheet of the kangaroo rat's water intake and outgo. At zero humidity its total intake on a diet of dry barley would be 54 grams, and the total loss would be 61 grams (14 in the urine, 3 in the feces and 44 by evaporation). At 50 per cent relative humidity at 75 degrees the intake would be 67 grams and the outgo only 43 grams. Thus it seems clear that the kangaroo rat cannot survive on a barley diet in completely dry air, for under those conditions it has a water deficit in spite of its marvelous mechanisms for water conservation. Actual tests showed that the minimum atmospheric conditions under which the animal can maintain its water balance on the dry diet is

10 to 20 per cent relative humidity at 75 degrees.

The desert atmosphere is often somewhat drier than this, but the explanation of the kangaroo rat's survival is that it is a night animal. During the day it stays in its burrow, where the air is always a little more humid than outside, even when the soil seems to be completely dry. To measure the temperature and humidity in the burrow we used a tiny instrument which included a humidity-sensitive hair hygrometer. The instrument makes a record on a smoked glass disk, which can be read afterward under a microscope. The recorder was tied to the rat's tail, and the animal dragged it into the burrow. It was secured by a thin wire so that the animal could not run away with the instrument. After 12 hours we opened the burrow and read the record. We found that in early summer in the Arizona desert the relative humidity inside the kangaroo rat's burrows ranged from 30 to 50 per cent, and the temperature from about 75 to 88 degrees. At night, in the desert outside the burrows, the relative humidity varied from 15 to 40 per cent and the temperature from about 60 to 75 degrees. The protection of the burrow by day provides just enough margin to enable the kangaroo rat to maintain its water balance and live in the driest of our deserts.

The same protection allows the kangaroo rat to survive the desert heat. A mammal such as man avoids overheating only by evaporating large quantities of water. The kangaroo rat cannot do this, nor can it tolerate a body temperature of much more than 100 degrees. It does not sweat or increase evaporation from its respiratory passages by panting, as a dog does. If it were exposed to the daytime heat in the desert, it would soon perish. The adaptations that permit it to thrive in the hot desert are its nocturnal habits and its extraordinary facilities for water conservation.

12

HORMONES AND GENES

ERIC H. DAVIDSON   June 1965

In the living cell the activities of life proceed under the direction of the genes. In a many-celled organism the cells are marshaled in tissues, and in order for each tissue to perform its role its cells must function in a cooperative manner. For more than a century biologists have studied the ways in which tissue functions are controlled, providing the organism with the flexibility it needs to adapt to a changing environment. Gradually it has become clear that among the primary controllers are the hormones. Thus whereas the genes control the activities of individual cells, these same cells constitute the tissues that respond to the influence of hormones.

New experimental evidence is now making it possible to complete this syllogism: it is being found that hormones can affect the activity of genes. Hormones of the most diverse sources, molecular structure and physiological influence appear able to rapidly alter the pattern of genetic activity in the cells responsive to them. The establishment of a link between hormones and gene action completes a conceptual bridge stretching from the molecular level to ecology and animal behavior.

In order to understand the nature of the link between hormones and genes it will be useful to review briefly what is known of how genes function in differentiated, or specialized, cells. One of the most striking examples of cell specialization in animals is the red blood cell, the protein content of which can be more than 90 percent hemoglobin. It has been shown that in man the ability to manufacture a given type of hemoglobin is inherited; this provides a clear case of a differentiated-cell function under genetic control. Hemoglobin also furnishes an example of another

principle that is fundamental to the study of differentiation: the specialized character of a cell depends on the type and quantity of proteins in it, and therefore the process of differentiation is basically the process of developing a specific pattern of protein synthesis. Some cells, such as red blood cells and the cells of the pancreas that produce digestive enzymes, specialize in synthesizing one kind of protein; other cells specialize in synthesizing an entire set of protein enzymes to manufacture nonprotein end products, for example glycogen, or animal starch (which is made by liver cells), and steroid hormones (which are made by cells of the adrenal cortex).

If one understood the means by which the type and quantity of protein made by cells was controlled, one would have taken a long step toward understanding the nature of the differentiated cell. Part of this objective has been attained: we now know something of how genes act and how proteins are synthesized. A protein owes its properties to the sequence of amino acid subunits in its chainlike molecule. The genes of most organisms consist of deoxyribonucleic acid (DNA), the chainlike molecules of which are made up of nucleotide subunits. The sequence of nucleotides in a single gene determines the sequence of amino acids in a single protein.

The protein is not assembled directly on the gene; instead the cell copies the sequence of nucleotides in the gene by synthesizing a molecule of ribonucleic acid (RNA). This "messenger" RNA moves away from the gene to the small bodies called ribosomes. On the ribosomes, which contain their own unique kind of RNA, the amino acids are assembled into protein. In the assembly

process each molecule of amino acid is identified and moved into position through its attachment to a specific molecule of a third kind of RNA: "transfer" RNA. It can therefore be said that the characteristics of the cell are determined at the level of "gene transcription"—the synthesis of messenger and ribosomal RNA.

Each differentiated cell in a many-celled organism contains a complete set of the organism's genes. It is obvious, however, that in such a cell only a small fraction of the genes are actually functioning; the gene for hemoglobin is not active in a skin cell and the assortment of genes active in a liver cell is not the same as the assortment active in an adrenal cell. The active genes release their information in the form of messenger RNA and the inactive genes do not. Exactly how the inactive genes are repressed is not clearly understood, but the repression seems to involve a chemical combination between DNA and the proteins called histones; it has been shown that histones inhibit the synthesis of messenger RNA in the isolated nuclei of calf-thymus cells, and similar results have been obtained with the nuclei of other kinds of cell. In any case it is clear that the characteristics of the cell are the result of variable gene activity. The prime question becomes: How are the genes selectively turned on or selectively repressed during the life of the cell?

Gene action is often closely linked to cell function in terms of time. It has been demonstrated that genes can exercise immediate control over the activities of differentiated cells—particularly very active or growing cells—and over cells that are going through some change of state. In many specialized cells at least part of the messenger RNA

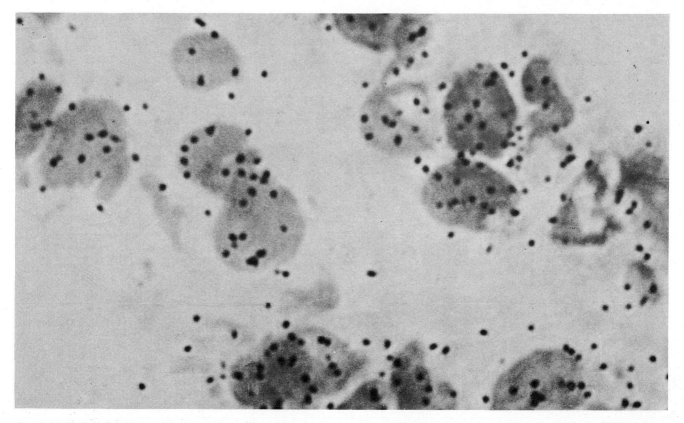

HORMONE IS LOCALIZED IN NUCLEI of cells in this radioautograph made by George A. Porter, Rita Bogoroch and Isidore S. Edelman of the University of California School of Medicine (San Francisco). The hormone aldosterone was radioactively labeled and administered to a preparation of toad bladder tissue. When the tissue was radioautographed, the hormone revealed its presence by black dots. The dots appear predominantly in the nuclei (*dark gray areas*) of the cells rather than in the cytoplasm (*light gray areas*).

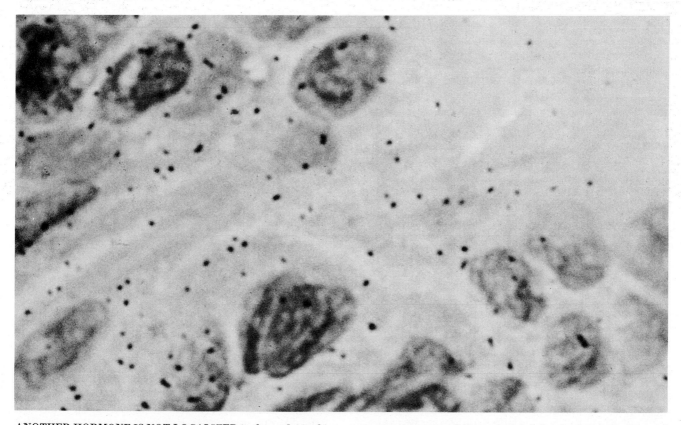

ANOTHER HORMONE IS NOT LOCALIZED in the nuclei in this radioautograph made by the same investigators. Here the hormone was progesterone, and it too was labeled and administered to toad bladder tissue. The dots are distributed more or less at random.

HORMONE	SOURCE		CHEMICAL NATURE	FUNCTION
ECDYSONE	INSECT PROTHORACIC GLAND		STEROID	Causes molting, initiation of adult development and puparium formation.
GLUCOCORTICOIDS (CORTISONE)	ADRENAL CORTEX		STEROID	Causes glycogen synthesis in liver. Causes redistribution of fat throughout organism. Alters nitrogen balance. Causes complete revision of white blood cell type frequencies. Is required for muscle function. Alters central nervous system excitation threshold. Affects connective tissue differentiation. Promotes healing. Induces appearance of new enzymes in liver. Affects almost all tissues.
INSULIN	PANCREAS (ISLETS OF LANGERHANS)		POLYPEPTIDE	Affects entry rate of carbohydrates, amino acids, cations and fatty acids into cells. Promotes protein synthesis. Affects glycogen synthetic activity. Stimulates fat synthesis. Stimulates acid mucopolysaccharide synthesis. Affects almost all tissues.
ESTROGEN	OVARY		STEROID	Promotes appearance of secondary sexual characteristics. Increases synthesis of contractile and other proteins in uterus. Increases synthesis of yolk proteins in fowl liver. Increases synthesis of polysaccharides. Affects rates of glycolysis, respiration and substrate uptake into cells. Probably affects almost all tissues.
ALDOSTERONE	ADRENAL CORTEX		STEROID	Controls sodium and potassium excretion and cation flux across many internal body membranes.
PITUITARY ACTH	ANTERIOR PITUITARY		POLYPEPTIDE	Stimulates glucocorticoid synthesis by adrenal cortex. Stimulates adrenal protein synthesis and glucose uptake. Inhibits protein synthesis in adipose tissue. Stimulates fat breakdown.
PITUITARY GH	ANTERIOR PITUITARY		PROTEIN	Stimulates all anabolic processes. Affects nitrogen balance, water balance, growth rate and all aspects of protein metabolism. Stimulates amino acid uptake and acid mucopolysaccharide synthesis. Affects fat metabolism. Probably affects all tissues.
THYROXIN	THYROID		THYRONINE DERIVATIVE	Affects metabolic rate, growth, water and ion excretion. Promotes protein synthesis. Is required for normal muscle function. Affects carbohydrate levels, transport and synthesis. Probably affects all tissues.

HORMONES DISCUSSED IN THIS ARTICLE are listed according to their source, their chemical nature and their effects, which are usually quite diverse. Pituitary GH is the pituitary growth hormone. The steroid hormones share a basic molecular skeleton consisting of adjoining four-ring structures. The polypeptide hormones and the protein hormones consist of chains of amino acid subunits.

produced by the active genes decays in a matter of hours, and therefore the genes must be continuously active for protein synthesis to continue normally. Other differentiated cells display the opposite characteristic, in that gene activity occurs at a time relatively remote from the time at which the messenger RNA acts. The very existence of this time element in gene control of cell function indicates how extensive that control is. Furthermore, certain genes can be alternately active and inactive over a short period; for example, if a leaf is bleached by being kept in the dark and is then exposed to light, it immediately begins to manufacture messenger RNA for the synthesis of chlorophyll.

The sum of such observations is that the patterns of gene activity in the living cell are in a state of continuous flux. For a cell in a many-celled organism, however, it is essential that the genetic apparatus be responsive to external conditions. The cell must be able to meet changing situations with altered metabolism, and if all the cells in a tissue are to alter their metabolism in a coordinated way, some kind of organized external control is needed. Evidence obtained from experiments with a number of biological systems suggests that such control is obtained by externally modulating the highly variable activity of the cellular genetic apparatus. The studies that will be reviewed here are cases of this general proposition; in these cases the external agents that alter the pattern of gene activity are hormones.

Many efforts have been made to explain the basis of hormone action. It has been suggested that hormones are coenzymes (that is, cofactors in enzymatic reactions), that they activate key enzymes, that they modify the outer membrane of cells and that they directly affect the physical state of structures within the cell. For each hypothesis there is evidence from studies of one or several hormones. As an example, experiments with the pituitary hormone vasopressin, which causes blood vessels to constrict and decreases the excretion of urine by the kidney, strongly support the conclusion that the hormone attaches itself to the outer membrane of the cells on which it acts.

To these hypotheses has been added the new one that hormones act by regulating the genetic apparatus, and many investigators have undertaken to study the effects of hormones on gene activity. It turns out that the gene-regulation

hypothesis is more successful than the others in explaining some of the most puzzling features of hormone activity, such as the time lag between the administration of some hormones and the initial appearance of their effects, and also the astonishing variety of these effects [see illustration on opposite page]. There can be no doubt that some hormone action is independent of gene activity, but it has now been shown that a wide variety of hormones can affect such activity. This conclusion is strongly supported by the fact that each of these same hormones is powerless to exert some or all of its characteristic effects when the genes of the cells on which it acts are prevented from functioning.

The genes can be blocked by the remarkably specific action of the antibiotic actinomycin D. The antibiotic penetrates the cell and forms a complex with the cell's DNA; once this has happened the DNA cannot participate in the synthesis of messenger RNA. The specificity of actinomycin is indicated by the fact that it does not affect other activities of the cell: protein synthesis, respiration and so on. These activities continue until the cellular machinery stops because it is starved for messenger RNA. In high concentrations actinomycin totally suppresses the synthesis of messenger RNA; in lower concentrations it depresses this synthesis and appears to prevent it from developing at new sites.

So far the greatest number of studies of the effects of hormones on genes have been concerned with the steroid hormones, particularly the estrogens produced by the ovaries. This work has been carried forward by many investigators in many laboratories. It has been found that when the ovaries are removed from an experimental animal and then estrogen is administered to the animal at a later date, the synthesis of protein by cells in the uterus of the animal increases by as much as 300 percent. The increase is detected by measuring the incorporation of radioactively labeled amino acids into uterine protein, or by testing the capacity for protein synthesis of homogenized uterine tissue removed from the animal at various times after the administration of estrogen. Added proof that these observations have to do with the synthesis of protein is provided by the fact that the stimulating effects of estrogen are blocked by the antibiotic puromycin, which specifically inhibits protein synthesis.

In these experiments the principal rise in protein synthesis is first observed between two and four hours after estrogen treatment. Less than 30 minutes after the treatment, however, there is a dramatic increase in the rate of RNA synthesis. When actinomycin is used to block the rise in RNA synthesis, the administration of estrogen has no effect on protein synthesis! What this means is that since the diverse metabolic changes brought about in uterine cells by estrogen are all mediated by protein enzymes, none of the changes can occur unless the estrogen has induced gene action. Among the changes are the increased synthesis of amino acids from glucose, the increased evolution of carbon dioxide and the increased synthesis of the fatty lipids and phospholipids. It is not surprising to find that none of these metabolic changes in uterine cells can be detected when estrogen is administered to an animal that has first been treated with actinomycin.

The effect of estrogen on the synthesis of RNA is not limited to messenger RNA. There is also an increase in the manufacture of the other two kinds of RNA: transfer RNA and ribosomal RNA. The administration of estrogen first stimulates the production of messenger RNA and transfer RNA. The genes responsible for the synthesis of ribosomal RNA become active somewhat later, and the number of ribosomes per cell increases. One of the earliest changes brought about by estrogen, however, is an increase in the activity of the enzyme RNA-DNA polymerase. This enzyme appears to be responsible for all RNA synthesis in such cells.

Two main conclusions can be drawn from these various observations. First, there can be no reasonable doubt that treatment with estrogenic hormones results in activation at the gene level, and that many of the well-known effects of estrogen on uterine cells result from this gene activation. Second, it is clear that a considerable number of genes must be activated in order to account for the many different responses of the cells to estrogen. Consider only the fact that estrogen stimulates the production of three different kinds of RNA. At least two different genes are known to be associated with the synthesis of ribosomal RNA, and each cell needs to manufacture perhaps as many as 60 species of transfer RNA. As for messenger RNA, the variety of the changes induced by estrogen implies that under such influences it too must be produced

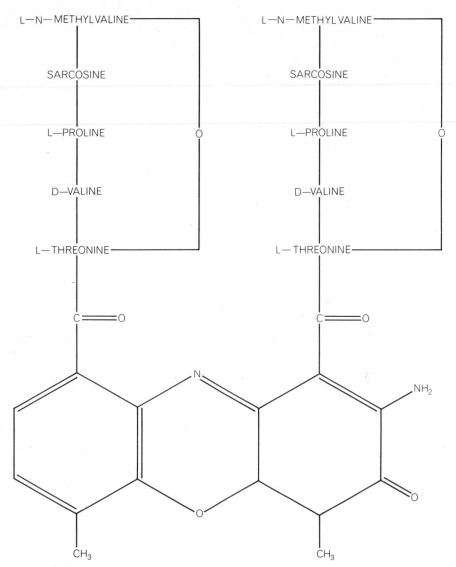

ANTIBIOTIC ACTINOMYCIN D has a complex chemical structure. The antibiotic blocks the participation of the genetic material in the synthesis of ribonucleic acid (RNA); thus it can be used in studies to determine whether or not a given hormone stimulates gene activity.

recognized that hormonal specificity resides less in the hormone than in the "target" cell. We are now, however, able to ask new questions: How are the sets of genes that are activated by a given hormone selected? Are these genes somehow preset for hormonal activation? How does the hormone interact not only with the gene itself but also with the cell's entire system of genetic regulation?

The male hormone testosterone has also been shown to operate by gene activation. Like the estrogens, the male sex hormones can give rise to dramatic increases of RNA synthesis in various cells. In experiments on male and female rats it has been found that the effect of testosterone on the liver cells of a female is somewhat different from that on the liver cells of a castrated male. In both cases the hormone causes an increase in the *amount* of messenger RNA produced, but in the female it also brings about the synthesis of a new *variety* of messenger RNA. This effect, like the ability of estrogen to stimulate a rooster's liver cells to produce egg-yolk proteins, provides a new approach for examining the whole question of sexual differentiation.

Apart from the sex hormones, the principal steroids in mammals are those secreted by the adrenal cortex. One group of adrenocortical hormones is typified by cortisone; this hormone and its relatives are known for their quite different effects in different tissues. Only a fraction of these effects have been studied from the standpoint of gene activation, and there is much evidence to indicate that some of them are not mediated by the genes. Some responses to cortisone, however, do appear to be the consequence of gene activation.

If the adrenal glands are removed from an experimental animal and cortisone is administered later, the hormone induces in the liver cells of the animal the production of a number of new proteins. Among these proteins are enzymes required for the synthesis of glucose (but not the breakdown of glucose) and enzymes involved in the metabolism of amino acids. Moreover, cortisone steps up the total production of protein by the liver cells. The effect of cortisone on the synthesis of messenger RNA is apparent as soon as five minutes after the hormone has been administered; within 30 minutes the amount of RNA produced has increased two to three times and probably includes not

in a number of molecular species. We are therefore confronted with a major mystery of gene regulation: How can a single hormone activate an entire set of functionally related but otherwise quite separate genes, and activate them in a specific sequence and to a specific degree?

The question can be sharpened somewhat by considering the effect of estrogen not on uterine cells but on the cells of the liver. When an egg is being formed in a hen, the estrogen produced by the hen's ovaries stimulates its liver to produce the yolk proteins lipovitellin and phosvitin. Obviously a rooster does not need to synthesize these proteins, but if it is treated with estrogen, its liver will make them in large amounts! A more unequivocal example of the

selective activation of repressed genes by a hormone could scarcely be imagined. What is more, experiments by E. N. Carlsen and his co-workers at the University of California School of Medicine (Los Angeles) have demonstrated that this gene-activating effect of estrogen is remarkably specific. Phosvitin is an unusual protein in that nearly half of its subunits are of one kind: they are residues of the amino acid serine. Carlsen and his colleagues found that estrogen most strongly stimulates liver cells to produce the particular species of transfer RNA that is associated with the incorporation of this amino acid into protein.

The effect of estrogen on liver cells is thus quite different from its effect on uterine cells. Indeed, it has long been

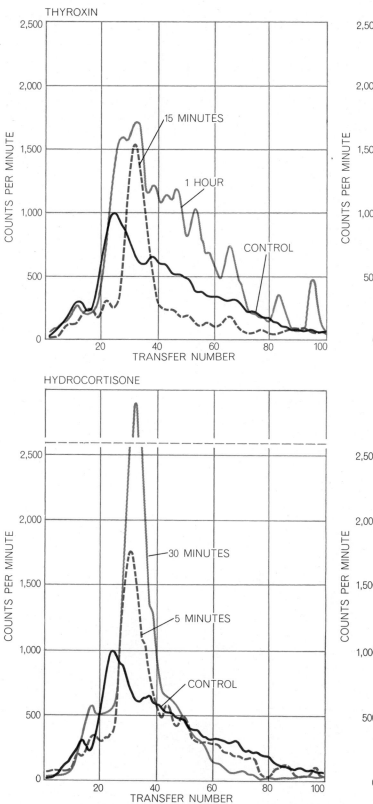

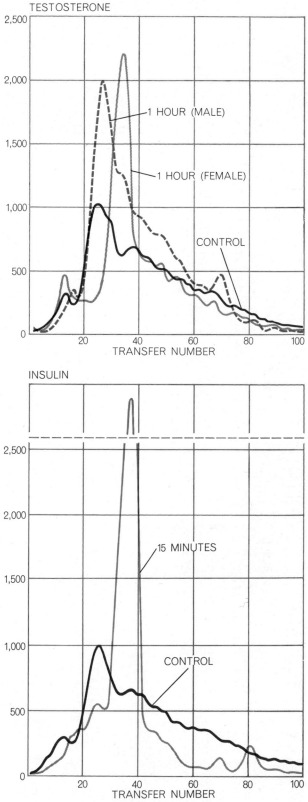

GENETIC ACTIVITY OF SEVERAL HORMONES is indicated by measurements made by Chev Kidson and K. S. Kirby of the Chester Beatty Research Institute in London. Their basic technique was first to administer to rats radioactively labeled orotic acid, which is a precursor of RNA. The tissues of the rat then incorporated the radioactive label into new RNA. Next liver tissue was removed from the rat and the species of RNA called "messenger" RNA was extracted from its cells. When the messenger RNA was analyzed by the method of countercurrent distribution, it gave rise to a characteristic curve (*black "Control" curve in each graph*); "Transfer number" refers to a stage of transfer in the countercurrent-distribution process and "Counts per minute" to the radioactivity of the solution at that point. Then, in separate measurements, rats were first given one of a number of hormones (*top left of each graph*) and shortly thereafter radioactively labeled orotic acid. The curves (*color*) of the messenger RNA obtained from such rats were entirely different, depending on the time that had elapsed before the administration of the orotic acid or on the sex of the animal (*top right*).

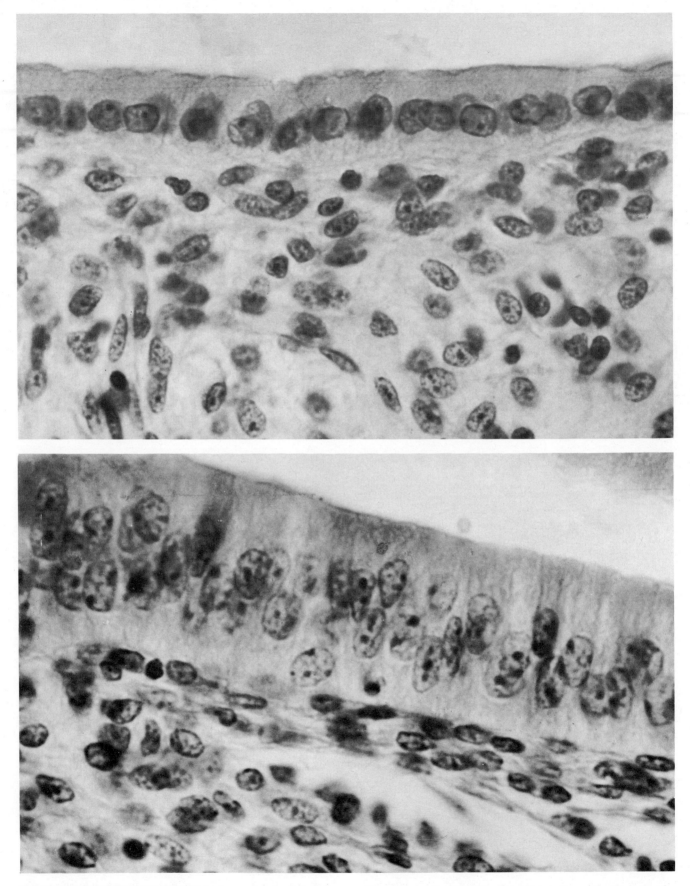

EFFECT OF ESTROGEN ON CELLS in the uterus of rats is demonstrated in these photomicrographs made by Sheldon J. Segal and G. P. Talwar of the Rockefeller Resitute. The photomicrograph at top shows uterine cells from a rat that had not been treated with estrogen; the layer of cells at the surface of the tissue is relatively thin. The photomicrograph at bottom shows uterine cells from a rat that had been treated with the hormone; the layer of cells is much thicker. The effect involves enhanced synthesis of protein.

only messenger RNA but also ribosomal RNA. These events are followed by the increase in enzyme activity. Olga Greengard and George Acs of the Institute for Muscle Disease in New York have shown that if the animal is treated with actinomycin before cortisone is administered, the new enzymes fail to appear in its liver cells.

Another clear case of the activation of genes by an adrenocortical hormone has been demonstrated by Isidore S. Edelman, Rita Bogoroch and George A. Porter of the University of California School of Medicine (San Francisco). They employed the hormone aldosterone, which regulates the passage through the cell membrane of sodium and potassium ions. Tracer studies with radioactively labeled aldosterone showed that when the bladder cells of a toad were exposed to the hormone, the molecules of hormone penetrated all the way into the nuclei of the cells [see illustrations on page 109]. About an hour and a half after the aldosterone has reached its peak concentration within the cells the movement of sodium ions across the cell membrane increases. It appears that this facilitation of sodium transport is brought about by proteins the cell is induced to make, because it will not occur if the cells have been treated beforehand with puromycin, the drug that blocks the synthesis of protein. Moreover, treatment of the cells with actinomycin will block the aldosterone-induced increase in sodium transport through the membrane. Thus the experiments indicate that aldosterone activates genes in the nucleus and gives rise to proteins—that is, enzymes—that speed up the passage of sodium ions across the membrane.

Ecdysone, a steroid hormone of insects, is also believed to be a gene activator. The evidence for this conclusion has been provided by Wolfgang Beermann and his colleagues at the Max Planck Institute for Biology in Tübingen [for further information, see "Chromosome Puffs," by Wolfgang Beermann and Ulrich Clever, Offprint #180]. If the larva of an insect lacks ecdysone, the development of the larva is indefinitely arrested at a stage preceding its metamorphosis into a pupa. Only when, in the course of normal development, the concentration of ecdysone in the tissues of the larva begins to rise does further differentiation take place; the larva then advances to metamorphosis. Ecdysone has been of especial interest to cell biologists because it has been observed

ROOSTER TREATED WITH ESTROGEN (*bottom*) is compared with a normal rooster (*top*). The signs of femaleness induced by estrogen include changes in comb and plumage.

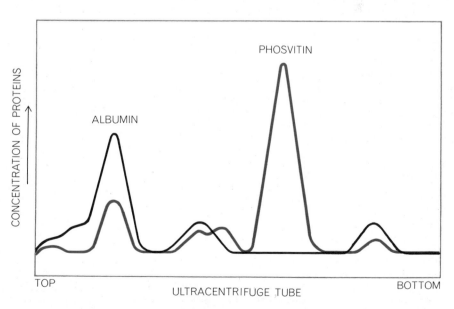

ULTRACENTRIFUGE PATTERNS show that phosvitin, a yolk protein found only in hens, is present in serum extracted from a bird that had been injected with estrogen (*colored curve*) but not in serum from a bird used as a control (*black curve*). Each curve gives the concentration of proteins as they are separated out of a mixture by an ultracentrifuge.

to cause startling changes in the chromosomes within the nuclei of the cells affected by it. Studies of this kind are possible in insects because the cells of certain insect tissues have giant chromosomes that can easily be examined in the microscope. These "polytene" chromosomes develop in many kinds of differentiated cell by means of a process in which the chromosomes repeatedly replicate but do not separate.

In some polytene chromosomes genetic loci, or specific regions, have a distended, diffuse appearance [see illustration below]. Biologists regard these regions, which have been named "puffs," as sites of intense gene activity. Evidence for this conclusion is provided by radioautograph studies, which show that the puffs are localized sites of intense RNA synthesis. In such studies a molecular precursor of RNA is radio-

actively labeled and after it has been incorporated into RNA reveals its presence as a black dot in the emulsion of the radioautograph. According to the view of differentiation presented in this article, different genes should be active in different types of cell, and this appears to be the case in insect cells with polytene chromosomes. In many different kinds of cell—salivary-gland cells, rectal-gland cells and excretory-tubule cells—the giant chromosomes have a different constellation of puffs; this suggests that different sets of genes are active, a given gene being active in one cell and quiescent in another.

On the polytene chromosomes of insect salivary-gland cells new puffs develop as metamorphosis begins. This is where ecdysone comes into the picture: the hormone seems to be capable of inducing the appearance of specific new puffs. When a minute amount of ecdysone is injected into an insect larva, a specific puff appears on one of its salivary-gland chromosomes; when a slightly larger amount of ecdysone is injected, a second puff materializes at a different chromosomal location. In the normal course of events the concentration of ecdysone increases as the larva nears metamorphosis; therefore there exists a mechanism whereby the more sensitive genetic locus can be aroused first. This example of hormone action at the gene level, which is directly visible to the investigator, seems to have provided some of the strongest evidence for the regulation of gene action by hormones. The effect of ecdysone, which is clearly needed for differentiation, appears to be to arouse quiescent genes to visible states of activity. In this way the specific patterns of gene activity required for differentiation are provided.

What about nonsteroid hormones? Here the overall picture is not as clearcut. The effects of some hormones are quite evidently due to gene activation, and yet other effects of the same hormones are not blocked by the administration of actinomycin; a small sample of these effects is listed in the illustration on the opposite page. As for the hormonal effects that are quite definitely not genetic, they fall into one of the following categories.

(1) Some hormones act on specific enzymes; for example, the thyroid hormone thyroxin promotes the dissociation of the enzyme glutamic dehydrogenase. (2) Other hormones, for instance insulin and vasopressin, act on systems that transport things through cell mem-

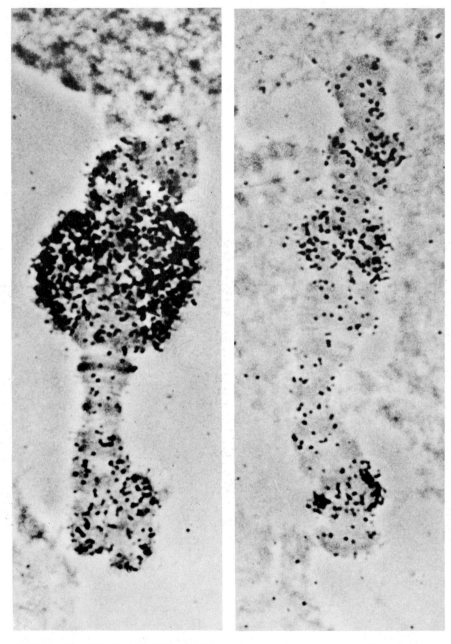

"PUFF" ON A GIANT CHROMOSOME from the salivary gland of the midge *Chironomus tentans* appears after administration of the insect hormone ecdysone. In the radioautograph at left the round area at top center is a puff. The black dots result from the fact that the midge was given radioactively labeled uridine, which is a precursor of RNA. The concentration of dots in the puff indicates that it is actively synthesizing RNA. In the radioautograph at right is a chromosome from a fly that had been treated with actinomycin before receiving ecdysone. No puff has occurred and RNA synthesis appears to be muted. The radioautographs were made by Claus Pelling of the Max Planck Institute for Biology in Tübingen.

HORMONE	EVIDENCE FOR HORMONAL ACTION BY GENE ACTIVATION.	EVIDENCE THAT HORMONAL ACTION IS CLEARLY INDEPENDENT OF IMMEDIATE GENE ACTIVATION.
PITUITARY GROWTH HORMONE	General stimulation of protein synthesis. Stimulation of rates of synthesis of ribosomal RNA, transfer RNA and messenger RNA within 90 minutes in liver. Effect blocked with actinomycin.	
PITUITARY ACTH	Stimulates adrenal protein synthesis. Messenger RNA and total RNA synthesis stimulated.	Steroid synthesis in isolated adrenal sections is independent of RNA synthesis and is insensitive to actinomycin D.
THYROXIN	Promotes new messenger RNA synthesis within 10 to 15 minutes of administration, promotes stimulation of all classes of RNA by 60 minutes. Promotes increase in RNA–DNA polymerase at 10 hours, later promotes general increase in protein synthesis.	Causes isolated, purified glutamic dehydrogenase to dissociate to the inactive form. Affects isolated mitochondria in vitro.
INSULIN	Promotes 100 percent increase in rate of RNA synthesis. Causes striking change in messenger RNA profile within 15 minutes of administration to rat diaphragm; effect blocked with actinomycin. Actinomycin-sensitive induction of glucokinase activity.	Actinomycin-insensitive increase in ATP synthesis and in glucose transport into cells; mechanism appears to involve insulin binding to cell membrane, occurs at 0 degrees C.
VASOPRESSIN		Actinomycin-insensitive promotion of water transport in isolated bladder preparation under same conditions in which aldosterone action is blocked by actinomycin.

SUMMARY OF EXPERIMENTAL EVIDENCE is given in table. Facts indicating that hormones activate the genes (*middle column*) are compared with facts suggesting that hormonal action does not entail the immediate activation of the genes (*column at right*).

branes; indeed, it is believed that both of these hormones attach themselves directly to the membranes whose function they affect. (3) Still other hormones rapidly activate a particular enzyme; phosphorylase, a key enzyme in determining the overall rate at which glycogen is broken down, is converted from an inactive form by several hormones, including epinephrine, glucagon and ACTH.

This does not alter the fact that many nonsteroid hormones operate at the gene level. Some of the best evidence for this statement is provided by studies of several hormones made by Chev Kidson and K. S. Kirby of the Chester Beatty Research Institute of the Royal Cancer Hospital in London. They separately injected rats with thyroxin, testosterone, cortisone and insulin and then mea-

sured the synthesis of messenger RNA by the rats' liver cells [*see illustration on page 113*]. The most striking aspect of their measurements is the extremely short time lag between the administration of the hormone and the change in the pattern of gene activity. The activation of genes in the nuclei of the affected cells occurs so quickly that one is tempted to assume that it is an initial effect of the hormone.

Here, however, we come face to face with a basic problem that must be solved in any attempt to explain the exact molecular mechanism of hormone action. The problem is simply that of identifying the initial site of reaction in a cell exposed to a hormone. Does a hormone move directly to the chromosome and exert its effect, so to speak, "in person"? As we have seen, aldoste-

rone does appear to enter the nucleus, but there is little real evidence that other hormones do so.

For many years biologists have been looking for the "receptor" substance of various hormones. The discovery that hormones ultimately act on genes makes this search all the more interesting. The evidence presented here only goes as far as to prove that an early stage in the operation of many hormones is the selective stimulation of genetic activity in the target cell. The molecules of the hormones range in size and structure from the tiny molecule of thyroxin to the unique multi-ring molecule of a steroid and the giant molecule of a protein; how these various molecules similarly affect the genetic apparatus of their target cells remains an intriguing mystery.

Part IV

SENSORY RECEPTORS

IV
Sensory Receptors

INTRODUCTION Much of the electrical activity that is eventually transformed by the central nervous system into an appropriate behavioral output begins in sensory organs. These range in complexity from scattered isolated cells in the skin that respond to mechanical deformation or to temperature change, to those that form highly complex, organized sensory structures in eyes, ears, or olfactory organs. The properties of sensation result from the abilities of receptor cells to change, or *transduce*, a variety of forms of stimulus energy into the common language of membrane depolarization and propagated electrical signals. The cells responsible for such transductions are, within a given category or *modality* of stimulus energy, quite similar. Most photoreceptor cells, for example, contain a series of internal membranes, stacked on top of one another; the membranes may be in the form of a stack of discs derived from cilia, as in vertebrate rods, or of a honeycomb of microtubules, as in the photoreceptors of arthropods, but the basic plan is the same.

Despite the common properties implied by the basic transduction process, we know that the performances of sensory systems, even within the same category, differ dramatically. Consider as an illustration the receptor systems that convert mechanical energy into impulses. In some of the touch receptors of human skin, a naked nerve ending is used for this purpose. In the Pacinian corpuscles found in joints and mesenteries and described in the first article of this section, the nerve ending is enclosed in a capsule; in the stretch receptors of muscles, it is wrapped around a specialized muscle fiber. In more complex receptor systems, the transducer may be a special cell derived from an epithelial layer. The neuromast cells of vertebrates have a crown of sensory cilia which, when subjected to a shearing force, somehow transmit the excitation to nerve fibers that connect with the base of the cell. Such receptors, which are coupled to the outside world in different ways, are employed for various purposes.

In the semicircular canals of the inner ear, the cilia of neuromast cells are embedded in a gelatinous cupula. The cupula, which is placed in a swelling of the fluid-filled canal, acts like a swinging door; acceleration in one direction in the plane of the canal twists the door, and produces a shearing force on the hairs so that the cells excite a higher rate of discharge in the nerve cells connected to them. Acceleration in the reverse direction lowers the rate of discharge. In the cochlea, the cilia of cells that are similar to the neuromast cells of the semicircular canals project to an overlying membrane, and traveling waves in the membrane that bears these cilia are transmitted from the eardrum through a series of bones in the middle ear. A shearing force similar to that which excites the cells of the semicircular canals excites the cells of the cochlea; but because the coupling system is so different, high-frequency airborne sound waves, rather than accelerations of the head, are the effective natural stimuli.

The properties of the cell membrane that are of significance in converting mechanical to electrical energy are discussed by Werner R. Loewenstein in "Biological Transducers." Loewenstein has based his analysis on one of the simpler types of primary receptor neuron, the Pacinian corpuscle. His results illustrate an important principle that applies to a variety of sensory events: the regions of membrane that transform mechanical energy into graded depolarization are separated from those that convert this graded depolarization into propagated, all-or-none nerve impulses. Moreover, the first events of energy transduction are localized in the membrane of the nerve ending itself, and do not depend upon the specialized encapsulation that covers it. This specialized encapsulation couples the site of transduction in an effective way to the environment.

The "special senses" are so called because of their special importance in man's own perceptual world. The special sense organs are composed of many thousands of such receptor units. These transduce various kinds of signals into the common language of nerve impulses; these signals, in turn, are filtered by a complex network of nerve cells that extracts information of special importance to the owner. The second article in this section is concerned with one of these special sense organs—a mechanoreceptor, but one much more complex than the Pacinian corpuscle. In "The Ear," Georg von Békésy, who was awarded the Nobel Prize in Physiology and Medicine for his research on the auditory system four years after the publication of this article, considers the total performance of this most remarkable of the mechanically sensitive receptors. Hearing is largely the result of the ingenious coupling of fundamentally simple receptor cells to the source of energy; the primary sensory cells are almost identical to those of the semicircular canals or those of the lateral line of fish, but perform very differently. Large-amplitude, low-energy sound waves that arrive at the eardrum are transformed to low-amplitude, high-energy events in the fluid-filled, helical cochlea of the inner ear. Within the cochlea an even more useful conversion takes place: the basilar membrane, upon which the sensory cells are borne, is vibrated in a series of traveling waves that change these mechanical forces into tension directed transversely against the hairs of the receptor elements, creating a shearing force. It is the task of the central nervous system to identify, in a barrage of incoming signals from many receptors, *which* receptors are most strongly excited. Each frequency of sound vibrates the basilar membrane maximally in a different place, but this excitatory locus is flanked by regions in which the movement, though less, is still high enough to generate significant impulse frequencies in the sensory cells located there. More recent experiments have shown that the specific detection of pitch is aided by the phenomenon of *lateral inhibition*. A nerve cell connected to receptors in a given place on the basilar membrane sends inhibitory signals to its neighbors. In this way a strongly excited cell can suppress a weakly responding adjacent one. Thus neurons that report from the focus of basilar membrane vibration dominate the signals that reach the conscious level of the central nervous system, giving the perceptual impression of a relatively pure tone.

The final two articles deal with a different kind of transduction, that of light into electrical energy, as it takes place in the most complex of sense organs. In "Eye and Camera," George Wald describes the optics and the photochemistry of eyes and pursues the analogy between them and similar man-made devices. Though this article was published in 1950, its discussion of the basic properties of vertebrate vision is as accurate today as it was then, and its comparisons between visual systems and optical devices are especially instructive. It is perhaps surprising that Wald only touches upon the biochemistry of visual pigments, about which he and his colleagues have gathered most of the basic information. The rod cells of the retina invest more than 40 percent of their dry weight in the visual pigment, rhodopsin, which consists of a protein coupled to a molecule which Wald discovered to be a derivative of vitamin A. This complex maximally absorbs light of wavelength 500 millimicrons; this accounts for the fact that the dark-adapted eye is most sensitive to the light of this wavelength, which is blue-green. When the complex absorbs light, the vitamin A derivative isomerizes, and then it dissociates from the protein component. The reunion of the two depends upon the ability of retinal tissue to reisomerize the vitamin A to its proper configuration. This explains why an eye that has just been exposed to bright light must remain in the dark for several minutes before it can perceive dim objects.

A similar chemical mechanism exists in the cones of the retina, which are responsible for vision in bright light. The cone pigments differ from those of rods, however, in that their proteins combine much more readily with the vitamin A derivative than do the proteins of rod pigments. This enables the cones to maintain their sensitivity in the very bright light the eye must operate in during the day. Although the biochemistry of these visual pigments was studied in the test tube by Wald and others, analysis of them in the living eye, where parallel physiological measurements of sensitivity could be made, seemed out of the question. W. A. H. Rushton made such analysis possible by the development of a remarkable technique of reflection photometry. The method, and its application to human color vision, are reported in "Visual Pigments in Man." Rushton describes the measurements used to determine the relation between the concentration of rhodopsin and rod sensitivity. He also provides direct evidence for the existence of separate visual pigments in man, and correlates their absence with the familiar types of color blindness.

13

BIOLOGICAL
TRANSDUCERS

WERNER R. LOEWENSTEIN   August 1960

Aristotle's maxim "Nothing is in the mind that did not pass through the senses" is questioned by some schools of philosophy. If "brain" is substituted for "mind," however, and the statement is made a physiological rather than a philosophical one, it then becomes literally true. For higher organisms the sensory receptor furnishes the only means of gaining information about the surrounding world.

This is not the case with primitive one-celled organisms; the cell is directly excited by the environmental stimulus and responds through movement, secretion and so on. But as organisms became more complex in the course of evolution, many of their body cells lost direct contact with the outside world. Certain cells appeared that specialize in the reception of external stimuli. They are, in general, attuned to a single type of stimulus: the rods and cones in the retina respond to light, the thermal receptors in the skin respond to heat and cold, and the mechanoreceptors in muscle respond to mechanical stimuli such as stretching and pressure. The specialized receptor is an outgrowth of a nerve cell, or is in intimate contact with one. Environmental stimuli of the appropriate kind excite the receptor, and the excitation is conveyed to other parts of the organism along nerve circuits of varying complexity.

From the physical point of view the sensory receptors are transducers, that is, they convert one form of energy into another. The various types of receptor convert the particular form of energy to which each is attuned into the electrical energy of the nerve impulse. One may compare them with the transducer devices which modern technology has developed in great variety for automatic control of machines and factories—de-

vices that measure temperature, pressure, rate of flow and so on, and feed their measurements into the artificial nerve-circuits of the control system. The biological transducers that nature took somewhat longer to develop have remarkable sensitivity and efficiency.

The mechanoreceptors were among the earliest to evolve; they enabled primitive marine animals to maintain their orientation with respect to gravity, to detect obstacles and to sense vibrations produced by other animals. The evolution of life on land brought the development of mechanoreceptors sensitive to vibrations of the air; with the growth of specialized internal organs and the need for fast regulatory mechanisms came the development of receptors sensitive to internal mechanical stimuli. Vertebrates possess mechanoreceptors in all organs in which active or passive movements occur, including the digestive tract, the lungs, the heart and the blood vessels, as well as the skin and the skeletal muscles. These receptors feed into the nervous system information about movement, tension and pressure.

The transducer mechanism in the mechanoreceptor was first demonstrated in 1950 by Bernhard Katz of University College London. He discovered that the stretching of a muscle spindle (a mechanoreceptor built into skeletal muscle) generates a local electric current. When the current reaches a certain intensity, he found, it triggers the firing of an impulse in the nerve fiber leading from the muscle spindle to the higher nerve-centers [see "The Nerve Impulse," by Bernhard Katz, Offprint #20, for further information].

This result did not come as a complete surprise. Since the time of Luigi Galvani and Alessandro Volta, physiolo-

gists have used electric currents to trigger impulses in nerve fibers. But the current that tripped the nerve impulse in Katz's experiment did not come from a battery or any other external power supply; it came from within the muscle spindle itself. Several other workers soon obtained similar results in experiments on mechanoreceptors from organisms as diverse as the crayfish and the cat.

In all these receptors mechanical stimulation produces a weak local current. It is known as the generator current, because it in turn triggers the nerve impulse. The generator current is the earliest detectable step in the transducer sequence. In a typical mammalian mechanoreceptor the current follows the stimulus within a thousandth of a second. Moreover, the generator current increases in direct relation with the increase in the energy of the stimulus. This relation between the input and output of energy resembles that of a good mechanoelectric transducer of the type represented by a carbon microphone. In the microphone mechanical deformation of the disk of carbon by the impinging sound wave reduces its electrical resistance, permitting the electric current to flow through it in strength proportional to the intensity of the sound. The output current from the biological and from the man-made transducer thus conveys a measurement of the strength of the stimulus and of the sound.

With this much known about the mechanoreceptor, the next problem was to discover which of its several structural components is the transducer element. I became interested in the question in 1955, and looked for an appropriate receptor with which to look into it. The Pacinian corpuscle—found in the skin, muscles, tendons and joints of mammals—occurs in an especially accessible

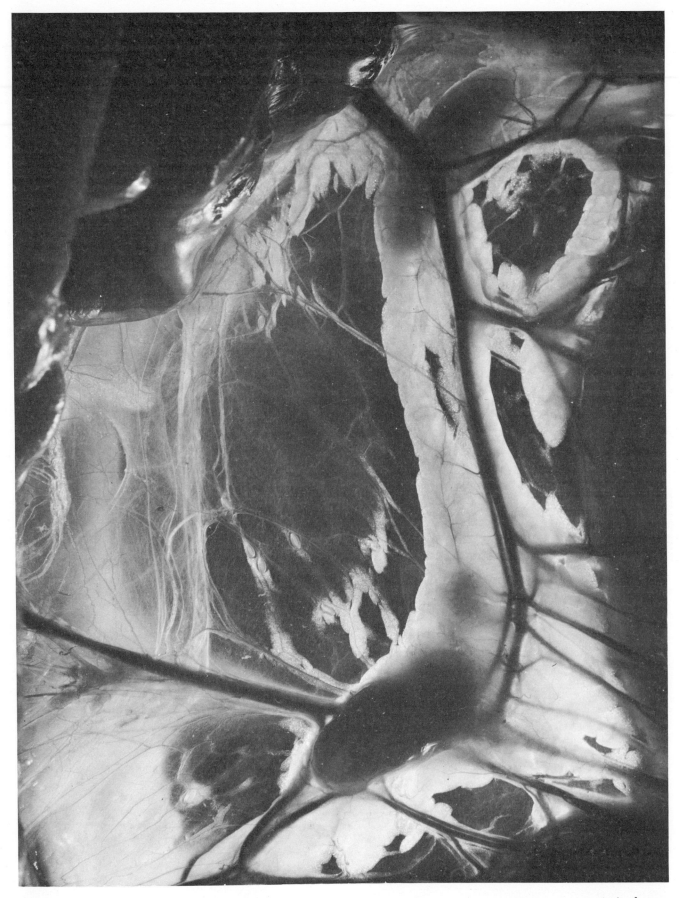

PACINIAN CORPUSCLES are small ellipsoidal bodies, three of which are visible at the lower left side of the dark area in the center of this photograph. Here the corpuscles are located in the mesentery (a thin membrane attached to the intestine) of a cat. They are enlarged some four diameters. In actual size each corpuscle is about one millimeter long and .6 millimeter thick.

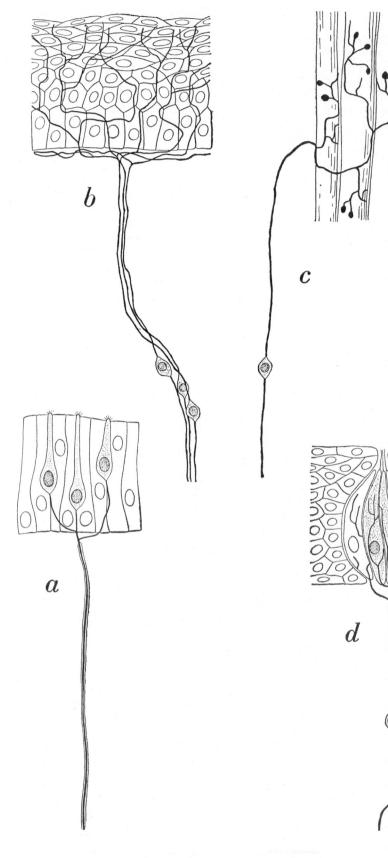

form in the mesentery of the cat, that is, the fold of tissue that connects the intestine to the rear wall of the abdomen. From the mesentery it can be easily removed and kept alive for hours in a suitable salt solution. Moreover, John Gray and his colleagues at University College London had already shown that the corpuscle can be stimulated mechanically and the resulting current recorded through an electrode attached to its nerve fiber. But the greatest experimental advantage of the Pacinian corpuscle is its large size. It is truly a giant among receptors, measuring almost one millimeter long and .6 millimeter thick. Under the microscope it looks like an onion: it consists of many concentric layers, or lamellae, ultimately enclosing a nerve ending. The nerve fiber leading from it can be kept functioning along with the corpuscle. This easily manageable unit became our experimental subject.

For mechanical stimulation of the corpuscle we used a piezoelectric crystal of the type employed in phonograph pickups. In its familiar applications such a crystal converts mechanical energy into electrical energy. But it can also be used to convert electrical energy into mechanical; it deflects when a voltage is applied across it, the deflection increasing linearly with voltage. In our experiment a glass stylus transmitted the deflection to the corpuscle as a readily measurable stimulus.

To locate the transducer site in the corpuscle, R. Rathkamp and I adopted a direct approach; we removed pieces of its structure in the hope that we might isolate the part essential to the transducer process. We peeled off the outer layers of the corpuscle, stimulating it after each step of dissection. It soon became clear that more than 99.9 per cent of the mass of the corpuscle could be dissected away without impairing the transducer function. A preparation consisting of the nerve ending surrounded only by a thin sheath of "inner core" was as good a transducer as the intact corpuscle. When stimulated mechanically, it produced generator currents which, if the outgoing nerve fiber was left intact, triggered the firing of nerve impulses in the normal manner.

Since the transducer mechanism had to be somewhere within the inner core, the nerve ending appeared to be the most likely site. We tried to strip away the inner core around the nerve ending, but did not succeed because the tissue here is only about .01 millimeter thick and is too intimately attached to the

SENSORY NERVES such as olfactory receptors of the nose (*a*) have their cell body at the periphery; pain receptors of the skin (*b*), mechanoreceptors of muscle (*c*) and taste receptors of the tongue (*d*) have their cell bodies buried in the organism. All of them act as biological transducers. The muscle receptors end in leaflike structures; the others are bare.

nerve. However, with a pair of micromanipulators we were able to tease off the outer layers, cut out a few pieces of the remaining layers and puncture the rest with a fine glass needle. This preparation, in which the nerve ending was the only intact structure, was still a good transducer, producing currents as in the intact organ.

Thus we could not yet say whether the core tissues or the nerve ending produced the current. Since the technique of microdissection could not completely free the nerve ending from the surrounding core material, we decided to try the opposite tack. We prepared an "endingless" core by severing the nerve fiber of the corpuscle in the living animal and allowing the nerve ending to degenerate. When this preparation was removed two or three days later and stimulated, it failed to produce a generator current, indicating that the nerve ending was indeed the transducer site.

Investigators have now begun to take a closer look at the mechanosensitive nerve-ending with the electron microscope. The two different types, the Pacinian corpuscle and the muscle spindle, that have been examined so far have three characteristics in common: (1) the absence of the insulating sheath of myelin found in nerve-circuit cells; (2) the presence, characteristic of cells that must produce large amounts of energy, of a relatively large number of mitochondria, the small bodies associated with metabolic activity; and (3) the presence of many small, round structures of unknown function that resemble certain structures found in motor-nerve endings.

The generator current produced by the nerve ending in the mechanoreceptor does not itself travel along the nerve fiber. It serves merely to trigger the nerve impulse which does propagate along the nerve circuit, often for considerable distances. Generator current and nerve impulse originate at different places in the corpuscle. Rathkamp and I located the site at which the impulse originates by blocking the activity of selected portions of the nerve fiber. The myelin sheath that covers this fiber extends well into the corpuscle, and is interrupted at intervals of about .25 millimeter by small gaps known as the nodes of Ranvier. In a dissected preparation several nodes and the nerve ending are visible under the microscope; the first node lies within the corpuscle. We applied pressure to the nodes with a wisp of glass about .004 millimeter in diameter, blocking the electric activity of each one in turn. The nerve continued to fire its impulse in response to mechanical stimulation of the nerve ending and to the resulting generator current until we blocked the first node. Plainly the first node is the point at which the nerve impulse starts.

The production of the nerve impulse could now be visualized as a two-step

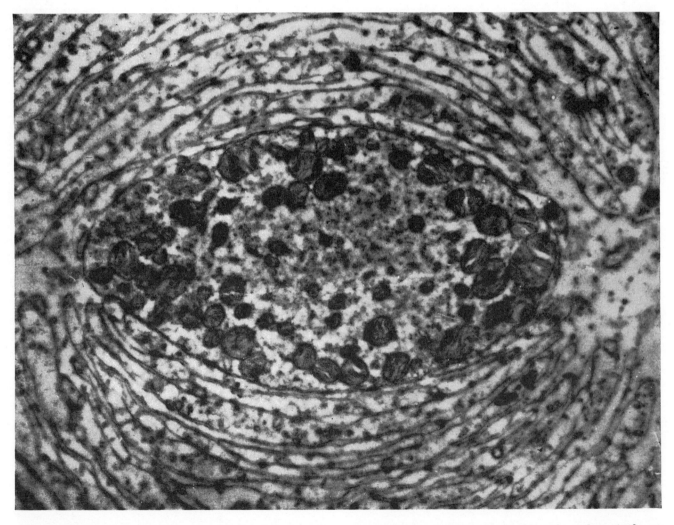

NERVE ENDING of the Pacinian corpuscle is enlarged some 20,000 diameters in this electron micrograph made by D. C. Pease and T. A. Quilliam of the University of California. The section cuts across the long axis of the ending, which is the oval area in the center. The round, dark bodies within this area are mitochondria. Around the area are the layers, or lamellae, of the corpuscle core.

process, each step related to a particular structure. Under resting conditions there is a potential across the "receptor" membrane or the nerve ending in the corpuscle; the inside of the ending is several tenths of a volt negative with respect to the outside. This potential appears to be produced and maintained by unequal concentrations of ions (that is, charged atoms or molecules) on the two sides of the membrane. The ending thus resembles other excitable—nerve and muscle—tissues. It differs markedly from such tissues, however, in its high sensitivity to mechanical stimulation: deformation of the receptor membrane leads to a drop in resting potential. Under resting conditions the membrane resistance is so high that no appreciable net ion current leaks through it. Distortion produces a decrease in resistance which allows ions to move along their concentration gradients across the membrane, causing the resting potential to drop. Mechanical stimulation thus results in a transfer of charges across the receptor membrane; this constitutes the generator current. Part of the generator current flows through the first node, where it triggers the nerve impulse [*see illustration on page 129*]. Apparently the generator current must be of a minimal intensity and must reach this intensity at a minimal rate in order to have this effect.

The nerve impulse has been far more thoroughly studied than has the generator current. It is the signal which in the sensory nerve fibers travels from the periphery to the nerve centers, conveying information about color, shape, texture, temperature, sound and so on; and in the motor and secretory fibers in the opposite direction, conveying orders for contraction of muscles or for secretion of glands. No basic qualitative differences are known to exist among these fibers. The nerve impulse is a pulse of current that under equal conditions and in any given fiber is of the same size and duration [for further information, see "The Nerve Impulse," by Bernhard Katz, Offprint #20]. In the case of the fiber that leads out of the Pacinian corpuscle, the generator current excites the membrane of the first node of Ranvier. The node responds with a change in permeability to certain ions, and the result is an abrupt surge of current through the membrane lasting for a thousandth of a second or so. From here on propagation of the nerve impulse follows the pattern of other myelin-insulated nerve fibers. Part of the current set up at the first node flows through the second node, and this current is more than sufficiently strong to trip off a current pulse of the same magnitude in the next node. In this manner the impulse regenerates itself at each node and propagates at full amplitude to the nerve centers. That amplitude bears no relation, however, to the intensity of the generator current that triggered it at the first node.

Here one encounters an apparent difficulty. The receptor chain begins with a mechanical stimulus of a given energy content; the stimulus is transformed into a generator current with an energy content proportional to that of the stimulus, and now the chain ends in a signal—the nerve impulse—with an energy content that bears no relationship to either of the preceding events. How can the all-or-none signal of the nerve impulse convey quantitative information about the strength of the stimulus along the nerve fiber?

A clue is furnished by the fact that many man-made information systems operate with all-or-none signals. Digital computers send and store information by all-or-none pulses, and the telegraph transmits messages of the most varied content with dots and dashes, two types of all-or-none pulses. Biological sensory systems operate on the same digital principle and use only dots. As long ago as 1926 E. D. Adrian of the University of Cambridge recorded the signals from skin and muscle receptors and made the far-reaching discovery that the frequency—not the amplitude—of the nerve impulses varies with the strength of the stimulus. Adrian and his colleagues soon broke the code of other sensory systems, and investigators in many parts of the world took up the task of decoding the rest. All turned out to be frequency-modulating systems which translate an increase in the intensity of the stimulus into an increase in the frequency of the nerve impulse. In fibers connected to mechanoreceptors it has been found that the impulse frequency varies with the intensity of the generator current and with the rate at which it increases.

For the student of biological transducers the most interesting and still not completely answered question is how the receptor produces a generator current that varies with the strength of the stimulus and serves to measure it. If the membrane were a carbon microphone, current would flow only in the region distorted by the stimulus, and this would explain the linear relationship between the flow of current and the strength of the stimulus. But physiologists deal with information systems in which signals are carried by ions through conduction lines composed chiefly of water and salts. In a

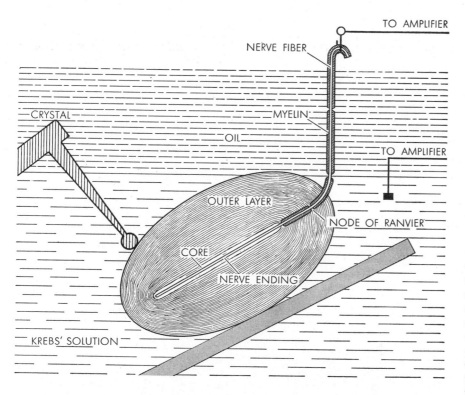

ISOLATED PACINIAN CORPUSCLE was stimulated by a rod attached to a vibrating phonograph crystal (*left*). The resulting nerve impulse was picked up by a pair of electrodes (*right*). This illustration also schematically depicts the corpuscle and its various parts.

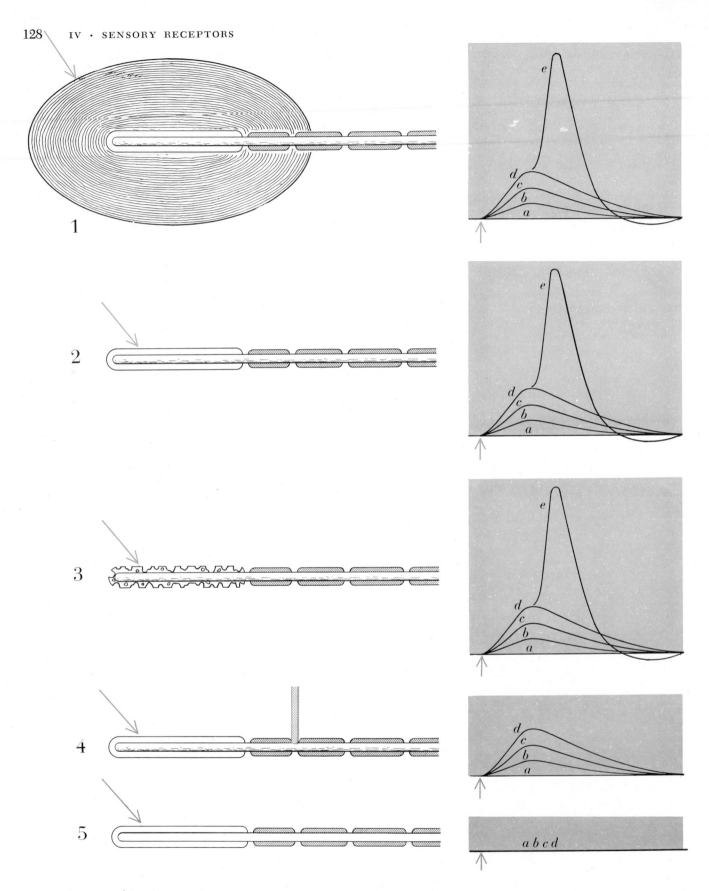

DISSECTION OF CORPUSCLE revealed the site of the transducer mechanism. Stimulation (*arrow*) of the corpuscle when intact (1), with outer layers removed (2) or after partial destruction of the core sheath (3) produced the same responses. A weak stimulus produced a weak generator current (*a*); progressively stronger stimuli produced correspondingly stronger generator current (*b* and *c*); the threshold stimulus (*d*) fired an all-or-none nerve impulse (*e*). When the first node of Ranvier was blocked (4), no all-or-none impulse could be induced. After degeneration of nerve ending (5), receptor did not respond at all to stimuli.

typical conductile nerve-fiber a region of membrane cannot remain unexcited for long next to an excited one. The excited region generates an electric current that is more than sufficient to excite the neighboring region; this region in turn excites the next, and so on. In this manner excitation sweeps in a wave over the entire membrane. Such is the behavior of the nerve fiber. Does the transducer membrane in the corpuscle act like a carbon microphone or like a nerve fiber? Or, to rephrase the question, is current generated throughout the membrane, or generated only in the mechanically distorted region?

A technique worked out in my laboratory at Columbia University provided an approach to this question. We applied a mechanical pulse to a tiny patch of membrane, about .03 millimeter in diameter, and measured the resulting generator currents at varying distances from the stimulated site. We found that the current decreases exponentially with the distance. This is precisely what one would expect if the excitation were restricted to the stimulated region of the membrane. Some of the generator cur-

rent leaks into the unexcited regions, but the unexcited membrane acts like a passive cable in which signals fade out with distance. The experiment showed clearly that excitation in the receptor membrane is confined to the mechanically distorted region.

I have recently succeeded in exciting two generator currents in the same nerve ending by simultaneously stimulating two spots on the membrane separated by about .5 millimeter. This experiment brought to light a most significant effect: Two such independently generated currents sum to produce a single large generator-current.

The summation of two or more currents, each generated at a separate active site on the membrane, thus promised to explain why the intensity of the generator current is proportional to the strength of the stimulus. To test this hypothesis we applied a series of mechanical stimuli of progressively increasing strength to part of the nerve ending and scanned the membrane with a microelectrode. We found that as the stimulus strength increased, deforming progressively more of the receptor mem-

brane, the excitation spread over a correspondingly greater area.

This opened the rather attractive possibility that the receptor membrane might contain a great number of tiny active sites that show the conventional all-or-none response to mechanical stimulation, and yet give rise to a continuously variable generator current which represents the sum of the currents generated by each of these sites. Theoretically this model can account for the entire input-output response of the mechanoreceptor, and it is alluring because of its simplicity. Unfortunately it must remain rather tentative because there may still be an intensity factor at work. A membrane node that fits the experimental results just as well is one that operates on the basis of spatial summation and in which each active site generates a variable current. We have as yet no way to distinguish between these two possibilities. The only evidence of excitation that can be traced at present is the flow of electric current through the membrane. Current flow is a good index, but a rather blurred one. Even the finest microelectrode is far too large to discrimi-

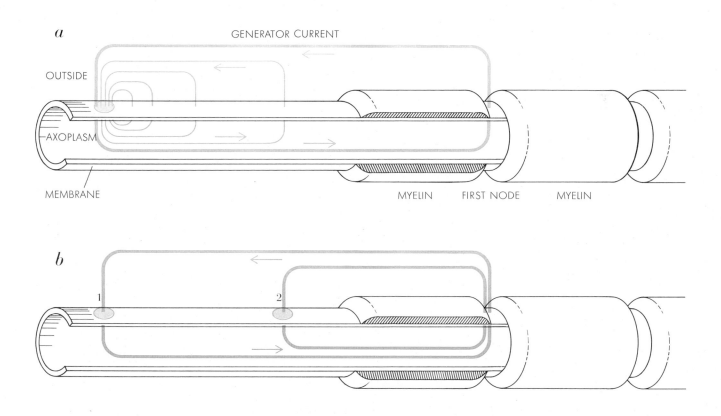

a

GENERATOR CURRENT

OUTSIDE

AXOPLASM

MEMBRANE   MYELIN   FIRST NODE   MYELIN

b

1   2

GENERATOR CURRENT arises in limited region of the receptor membrane of the nerve ending (*at left in top diagram*) in response to mechanical stimulation of that area. This current dies out quickly over the nonstimulated area of the membrane, but if sufficient current reaches the first node of the conducting fiber (*at right*) it triggers a nerve impulse. Two or more generator currents produced by stimulation of separate membrane regions (*bottom diagram*) sum to produce a strong current at the first node.

nate changes in the molecular structure of the membrane which apparently account for the flow of electric charges.

The finding that current flow increases with the area of membrane deformed by the stimulus suggested that the excitation might be a statistical process. In other words, the deformation of a given area of membrane might be expected to excite a statistically fluctuating number of active sites, producing a statistically fluctuating generator current. The fluctuations proved, upon measurement, to be large. My colleague Nobusada Ishiko and I were able to show that a constant mechanical stimulus elicits a generator current that fluctuates at random, and that these fluctuations increase with stimulus strength as predicted by the spatial-summation model.

One may now perhaps picture the receptor as a membrane in which there is a number of tiny holes. In the resting state the holes are too small for certain ions to pass. Mechanical deformation of a given area opens (excites) a statistically determined number of holes, and the ions move through these, setting up the generator current. The opening may occur directly—through stretching, for example—or indirectly, through some biochemical process. As the stimulus strength increases, an increasing number of holes opens up and a correspondingly increasing number of ions passes through the membrane.

A glance at an electron micrograph of the receptor suggests that the number of ions available for transfer must be limited. The lamellae of the receptor core, which are formidable barriers for ion diffusion, are tightly wrapped around the ending, leaving little fluid space between the receptor membrane and the first lamella [see illustration on page 126]. This prompted Stanley Cohen and me to see whether the receptor could be "depleted" of ions by repeated stimulation. We found that the reduction of responsiveness is considerable. For example, a stimulus that produces a generator current of 100 units in the fully rested receptor produces a current of barely 10 units after the application of 5,000 stimuli (at the rate of 500 per second) and none at all after 7,000. The effect is now being studied in our laboratory by Sidney J. Socolar and Masayasu Sato. Preliminary results suggest that the transfer of charges across the membrane depends on the interplay of two competing processes: the depletion and the restoration of something, or the inactivation and reactivation of something. But what this something is—whether it is ions or some

chemical precursor—is not yet clear.

The linear relationship between stimulus and response observed in the mechanoreceptor is characteristic of all sensory receptors, and in all sensory circuits the rise in the intensity of the generator current increases the frequency of the nerve impulses dispatched to the higher

nerve centers. Another important factor reinforcing the correspondence between input and output in the sensory system is the fact that the receptors occur in groups. Thus a weight pressing on the skin excites many Pacinian corpuscles; light shining into the eye excites a large number of photoreceptors. The greater

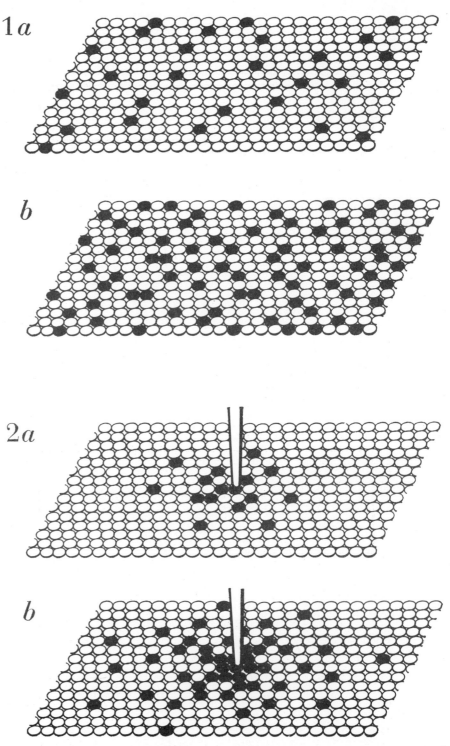

EXCITED REGIONS (black) on receptor membranes increase in number as strength of stimulus increases. A possible pattern of spread with increase in stimulus strength is shown for the case in which the stimulus is distributed rather uniformly over the entire membrane, as probably occurs in the intact Pacinian corpuscle (1a and b), and for the case in which a small area of membrane is stimulated by means of a fine stylus (2a and b).

the strength of the stimulus, the greater the number of excited receptors; if the light is brightened or the pressure on the skin is increased, more receptors are excited, and hence more parallel nerve-fibers fire off impulses to higher centers. Moreover, since several receptor endings are generally the twigs of a single nerve-fiber, there will be considerable con-

vergence of impulses in the common fiber. Thus when a strong stimulus increases the number of activated receptors, this also increases the frequency of impulses traveling along the fiber.

The higher nerve-centers may decipher a receptor message and estimate the strength of the stimulus in at

least two ways: by counting the number of parallel information channels engaged in impulse traffic, and by gauging the frequency and sequence patterns of impulses transmitted in each information channel. However, the nature of the mechanisms that decode and store such information is still one of the many open questions on the brain-mind frontier.

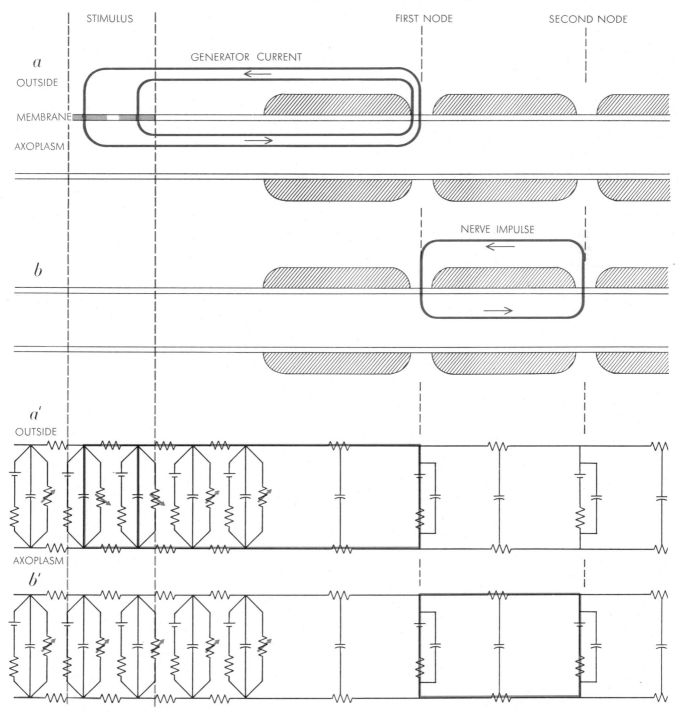

MECHANISM OF TRANSDUCER can be explained by analogy with an electrical circuit. Stimulating a portion of the receptor membrane (*colored area*) causes a drop in the resistance of this membrane region to ion movement. This leads to a transfer of charge, a drop in membrane potential and the generation of current (*colored loops*) in that region. The current flows through the first node of Ranvier and triggers a nerve impulse, the current of which in turn excites the second node (*b*) and so on. In the electrical equivalent (*a'* and *b'*), membrane potential is represented by battery units, and membrane resistance and capacitance is distributed uniformly over a large number of units. (Only five are shown here.) Excitation of a unit causes its resistance to drop (*colored arrows*) and starts events indicated by colored loops. Current discharging into inactive receptor membrane is omitted.

14

THE EAR

GEORG VON BÉKÉSY   August 1957

Even in our era of technological wonders, the performances of our most amazing machines are still put in the shade by the sense organs of the human body. Consider the accomplishments of the ear. It is so sensitive that it can almost hear the random rain of air molecules bouncing against the eardrum. Yet in spite of its extraordinary sensitivity the ear can withstand the pounding of sound waves strong enough to set the body vibrating. The ear is equipped, moreover, with a truly impressive selectivity. In a room crowded with people talking, it can suppress most of the noise and concentrate on one speaker. From the blended sounds of a symphony orchestra the ear of the conductor can single out the one instrument that is not performing to his satisfaction.

In structure and in operation the ear is extraordinarily delicate. One measure of its fineness is the tiny vibrations to which it will respond. At some sound frequencies the vibrations of the eardrum are as small as one billionth of a centimeter—about one tenth the diameter of the hydrogen atom! And the vibrations of the very fine membrane in the inner ear which transmits this stimulation to the auditory nerve are nearly 100 times smaller in amplitude. This fact alone is enough to explain why hearing has so long been one of the mysteries of physiology. Even today we do not know how these minute vibrations stimulate the nerve endings. But thanks to refined electro-acoustical instruments we do know quite a bit now about how the ear functions.

What are the ear's abilities? We can get a quick picture of the working condition of an ear by taking an audiogram, which is a measure of the threshold of hearing at the various sound frequencies. The hearing is tested with pure tones at various frequencies, and the audiogram tells how much sound pressure on the eardrum (*i.e.*, what intensity of sound) is necessary for the sound at each frequency to be just barely audible. Curiously, the audiogram curve often is very much the same for the various members of a family; possibly this is connected in some way with the similarity in the shape of the face.

The ear is least sensitive at the low frequencies: for instance, its sensitivity for a tone of 100 cycles per second is 1,000 times lower than for one at 1,000 cycles per second. This comparative insensitivity to the slower vibrations is an obvious physical necessity, because otherwise we would hear all the vibrations of our own bodies. If you stick a finger in each ear, closing it to air-borne sounds, you hear a very low, irregular tone, produced by the contractions of the muscles of the arm and finger. It is interesting that the ear is just insensitive enough to low frequencies to avoid the disturbing effect of the noises produced by muscles, bodily movements, etc. If it were any more sensitive to these frequencies than it is, we would even hear the vibrations of the head that are produced by the shock of every step we take when walking.

On the high-frequency side the range that the ear covers is remarkable. In childhood some of us can hear well at frequencies as high as 40,000 cycles per second. But with age our acuteness of hearing in the high-frequency range steadily falls. Normally the drop is almost as regular as clockwork: testing several persons in their 40s with tones at a fixed level of intensity, we found that over a period of five years their upper limit dropped about 80 cycles per second every six months. (The experiment was quite depressing to most of the participants.) The aging of the ear is not difficult to understand if we assume that the elasticity of the tissues in the inner ear declines in the same way as that of the skin: it is well known that the skin becomes less resilient as we grow old—a

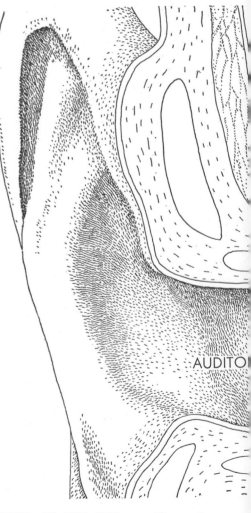

PARTS OF THE EAR are illustrated in somewhat simplified cross section. Be-

phenomenon anyone can test by lifting the skin on the back of his hand and measuring the time it takes to fall back.

However, the loss of hearing sensitivity with age may also be due to nerve deterioration. Damage to the auditory nervous system by extremely loud noises, by drugs or by inflammation of the inner ear can impair hearing. Sometimes after such damage the hearing improves with time; sometimes (*e.g.*, when the damaging agent is streptomycin) the loss is permanent. Unfortunately a physician cannot predict the prospects for recovery of hearing loss, because they vary from person to person.

Psychological factors seem to be involved. Occasionally, especially after an ear operation, a patient appears to improve in hearing only to relapse after a short time. Some reports have even suggested that operating on one ear has improved the unoperated ear as well. Since such an interaction between the two ears would be of considerable neuro-

logical interest, I have investigated the matter, but I have never found an improvement in the untreated ear that could be validated by an objective test.

Structure of the Ear

To understand how the ear achieves its sensitivity, we must take a look at the anatomy of the middle and the inner ear. When sound waves start the eardrum (tympanic membrane) vibrating, the vibrations are transmitted via certain small bones (ossicles) to the fluid of the inner ear. One of the ossicles, the tiny stirrup (weighing only about 1.2 milligrams), acts on the fluid like a piston, driving it back and forth in the rhythm of the sound pressure. These movements of the fluid force into vibration a thin membrane, called the basilar membrane. The latter in turn finally transmits the stimulus to the organ of Corti, a complex structure which contains the endings of the auditory nerves. The question im-

mediately comes up: Why is this long and complicated chain of transmission necessary?

The reason is that we have a formidable mechanical problem if we are to extract the utmost energy from the sound waves striking the eardrum. Usually when a sound hits a solid surface, most of its energy is reflected away. The problem the ear has to solve is to absorb this energy. To do so it has to act as a kind of mechanical transformer, converting the large amplitude of the sound pressure waves in the air into more forceful vibrations of smaller amplitude. A hydraulic press is such a transformer: it multiplies the pressure acting on the surface of a piston by concentrating the force of the pressure upon a second piston of smaller area. The middle ear acts exactly like a hydraulic press: the tiny footplate of the stirrup transforms the small pressure on the surface of the eardrum into a 22-fold greater pressure on the fluid of the inner ear. In this way the

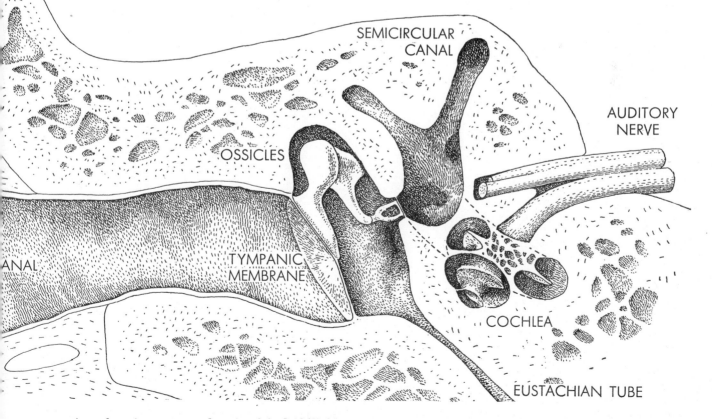

tween the eardrum (tympanic membrane) and the fluid-filled inner ear are the three small bones (ossicles) of the middle ear. The audi-

tory nerve endings are in an organ (*not shown*) between the plate of bone which spirals up the cochlea and the outer wall of the cochlea.

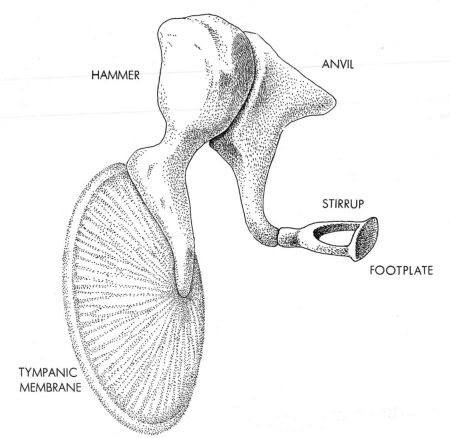

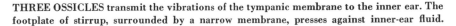

THREE OSSICLES transmit the vibrations of the tympanic membrane to the inner ear. The footplate of stirrup, surrounded by a narrow membrane, presses against inner-ear fluid.

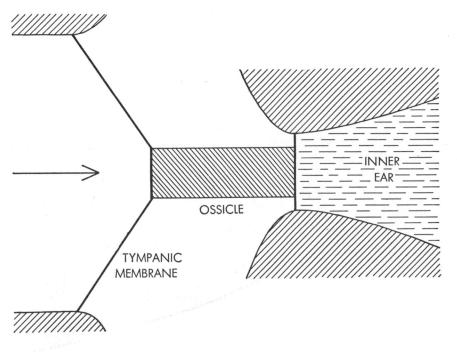

HOW OSSICLES ACT as a piston pressing against the fluid of the inner ear is indicated by this drawing. Pressure of the vibrations of tympanic membrane are amplified 22 times.

ear absorbs the greater part of the sound energy and transmits it to the inner ear without much loss.

But it needs another transformer to amplify the pressure of the fluid into a still larger force upon the tissues to which the nerves are attached. I think the ear's mechanism for this purpose is very ingenious indeed. It is based on the fact that a flat membrane, stretched to cover the opening of a tube, has a lateral tension along its surface. This tension can be increased tremendously if pressure is applied to one side of the membrane. And that is the function of the organ of Corti. It is constructed in such a way that pressure on the basilar membrane is transformed into shearing forces many times larger on the other side of the organ [*see diagram at bottom of opposite page*]. The enhanced shearing forces rub upon extremely sensitive cells attached to the nerve endings.

The eardrum is not by any means the only avenue through which we hear. We also hear through our skull, which is to say, by bone conduction. When we click our teeth or chew a cracker, the sounds come mainly by way of vibrations of the skull. Some of the vibrations are transmitted directly to the inner ear, by-passing the middle ear. This fact helps in the diagnosis of hearing difficulties. If a person can hear bone-conducted sounds but is comparatively deaf to air-borne sounds, we know that the trouble lies in the middle ear. But if he hears no sound by bone conduction, then his auditory nerves are gone, and there is no cure for his deafness. This is an old test, long used by deaf musicians. If a violin player cannot hear his violin even when he touches his teeth to the vibrating instrument, then he knows he suffers from nerve deafness, and there is no cure.

Speaking and Hearing

Hearing by bone conduction plays an important role in the process of speaking. The vibrations of our vocal cords not only produce sounds which go to our ears via the air but also cause the body to vibrate, and the vibration of the jawbone is transmitted to the ear canal. When you hum with closed lips, the sounds you hear are to a large degree heard by bone conduction. (If you stop your ears with your fingers, the hum sounds much louder.) During speaking and singing, therefore, you hear two different sounds—one by bone conduction and the other by air conduction. Of course another listener hears only the air-conducted sounds. In these sounds

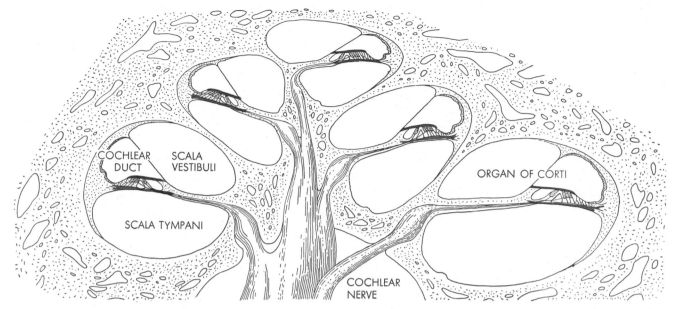

TUBE OF THE COCHLEA, coiled like the shell of a snail, is depicted in cross section. The plate of bone which appears in the cross section on pages 132 and 133 juts from the inside of the tube. Between it and the outside of the tube is the sensitive organ of Corti.

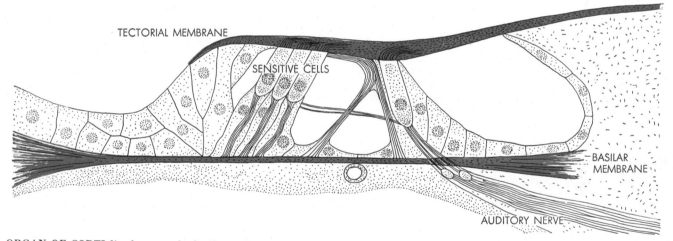

ORGAN OF CORTI lies between the basilar and tectorial membranes. Within it are sensitive cells which are attached to a branch of the auditory nerve (*lower right*). When fluid in scala tympani (*see drawing at top of page*) vibrates, these cells are stimulated.

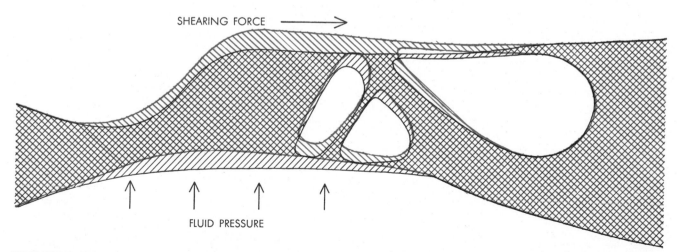

HOW VIBRATION FORCES ARE AMPLIFIED by the organ of Corti is indicated by this drawing. When the vibration of the fluid in the scala tympani exerts a force on the basilar membrane, a larger shearing force is brought to bear on tectorial membrane.

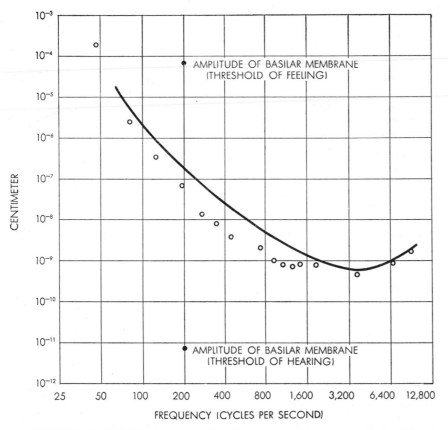

SENSITIVITY OF THE EAR is indicated by this curve, in which the amplitude of the vibrations of the tympanic membrane in fractions of a centimeter is plotted against the frequency of sound impinging on the membrane. Diameter of hydrogen atom is 10^{-8} centimeter.

cording of our voice may strike us as very thin and disappointing. From this point of view we have to admire the astonishing performance of an opera singer. The singer and the audience hear rather different sounds, and it is a miracle to me that they understand each other so well. Perhaps young singers would progress faster if during their training they spent more time studying recordings of their voices.

Feedback to the Voice

The control of speaking and singing involves a complicated feedback system. Just as feedback between the eyes and the muscles guides the hand when it moves to pick up an object, so feedback continually adjusts and corrects the voice as we speak or sing. When we start to sing, the beginning of the sound tells us the pitch, and we immediately adjust the tension of the vocal cords if the pitch is wrong. This feedback requires an exceedingly elaborate and rapid mechanism. How it works is not yet entirely understood. But it is small wonder that it takes a child years to learn to speak, or that it is almost impossible for an adult to learn to speak a foreign language with the native accents.

Any disturbance in the feedback immediately disturbs the speech. For instance, if, while a person is speaking, his speech is fed back to him with a time delay by means of a microphone and receivers at his ears, his pronunciation and accent will change, and if the delay interval is made long enough, he will find it impossible to speak at all.

some of the low-frequency components of the vocal cords' vibrations are lost. This explains why one can hardly recognize his own voice when he listens to a recording of his speech. As we normally hear ourselves, the low-frequency vibra-

tions of our vocal cords, conducted to our own ears by the bones, make our speech sound much more powerful and dynamic than the pure sound waves heard by a second person or through a recording system. Consequently the re-

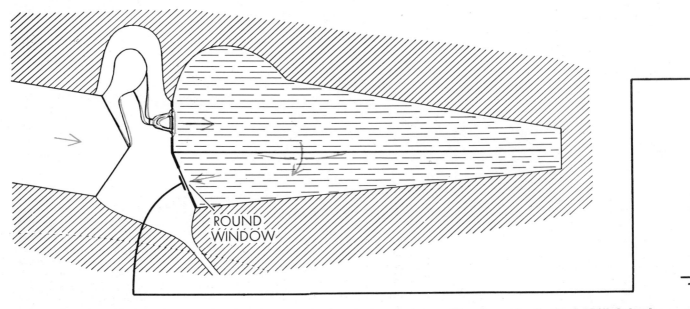

ELECTRICAL POTENTIALS of the microphonic type generated by the inner ear of an experimental animal can be detected by this arrangement. At left is a highly schematic diagram of the ear, the

cochlea is represented in cross section by the fluid-filled chamber and the organ of Corti by the horizontal line in this chamber. When the vibrations of the eardrum are transmitted to the organ of Corti,

This phenomenon affords an easy test for exposing pretended deafness. If the subject can continue speaking normally in the face of a delayed feedback through the machine to his ears, we can be sure that he is really deaf.

The same technique can be used to assess the skill of a pianist. A piano player generally adjusts his touch to the acoustics of the room: if the room is very reverberant, so that the music sounds too loud, he uses a lighter touch; if the sound is damped by the walls, he strengthens his touch. We had a number of pianists play in a room where the damping could be varied, and recorded the amplitude of the vibrations of the piano's sounding board while the musicians played various pieces. When they played an easy piece, their adjustment to the acoustics was very clear: as the sound absorption of the room was increased, the pianist played more loudly, and when the damping on the walls was taken away, the pianist's touch became lighter. But when the piece was difficult, many of the pianists concentrated so hard on the problems of the music that they failed to adjust to the feedback of the room. A master musician, however, was not lost to the sound effects. Taking the technical difficulties of the music in stride, he was able to adjust the sound level to the damping of the room with the same accuracy as for an easy piece. Our rating of the pianists by this test closely matched their reputation among musical experts.

In connection with room acoustics, I should like to mention one of the ear's most amazing performances. How is it that we can locate a speaker, even without seeing him, in a bare-walled room where reflections of his voice come at us from every side? This is an almost unbelievable performance by the ear. It is as if, looking into a room completely lined with mirrors, we saw only the real figure and none of the hundreds of reflected images. The eye cannot suppress the reflections, but the ear can. The ear is able to ignore all the sounds except the first that strikes it. It has a built-in inhibitory mechanism.

Suppressed Sounds

One of the most important factors that subordinate the reflected sounds is the delay in their arrival; necessarily they come to the ear only after the sound that has traveled directly from the speaker to the listener. The reflected sounds reinforce the loudness and tone volume of the direct sound, and perhaps even modify its localization, but by and large, they are not distinguishable from it. Only when the delay is appreciable does a reflected sound appear as a separate unit—an echo. Echoes often are heard in a large church, where reflections may lag more than half a second behind the direct sound. They are apt to be a problem in a concert hall. Dead walls are not desirable, because the music would sound weak. For every size of concert room there is an optimal compromise on wall reflectivity which will give amplification to the music but prevent disturbing echoes.

In addition to time delay, there are other factors that act to inhibit some sounds and favor others. Strong sounds generally suppress weaker ones. Sounds in which we are interested take precedence over those that concern us less, as I pointed out in the examples of the speaker in a noisy room and the orchestra conductor detecting an errant instrument. This brings us to the intimate collaboration between the ear and the nervous system.

Any stimulation of the ear (*e.g.*, any change in pressure) is translated into electrical messages to the brain via the nerves. We can therefore draw information about the ear from an analysis of these electrical impulses, now made possible by electronic instruments. There are two principal types of electric potential that carry the messages. One is a continuous, wavelike potential which has been given the name microphonic. In experimental animals such as guinea pigs and cats the microphonics are large enough to be easily measured (they range up to about half a millivolt). It

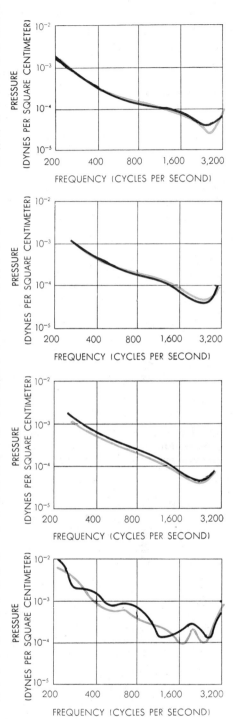

AUDIOGRAMS plot the threshold of hearing (in terms of pressure on the tympanic membrane) against the frequency of sound. The first three audiograms show the threshold for three members of the same family; the fourth, the threshold for an unrelated person. The black curves represent the threshold for one ear of the subject; the colored curves, for the other ear of the same subject. The audiogram curves indicate that in normal hearing the threshold in both ears, and the threshold in members of the same family, are remarkably similar.

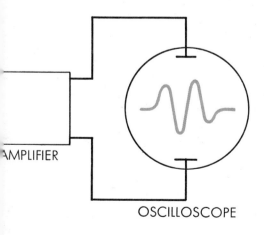

NDIFFERENT ELECTRODE N THE MUSCLE

its microphonic potentials can be picked up at the round window of the cochlea and displayed on the face of an oscilloscope (**right**).

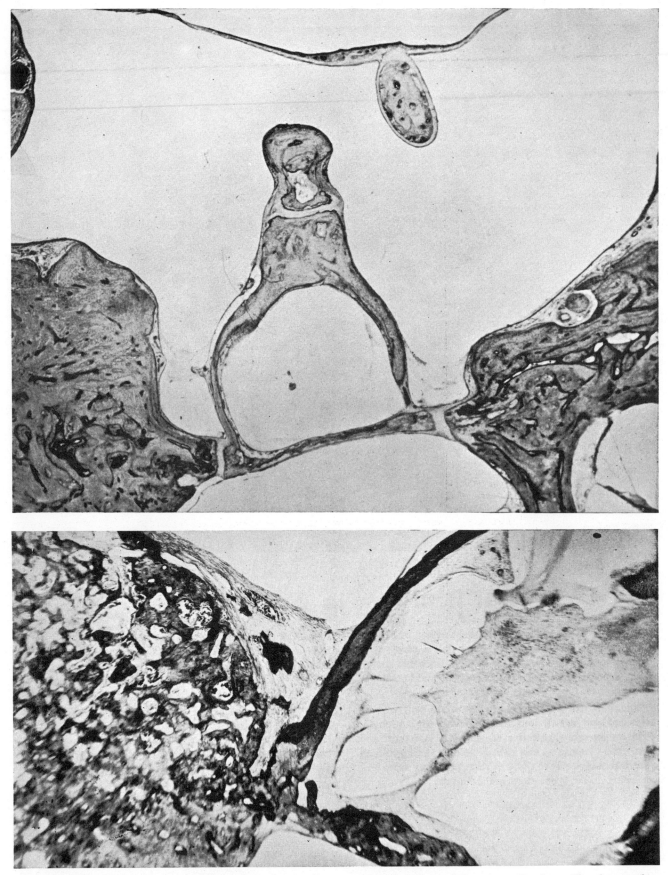

STIRRUP of the normal human ear is enlarged 19 times in the photograph at the top of this page. The thin line at the top of the photograph is the tympanic membrane seen in cross section. The hammer and anvil do not appear. The narrow membrane around the footplate of the stirrup may be seen as a translucent area between the footplate and the surrounding bone. The photograph at the bottom shows the immobilized footplate of an otosclerotic ear. In this photograph only the left side of the stirrup appears; the footplate is the dark area at the bottom center. The membrane around the footplate has been converted into a rigid bony growth.

has turned out that the magnitude of the microphonics produced in the inner ear is directly proportional to the displacements of the stirrup footplate that set the fluid in the inner ear in motion. The microphonics therefore permit us to determine directly to what extent the sound pressure applied to the eardrum is transmitted to the inner ear, and they have become one of the most useful tools for exploring sound transmission in the middle ear. For instance, there used to be endless discussion of the simple question: Just how much does perforation of the eardrum affect hearing? The question has now been answered with mathematical precision by experiments on animals. A hole of precisely measured size is drilled in the eardrum, and the amount of hearing loss is determined by the change in the microphonics. This type of observation on cats has shown that a perforation about one millimeter in diameter destroys hearing at the frequencies below 100 cycles per second but causes almost no impairment of hearing in the range of frequencies above 1,000 cycles per second. From studies of the physical properties of the human ear we can judge that the findings on animals apply fairly closely to man also.

The second type of electric potential takes the form of sharp pulses, which appear as spikes in the recording instrument. The sound of a sharp click produces a series of brief spikes; a pure tone generates volleys of spikes, generally in the rhythm of the period of the tone. We can follow the spikes along the nerve pathways all the way from the inner ear up to the cortex of the brain. And when we do, we find that stimulation of specific spots on the membrane of the inner ear seems to be projected to corresponding spots in the auditory area of the cortex. This is reminiscent of the projection of images on the retina of the eye to the visual area of the brain. But in the case of the ear the situation must be more complex, because there are nerve branches leading to the opposite ear and there seem to be several auditory projection areas on the surface of the brain. At the moment research is going on to find out how the secondary areas function and what their purpose is.

Detecting Pitch

The orderly projection of the sensitive area of the inner ear onto the higher brain levels is probably connected with the resolution of pitch. The ear itself can analyze sounds and separate one tone from another. There are limits to this

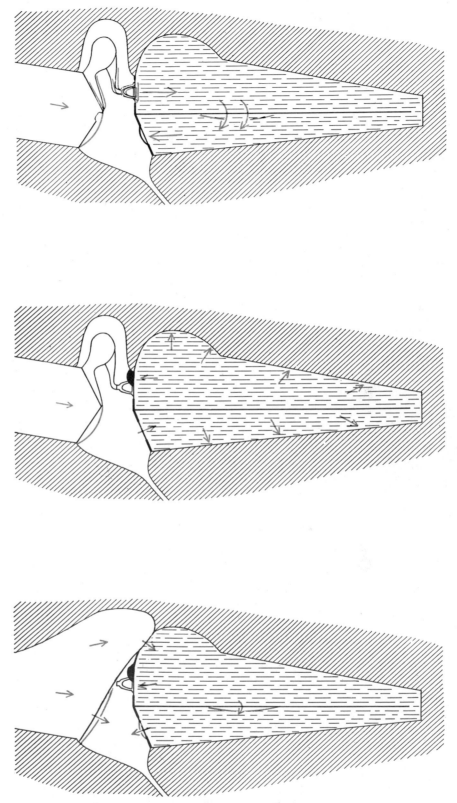

FENESTRATION OPERATION can alleviate the effects of otosclerosis. The drawing at the top schematically depicts the normal human ear as described in the caption for the illustration on pages 136 and 137. The pressure on the components of the ear is indicated by the colored arrows. The drawing in the middle shows an otosclerotic ear; the otosclerotic growth is represented as a black protuberance. Because the stirrup cannot move, the pressure on the tympanic membrane is transmitted to the organ of Corti only through the round window of the cochlea; and because the fluid in the cochlea is incompressible, the organ of Corti cannot vibrate. The drawing at the bottom shows how the fenestration operation makes a new window into the cochlea to permit the organ of Corti to vibrate freely.

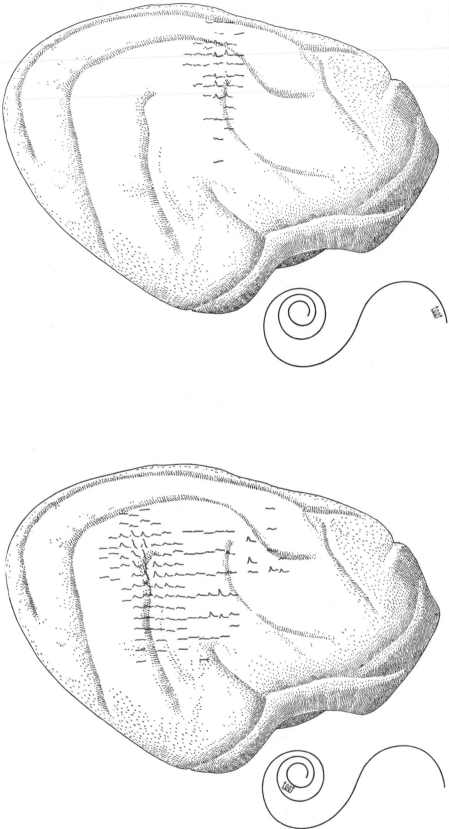

ability, but if the frequencies of the tones presented are not too close together, they are discriminated pretty well. Long ago this raised the question: How is the ear able to discriminate the pitch of a tone? Many theories have been argued, but only within the last decade has it been possible to plan pertinent experiments.

In the low-frequency range up to 60 cycles per second the vibration of the basilar membrane produces in the auditory nerve volleys of electric spikes synchronous with the rhythm of the sound. As the sound pressure increases, the number of spikes packed into each period increases. Thus two variables are transmitted to the cortex: (1) the number of spikes and (2) their rhythm. These two variables alone convey the loudness and the pitch of the sound.

Above 60 cycles per second a new phenomenon comes in. The basilar membrane now begins to vibrate unequally over its area: each tone produces a maximal vibration in a different area of the membrane. Gradually this selectivity takes over the determination of pitch, for the rhythm of the spikes, which indicates the pitch at low frequencies, becomes irregular at the higher ones. Above 4,000 cycles per second pitch is determined entirely by the location of the maximal vibration amplitude along the basilar membrane. Apparently there is an inhibitory mechanism which suppresses the weaker stimuli and thus sharpens considerably the sensation around the maximum. This type of inhibition can also operate in sense organs such as the skin and the eye. In order to see sharply we need not only a sharp image of the object on the retina but also an inhibitory system to suppress stray light entering the eye. Otherwise we would see the object surrounded by a halo. The ear is much the same. Without inhibitory effects a tone would sound like a noise of a certain pitch but not like a pure tone.

We can sum up by saying that the basilar membrane makes a rough, mechanical frequency analysis, and the auditory nervous system sharpens the analysis in some manner not yet understood. It is a part of the general functioning of the higher nerve centers, and it will be understood only when we know more about the functioning of these centers. If the answer is found for the ear, it will probably apply to the other sense organs as well.

Deafness

Now let us run briefly over some of

NERVE IMPULSES due to the electrical stimulation of the organ of Corti were localized on the surface of the brain of a cat. The spirals below each of these drawings of a cat's brain represent the full length of the organ of Corti. The pairs of colored arrows on each spiral indicate the point at which the organ was stimulated. The colored peaks superimposed on the brains represent the electrical potentials detected by an electrode placed at that point.

the types of hearing disorders, which have become much more understandable as a result of recent experimental researches.

Infections of the ear used to be responsible for the overwhelming majority of the cases of deafness. Ten years ago in a large city hospital there was a death almost every day from such infections. Thanks to antibiotics, they can now be arrested, and, if treated in time, an ear infection is seldom either fatal or destructive of hearing, though occasionally an operation is necessary to scoop out the diseased part of the mastoid bone.

The two other principal types of deafness are those caused by destruction of the auditory nerves and by otosclerosis (a tumorous bone growth). Nerve deafness cannot be cured: no drug or mechanical manipulation or operation can restore the victim's hearing. But the impairment of hearing caused by otosclerosis can usually be repaired, at least in part.

Otosclerosis is an abnormal but painless growth in a temporal bone (*i.e.*, at the side of the skull, near the middle ear). If it does not invade a part of the ear that participates in the transmission of sound, no harm is done to the hearing. But if the growth happens to involve the stirrup footplate, it will reduce or even completely freeze the footplate's ability to make its piston-like movements; the vibrations of the eardrum then can no longer be transmitted to the inner ear. An otosclerotic growth can occur at any age, may slow down for many years, and may suddenly start up again. It is found more often in women than in men and seems to be accelerated by pregnancy.

Immobilization of the stirrup blocks the hearing of air-borne sound but leaves hearing by bone conduction unimpaired. This fact is used for diagnosis. A patient who has lost part of his hearing ability because of otosclerosis does not find noise disturbing to his understanding of speech; in fact, noise may even improve his discrimination of speech. There is an old story about a somewhat deaf English earl (in France it is a count) who trained his servant to beat a drum whenever someone else spoke, so that he could understand the speaker better. The noise of the drum made the speaker

raise his voice to the earl's hearing range. For the hard-of-hearing earl the noise of the drum was tolerable, but for other listeners it masked what the speaker was saying, so that the earl enjoyed exclusive rights to his conversation.

Difficulty in hearing air-borne sound can be corrected by a hearing aid. Theoretically it should be possible to compensate almost any amount of such hearing loss, because techniques for amplifying sound are highly developed, particularly now with the help of the transistor. But there is a physiological limit to the amount of pressure amplification that the ear will stand. Heightening of the pressure eventually produces an unpleasant tickling sensation through its effect on skin tissue in the middle ear. The sensation can be avoided by using a bone-conduction earphone, pressed firmly against the surface of the skull, but this constant pressure is unpleasant to many people.

Operations

As is widely known, there are now operations (*e.g.*, "fenestration") which can cure otosclerotic deafness. In the 19th century physicians realized that if they could somehow dislodge or loosen the immobilized stirrup footplate, they might restore hearing. Experimenters in France found that they could sometimes free the footplate sufficiently merely by pressing a blunt needle against the right spot on the stirrup. Although it works only occasionally, the procedure seems so simple that it has recently had a revival of popularity in the U. S. If the maneuver is successful (and I am told that 30 per cent of these operations are) the hearing improves immediately. But unfortunately the surgeon cannot get a clear look at the scene of the operation and must apply the pushing force at random. This makes the operation something of a gamble, and the patient's hearing may not only fail to be improved but may even be reduced. Moreover, the operation is bound to be ineffectual when a large portion of the footplate is fixed. There are other important objections to the operation. After all, it involves the breaking of bone, to free the adhering part of the stirrup. I do not think that

bone-breaking can be improved to a standard procedure. In any case, precision cutting seems to me always superior to breaking, in surgery as in mechanics. This brings us to the operation called fenestration.

For many decades it has been known that drilling a small opening, even the size of a pinhead, in the bony wall of the inner ear on the footplate side can produce a remarkable improvement in hearing. The reason, now well understood, is quite simple. If a hole is made in the bone and then covered again with a flexible membrane, movements of the fluid in, for instance, the lateral canal of the vestibular organ can be transmitted to the fluid of the inner ear, and so vibrations are once again communicable from the middle to the inner ear. In the typical present fenestration operation the surgeon bores a small hole in the canal wall with a dental drill and then covers the hole with a flap of skin. The operation today is a straightforward surgical procedure, and all its steps are under accurate control.

Hazards to Hearing

I want to conclude by mentioning the problem of nerve deafness. Many cases of nerve deafness are produced by intense noise, especially noise with high-frequency components. Since there is no cure, it behooves us to look out for such exposures. Nerve deafness creeps up on us slowly, and we are not as careful as we should be to avoid exposure to intense noise. We should also be more vigilant about other hazards capable of producing nerve deafness, notably certain drugs and certain diseases.

We could do much to ameliorate the tragedy of deafness if we changed some of our attitudes toward it. Blindness evokes our instant sympathy, and we go out of our way to help the blind person. But deafness often goes unrecognized. If a deaf person misunderstands what we say, we are apt to attribute it to lack of intelligence instead of to faulty hearing. Very few people have the patience to help the deafened. To a deaf man the outside world appears unfriendly. He tries to hide his deafness, and this only brings on more problems.

15

EYE AND CAMERA

GEORGE WALD   August 1950

OF all the instruments made by man, none resembles a part of his body more than a camera does the eye. Yet this is not by design. A camera is no more a copy of an eye than the wing of a bird is a copy of that of an insect. Each is the product of an independent evolution; and if this has brought the camera and the eye together, it is not because one has mimicked the other, but because both have had to meet the same problems, and frequently have done so in much the same way. This is the type of phenomenon that biologists call convergent evolution, yet peculiar in that the one evolution is organic, the other technological.

Over the centuries much has been learned about vision from the camera, but little about photography from the eye. The camera made its first appearance not as an instrument for making pictures but as the *camera obscura* or dark chamber, a device that attempted no more than to project an inverted image upon a screen. Long after the optics of the camera obscura was well understood, the workings of the eye remained mysterious.

In part this was because men found it difficult to think in simple terms about the eye. It is possible for contempt to breed familiarity, but awe does not help one to understand anything. Men have often approached light and the eye in a spirit close to awe, probably because they were always aware that vision provides their closest link with the external

world. Stubborn misconceptions held back their understanding of the eye for many centuries. Two notions were particularly troublesome. One was that radiation shines out of the eye; the other, that an inverted image on the retina is somehow incompatible with seeing right side up.

I am sure that many people are still not clear on either matter. I note, for example, that the X-ray vision of the comic-strip hero Superman, while regarded with skepticism by many adults, is not rejected on the ground that there are no X-rays about us with which to see. Clearly Superman's eyes supply the X-rays, and by directing them here and there he not only can see through opaque objects, but can on occasion shatter a brick wall or melt gold. As for the inverted image on the retina, most people who learn of it concede that it presents a problem, but comfort themselves with the thought that the brain somehow compensates for it. But of course there is no problem, and hence no compensation. We learn early in infancy to associate certain spatial relations in the outside world with certain patterns of nervous activity stimulated through the eyes. The spatial arrangements of the nervous activity itself are altogether irrelevant.

It was not until the 17th century that the gross optics of image formation in the eye was clearly expressed. This was accomplished by Johannes Kepler in 1611, and again by René Descartes in

1664. By the end of the century the first treatise on optics in English, written by William Molyneux of Dublin, contained several clear and simple diagrams comparing the projection of a real inverted image in a "pinhole" camera, in a camera obscura equipped with a lens and in an eye.

Today every schoolboy knows that the eye is like a camera. In both instruments a lens projects an inverted image of the surroundings upon a light-sensitive surface: the film in the camera and the retina in the eye. In both the opening of the lens is regulated by an iris. In both the inside of the chamber is lined with a coating of black material which absorbs stray light that would otherwise be reflected back and forth and obscure the image. Almost every schoolboy also knows a difference between the camera and the eye. A camera is focused by moving the lens toward or away from the film; in the eye the distance between the lens and the retina is fixed, and focusing is accomplished by changing the thickness of the lens.

The usual fate of such comparisons is that on closer examination they are exposed as trivial. In this case, however, just the opposite has occurred. The more we have come to know about the mechanism of vision, the more pointed and fruitful has become its comparison with photography. By now it is clear that the relationship between the eye and the camera goes far beyond simple optics, and has come to involve much of the

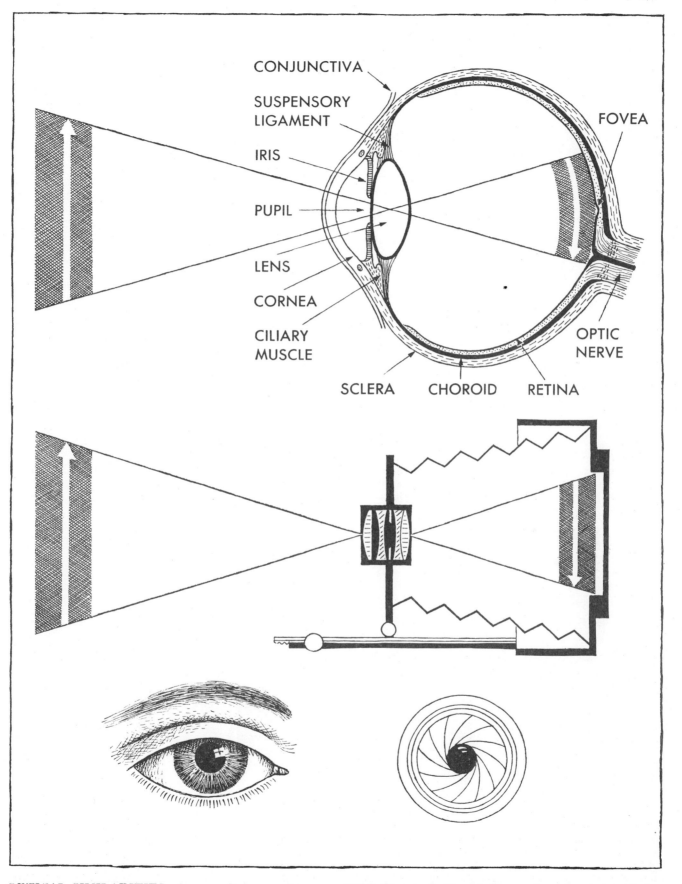

CONJUNCTIVA
SUSPENSORY LIGAMENT
IRIS
PUPIL
LENS
CORNEA
CILIARY MUSCLE
SCLERA CHOROID RETINA
FOVEA
OPTIC NERVE

OPTICAL SIMILARITIES of eye and camera are apparent in their cross sections. Both utilize a lens to focus an inverted image on a light-sensitive surface. Both possess an iris to adjust to various intensities of light. The single lens of the eye, however, cannot bring light of all colors to a focus at the same point. The compound lens of the camera is better corrected for color because it is composed of two kinds of glass.

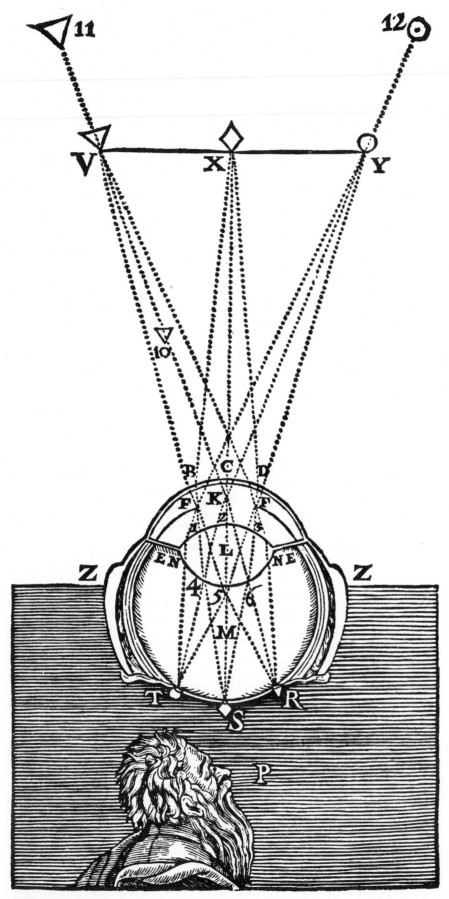

essential physics and chemistry of both devices.

Bright and Dim Light

A photographer making an exposure in dim light opens the iris of his camera. The pupil of the eye also opens in dim light, to an extent governed by the activity of the retina. Both adjustments have the obvious effect of admitting more light through the lens. This is accomplished at some cost to the quality of the image, for the open lens usually defines the image less sharply, and has less depth of focus.

When further pressed for light, the photographer changes to a more sensitive film. This ordinarily involves a further loss in the sharpness of the picture. With any single type of emulsion the more sensitive film is coarser in grain, and thus the image cast upon it is resolved less accurately.

The retina of the eye is grainy just as is photographic film. In film the grain is composed of crystals of silver bromide embedded in gelatin. In the retina it is made up of the receptor cells, lying side by side to form a mosaic of light-sensitive elements.

There are two kinds of receptors in the retinas of man and most vertebrates: rods and cones. Each is composed of an inner segment much like an ordinary nerve cell, and a rod- or cone-shaped outer segment, the special portion of the cell that is sensitive to light. The cones are the organs of vision in bright light, and also of color vision. The rods provide a special apparatus for vision in dim light, and their excitation yields only neutral gray sensations. This is why at night all cats are gray.

The change from cone to rod vision, like that from slow to fast film, involves a change from a fine- to a coarse-grained mosaic. It is not that the cones are smaller than the rods, but that the cones act individually while the rods act in large clumps. Each cone is usually connected with the brain by a single fiber of the optic nerve. In contrast large clusters of rods are connected by single optic nerve fibers. The capacity of rods for image vision is correspondingly coarse. It is not only true that at night all cats are gray, but it is difficult to be sure that they are cats.

Vision in very dim light, such as starlight or most moonlight, involves only the rods. The relatively insensitive cones are not stimulated at all. At moderately low intensities of light, about 1,000 times greater than the lowest intensity to which the eye responds, the cones begin to function. Their entrance is marked by dilute sensations of color. Over an intermediate range of intensities rods and cones function together, but as the brightness increases, the cones come to dominate vision. We do not know that

FORMATION OF AN IMAGE on the retina of the human eye was diagrammed by Rene Descartes in 1664. This diagram is from Descartes' *Dioptrics.*

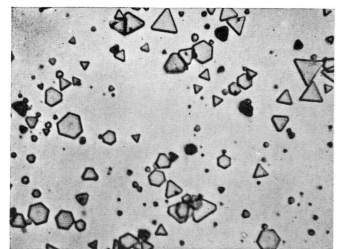

GRAIN of the photographic emulsion, magnified 2,500 times, is made up of silver-bromide crystals in gelatin.

"GRAIN" of the human retina is made up of cones and rods (*dots at far right*). Semicircle indicates fovea.

the rods actually stop functioning at even the highest intensities, but in bright light their relative contribution to vision falls to so low a level as to be almost negligible.

To this general transfer of vision from rods to cones certain cold-blooded animals add a special anatomical device. The light-sensitive outer segments of the rods and cones are carried at the ends of fine stalks called myoids, which can shorten and lengthen. In dim light the rod myoids contract while the cone myoids relax. The entire field of rods is thus pulled forward toward the light, while the cones are pushed into the background. In bright light the reverse occurs: the cones are pulled forward and the rods pushed back. One could scarcely imagine a closer approach to the change from fast to slow film in a camera.

The rods and cones share with the grains of the photographic plate another deeply significant property. It has long been known that in a film exposed to light each grain of silver bromide given enough developer blackens either completely or not at all, and that a grain is made susceptible to development by the absorption of one or at most a few quanta of light. It appears to be equally true that a cone or rod is excited by light to yield either its maximal response or none at all. This is certainly true of the nerve fibers to which the rods and cones are connected, and we now know that to produce this effect in a rod—and possibly also in a cone—only one quantum of light need be absorbed.

It is a basic tenet of photochemistry that one quantum of light is absorbed by, and in general can activate, only one molecule or atom. We must attempt to understand how such a small beginning can bring about such a large result as the development of a photographic grain or the discharge of a retinal receptor. In the photographic process the answer to this question seems to be that the ab-

sorption of a quantum of light causes the oxidation of a silver ion to an atom of metallic silver, which then serves as a catalytic center for the development of the entire grain. It is possible that a similar mechanism operates in a rod or a cone. The absorption of a quantum of light by a light-sensitive molecule in either structure might convert it into a biological catalyst, or an enzyme, which could then promote the further reactions that discharge the receptor cell. One wonders whether such a mechanism could possibly be rapid enough. A rod or a cone responds to light within a small fraction of a second; the mechanism would therefore have to complete its work within this small interval.

One of the strangest characteristics of the eye in dim light follows from some of these various phenomena. In focusing the eye is guided by its evaluation of the sharpness of the image on the retina. As the image deteriorates with the opening of the pupil in dim light, and as the retinal capacity to resolve the image falls with the shift from cones to rods, the ability to focus declines also. In very dim light the eye virtually ceases to adjust its focus at all. It has come to resemble a very cheap camera, a fixed-focus instrument.

In all that concerns its function, therefore, the eye is one device in bright light and another in dim. At low intensities all its resources are concentrated upon sensitivity, at whatever sacrifice of form; it is predominantly an instrument for seeing light, not pattern. In bright light all this changes. By narrowing the pupil, shifting from rods to cones, and other stratagems still to be described, the eye sacrifices light in order to achieve the utmost in pattern vision.

Images

In the course of evolution animals have used almost every known device

for forming or evaluating an image. There is one notable exception: no animal has yet developed an eye based upon the use of a concave mirror. An eye made like a pinhole camera, however, is found in Nautilus, a cephalopod mollusk related to the octopus and squid. The compound eye of insects and crabs forms an image which is an upright patchwork of responses of individual "eyes" or ommatidia, each of which records only a spot of light or shade. The eye of the tiny arthropod Copilia possesses a large and beautiful lens but only one light receptor attached to a thin strand of muscle. It is said that the muscle moves the receptor rapidly back and forth in the focal plane of the lens, scanning the image in much the same way as it is scanned by the light-sensitive tube of a television camera.

Each of these eyes, like the lens eye of vertebrates, represents some close compromise of advantages and limitations. The pinhole eye is in focus at all distances, yet to form clear images it must use a small hole admitting very little light. The compound eye works well at distances of a few millimeters, yet it is relatively coarse in pattern resolution. The vertebrate eye is a long-range, high-acuity instrument useless in the short distances at which the insect eye resolves the greatest detail.

These properties of the vertebrate eye are of course shared by the camera. The use of a lens to project an image, however, has created for both devices a special group of problems. All simple lenses are subject to serious errors in image formation: the lens aberrations.

Spherical aberration is found in all lenses bounded by spherical surfaces. The marginal portions of the lens bring rays of light to a shorter focus than the central region. The image of a point in space is therefore not a point, but a little "blur circle." The cost of a camera is largely determined by the extent to

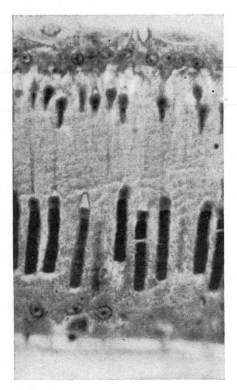

CONES of the catfish *Ameiurus* are pulled toward the surface of the retina (*top*) in bright light. The rods remain in a layer below the surface.

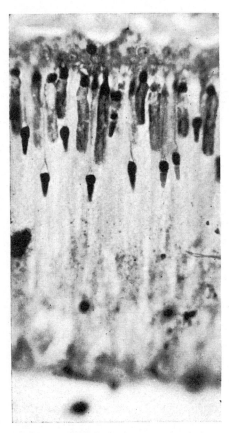

RODS advance and cones retreat in dim light. This retinal feature is not possessed by mammals. It is peculiar to some of the cold-blooded animals.

which this aberration is corrected by modifying the lens.

The human eye is astonishingly well corrected—often slightly overcorrected—for spherical aberration. This is accomplished in two ways. The cornea, which is the principal refracting surface of the eye, has a flatter curvature at its margin than at its center. This compensates in part for the tendency of a spherical surface to refract light more strongly at its margin. More important still, the lens is denser and hence refracts light more strongly at its core than in its outer layers.

A second major lens error, however, remains almost uncorrected in the human eye. This is chromatic aberration, or color error. All single lenses made of one material refract rays of short wavelength more strongly than those of longer wavelength, and so bring blue light to a shorter focus than red. The result is that the image of a point of white light is not a white point, but a blur circle fringed with color. Since this seriously disturbs the image, even the lenses of inexpensive cameras are corrected for chromatic aberration.

It has been known since the time of Isaac Newton, however, that the human eye has a large chromatic aberration. Its lens system seems to be entirely uncorrected for this defect. Indeed, living organisms are probably unable to manufacture two transparent materials of such widely different refraction and dispersion as the crown and flint glasses from which color-corrected lenses are constructed.

The large color error of the human eye could make serious difficulties for image vision. Actually the error is moderate between the red end of the spectrum and the blue-green, but it increases rapidly at shorter wavelengths: the blue, violet and ultraviolet. These latter parts of the spectrum present the most serious problem. It is a problem for both the eye and the camera, but one for which the eye must find a special solution.

The first device that opposes the color error of the human eye is the yellow lens. The human lens is not only a lens but a color filter. It passes what we ordinarily consider to be the visible spectrum, but sharply cuts off the far edge of the violet, in the region of wavelength 400 millimicrons. It is this action of the lens, and not any intrinsic lack of sensitivity of the rods and cones, that keeps us from seeing in the near ultraviolet. Indeed, persons who have lost their lenses in the operation for cataract and have had them replaced by clear glass lenses, have excellent vision in the ultraviolet. They are able to read an optician's chart from top to bottom in ultraviolet light which leaves ordinary people in complete darkness.

The lens therefore solves the problem of the near ultraviolet, the region of the spectrum in which the color error is greatest, simply by eliminating the region from human vision. This boon is distributed over one's lifetime, for the lens becomes a deeper yellow and makes more of the ordinary violet and blue invisible as one grows older. I have heard it said that for this reason aging artists tend to use less blue and violet in their paintings.

The lens filters out the ultraviolet for the eye as a whole. The remaining devices which counteract chromatic aberration are concentrated upon vision in bright light, upon cone vision. This is good economy, for the rods provide such a coarse-grained receptive surface that they would be unable in any case to evaluate a sharp image on the retina.

As one goes from dim to bright light, from rod to cone vision, the sensitivity of the eye shifts toward the red end of the spectrum. This phenomenon was described in 1825 by the Czech physiologist Johannes Purkinje. He had noticed that with the first light of dawn blue objects tend to look relatively bright compared with red, but that they come to look relatively dim as the morning advances. The basis of this change is a large difference in spectral sensitivity between rods and cones. Rods have their maximal sensitivity in the blue-green at about 500 millimicrons; the entire spectral sensitivity of the cones is transposed toward the red, the maximum lying in the yellow-green at about 562 millimicrons. The point of this difference for our present argument is that as one goes from dim light, in which pattern vision is poor in any case, to bright light, in which it becomes acute, the sensitivity of the eye moves away from the region of the spectrum in which the chromatic aberration is large toward the part of the spectrum in which it is least.

The color correction of the eye is completed by a third dispensation. Toward the center of the human retina there is a small, shallow depression called the fovea, which contains only cones. While the retina as a whole sweeps through a visual angle of some 240 degrees, the fovea subtends an angle of only about 1.7 degrees. The fovea is considerably smaller than the head of a pin, yet with this tiny patch of retina the eye accomplishes all its most detailed vision.

The fovea also includes the fixation point of the eye. To look directly at something is to turn one's eye so that its image falls upon the fovea. Beyond the boundary of the fovea rods appear, and they become more and more numerous as the distance from the fovea increases. The apparatus for vision in bright light is thus concentrated toward the center of the retina, that for dim light toward its periphery. In very dim light, too dim to excite the cones, the fovea is blind. One can see objects then only by looking at them slightly askance to catch their images on areas rich in rods.

In man, apes and monkeys, alone of all known mammals, the fovea and the region of retina just around it is colored yellow. This area is called the yellow patch, or *macula lutea*. Its pigmentation lies as a yellow screen over the light receptors of the central retina, subtending a visual angle some five to 10 degrees in diameter.

Several years ago in our laboratory at Harvard University we measured the color transmission of this pigment in the living human eye by comparing the spectral sensitivities of cones in the yellow patch with those in a colorless peripheral area. The yellow pigment was also extracted from a small number of human maculae, and was found to be xanthophyll, a carotenoid pigment that occurs also in all green leaves. This pigment in the yellow patch takes up the absorption of light in the violet and blue regions of the spectrum just where absorption by the lens falls to very low values. In this way the yellow patch removes for the central retina the remaining regions of the spectrum for which the color error is high.

So the human eye, unable to correct its color error otherwise, throws away those portions of the spectrum that would make the most trouble. The yellow lens removes the near ultraviolet for the eye as a whole, the macular pigment eliminates most of the violet and blue for the central retina, and the shift from rods to cones displaces vision in bright light bodily toward the red. By these three devices the apparatus of most acute vision avoids the entire range of the spectrum in which the chromatic aberration is large.

Photography with Living Eyes

In 1876 Franz Boll of the University of Rome discovered in the rods of the frog retina a brilliant red pigment. This bleached in the light and was resynthesized in the dark, and so fulfilled the elementary requirements of a visual pigment. He called this substance visual red; later it was renamed visual purple or rhodopsin. This pigment marks the point of attack by light on the rods: the absorption of light by rhodopsin initiates the train of reactions that end in·rod vision.

Boll had scarcely announced his discovery when Willy Kühne, professor of physiology at Heidelberg, took up the study of rhodopsin, and in one extraordinary year learned almost everything about it that was known until recently. In his first paper on retinal chemistry Kühne said: "Bound together with the pigment epithelium, the retina behaves not merely like a photographic plate, but like an entire photographic workshop, in which the workman continually renews

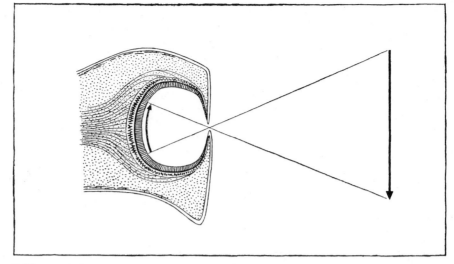

PINHOLE-CAMERA EYE is found in Nautilus, the spiral-shelled mollusk which is related to the octopus and the squid. This eye has the advantage of being in focus at all distances from the object that is viewed. It has the serious disadvantage, however, of admitting very little light to the retina.

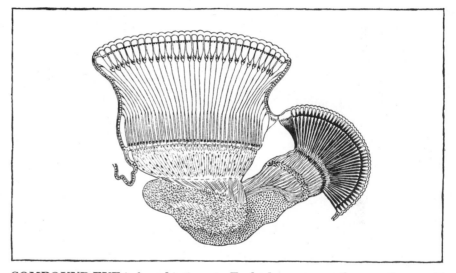

COMPOUND EYE is found in insects. Each element contributes only a small patch of light or shade to make up the whole mosaic image. This double compound eye is found in the mayfly *Chloeon*. The segment at the top provides detailed vision; the segment at the right, coarse, wide-angled vision.

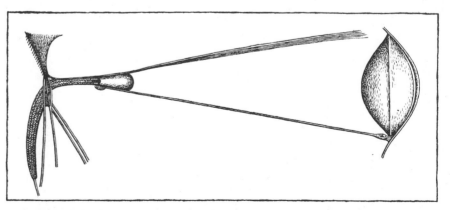

SCANNING EYE is found in the arthropod Copilia. It possesses a large lens (*right*) but only one receptor element (*left*). Attached to the receptor are the optic nerve and a strand of muscle. The latter is reported to move the receptor back and forth so that it scans the image formed by the lens.

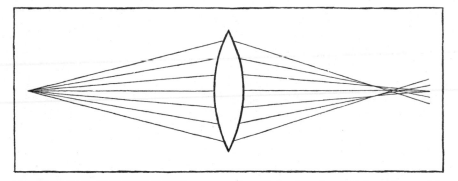

SPHERICAL ABERRATION occurs when light is refracted by a lens with spherical surfaces. The light which passes through the edge of the lens is brought to a shorter focus than that which passes through the center. The result of this is that the image of a point is not a point but a "blur circle."

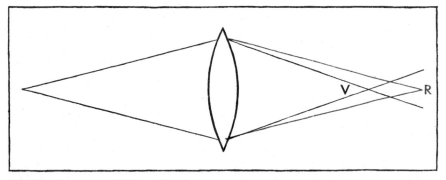

CHROMATIC ABERRATION occurs when light of various colors is refracted by a lens made of one material. The light of shorter wavelength is refracted more than that of longer wavelength, i.e., violet is brought to a shorter focus than red. The image of a white point is a colored blur circle.

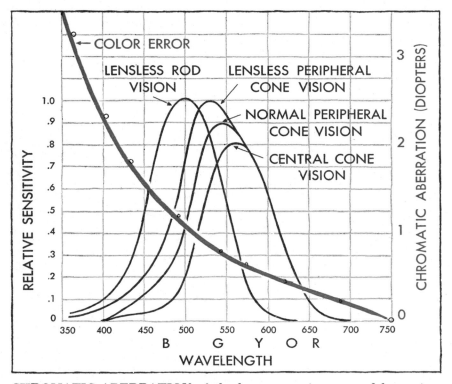

CHROMATIC ABERRATION of the human eye is corrected by various stratagems which withdraw the cones from the region of maximum aberration, i.e., the shorter wavelengths. The horizontal coordinate of this diagram is wavelength in millimicrons; the colors are indicated by initial letters.

the plate by laying on new light-sensitive material, while simultaneously erasing the old image."

Kühne saw at once that with this pigment which was bleached by light it might be possible to take a picture with the living eye. He set about devising methods for carrying out such a process, and succeeded after many discouraging failures. He called the process optography and its products optograms.

One of Kühne's early optograms was made as follows. An albino rabbit was fastened with its head facing a barred window. From this position the rabbit could see only a gray and clouded sky. The animal's head was covered for several minutes with a cloth to adapt its eyes to the dark, that is to let rhodopsin accumulate in its rods. Then the animal was exposed for three minutes to the light. It was immediately decapitated, the eye removed and cut open along the equator, and the rear half of the eyeball containing the retina laid in a solution of alum for fixation. The next day Kühne saw, printed upon the retina in bleached and unaltered rhodopsin, a picture of the window with the clear pattern of its bars.

I remember reading as a boy a detective story in which at one point the detective enters a dimly lighted room, on the floor of which a corpse is lying. Working carefully in the semidarkness, the detective raises one eyelid of the victim and snaps a picture of the open eye. Upon developing this in his darkroom he finds that he has an optogram of the last scene viewed by the victim, including of course an excellent likeness of the murderer. So far as I know Kühne's optograms mark the closest approach to fulfilling this legend.

The legend itself has nonetheless flourished for more than 60 years, and all of my readers have probably seen or heard some version of it. It began with Kühne's first intimation that the eye resembles a photographic workshop, even before he had succeeded in producing his first primitive optogram, and it spread rapidly over the entire world. In the paper that announces his first success in optography, Kühne refers to this story with some bitterness. He says: "I disregard all the journalistic potentialities of this subject, and willingly surrender it in advance to all the claims of fancy-free coroners on both sides of the ocean, for it certainly is not pleasant to deal with a serious problem in such company. Much that I could say about this had better be suppressed, and turned rather to the hope that no one will expect from me any corroboration of announcements that have not been authorized with my name."

Despite these admirable sentiments we find Kühne shortly afterward engaged in a curious adventure. In the nearby town of Bruchsal on November 16, 1880, a young man was beheaded by guillotine. Kühne had made arrangements to receive the corpse. He had prepared a dimly lighted room screened with red and yellow glass to keep any rhodopsin left in the eyes from bleaching further. Ten minutes after the knife had fallen he obtained the whole retina from the left eye, and had the satisfaction of seeing and showing to several colleagues a sharply demarcated optogram printed upon its surface. Kühne's drawing of it is reproduced at the bottom of the next page. To my knowledge it is the only human optogram on record.

Kühne went to great pains to determine what this optogram represented. He says: "A search for the object which served as source for this optogram remained fruitless, in spite of a thorough inventory of all the surroundings and reports from many witnesses. The delinquent had spent the night awake by the light of a tallow candle; he had slept

human eye as did the original subject of the picture.

How the human eye resolves colors is not known. Normal human color vision seems to be compounded of three kinds of responses; we therefore speak of it as trichromatic or three-color vision. The three kinds of response call for at least three kinds of cone differing from one another in their sensitivity to the various regions of the spectrum. We can only guess at what regulates these differences. The simplest assumption is that the human cones contain three different light-sensitive pigments, but this is still a matter of surmise.

There exist retinas, however, in which one can approach the problem of color vision more directly. The eyes of certain turtles and of certain birds such as chickens and pigeons contain a great predominance of cones. Since cones are the organs of vision in bright light as well as

In a paper published in 1907 the German ophthalmologist Siegfried Garten remarked that he was led by such retinal color filters to invent a system of color photography based upon the same principle. This might have been the first instance in which an eye had directly inspired a development in photography. Unfortunately, however, in 1906 the French chemist Louis Lumière, apparently without benefit of chicken retinas, had brought out his autochrome process for color photography based upon exactly this principle.

To make his autochrome plates Lumiere used suspensions of starch grains from rice, which he dyed red, green and blue. These were mixed in roughly equal proportions, and the mixture was strewn over the surface of an ordinary photographic plate. The granules were then squashed flat and the interstices were filled with particles of carbon. Each dyed granule served as a color filter for the patch of silver-bromide emulsion that lay just under it.

Just as the autochrome plate can accomplish color photography with a single light-sensitive substance, so the cones of the chicken retina should require no more than one light-sensitive pigment. We extracted such a pigment from the chicken retina in 1937. It is violet in color, and has therefore been named iodopsin from *ion*, the Greek word for violet. All three pigments of the colored oil globules have also been isolated and crystallized. Like the pigment of the human macula, they are all carotenoids: a greenish-yellow carotene; the golden mixture of xanthophylls found in chicken egg yolk; and red astaxanthin, the pigment of the boiled lobster.

Controversy thrives on ignorance, and we have had many years of disputation regarding the number of kinds of cone concerned in human color vision. Many investigators prefer three, some four, and at least one of my English colleagues seven. I myself incline toward three. It is a good number, and sufficient unto the day.

The appearance of three colors of oil globule in the cones of birds and turtles might be thought to provide strong support for trichromatic theories of color vision. The trouble is that these retinas do in fact contain a fourth class of globule which is colorless. Colorless globules have all the effect of a fourth color; there is no doubt that if we include them, bird and turtle retinas possess the basis for four-color vision.

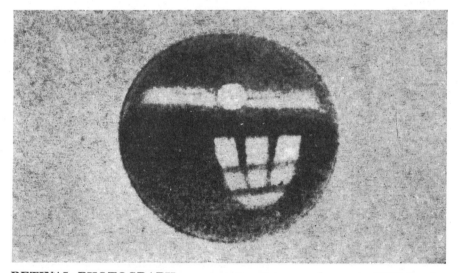

RETINAL PHOTOGRAPH, or an optogram, was drawn in 1878 by the German investigator Willy Kühne. He had exposed the eye of a living rabbit to a barred window, killed the rabbit, removed its retina and fixed it in alum.

from four to five o'clock in the morning; and had read and written, first by candlelight until dawn, then by feeble daylight until eight o'clock. When he emerged in the open, the sun came out for an instant, according to a reliable observer, and the sky became somewhat brighter during the seven minutes prior to the bandaging of his eyes and his execution, which followed immediately. The delinquent, however, raised his eyes only rarely."

Color

One of the triumphs of modern photography is its success in recording color. For this it is necessary not only to graft some system of color differentiation and rendition upon the photographic process; the finished product must then fulfill the very exacting requirement that it excite the same sensations of color in the

of color vision, these animals necessarily function only at high light intensities. They are permanently night-blind, due to a poverty or complete absence of rods. It is for this reason that chickens must roost at sundown.

In the cones of these animals we find a system of brilliantly colored oil globules, one in each cone. The globule is situated at the joint between the inner and outer segments of the cone, so that light must pass through it just before entering the light-sensitive element. The globules therefore lie in the cones in the position of little individual color filters.

One has only to remove the retina from a chicken or a turtle and spread it on the stage of a microscope to see that the globules are of three colors: red, orange and greenish yellow. It was suggested many years ago that they provide the basis of color differentiation in the animals that possess them.

Latent Images

Recent experiments have exposed a wholly unexpected parallel between vision and photography. Many years ago Kühne showed that rhodopsin can be extracted from the retinal rods into clear water solution. When such solutions are

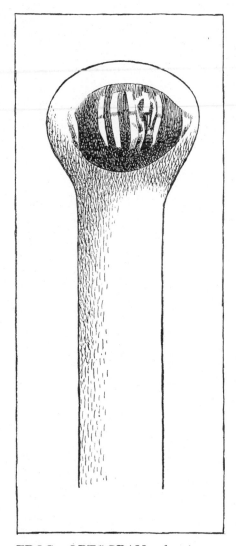

FROG OPTOGRAM showing a barred pattern was made by the German ophthalmologist Siegfried Garten. The retina is mounted on a rod.

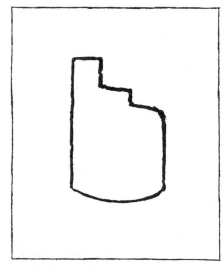

HUMAN OPTOGRAM was drawn by Kühne after he had removed the retina of a beheaded criminal. Kühne could not determine what it showed.

exposed to light, the rhodopsin bleaches just as it does in the retina.

It has been known for some time that the bleaching of rhodopsin in solution is not entirely accomplished by light. It is started by light, but then goes on in the dark for as long as an hour at room temperature. Bleaching is therefore a composite process. It is ushered in by a light reaction that converts rhodopsin to a highly unstable product; this then decomposes by ordinary chemical reactions—"dark" reactions in the sense that they do not require light.

Since great interest attaches to the initial unstable product of the light reaction, many attempts were made in our laboratory and at other laboratories to seize upon this substance and learn its properties. It has such a fleeting existence, however, that for some time nothing satisfactory was achieved.

In 1941, however, two English workers, E. E. Broda and C. F. Goodeve, succeeded in isolating the light reaction by irradiating rhodopsin solutions at about −73 degrees Celsius, roughly the temperature of dry ice. In such extreme cold, light reactions are unhindered, but ordinary dark processes cannot occur. Broda and Goodeve found that an exhaustive exposure of rhodopsin to light under these conditions produced only a very small change in its color, so small that though it could be measured one might not have been certain merely by looking at these solutions that any change had occurred at all. Yet the light reaction had been completed, and when such solutions were allowed to warm up to room temperature they bleached *in the dark*. We have recently repeated such experiments in our laboratory. With some differences which need not be discussed, the results were qualitatively as the English workers had described them.

These observations led us to re-examine certain early experiments of Kühne's. Kühne had found that if the retina of a frog or rabbit was thoroughly dried over sulfuric acid, it could be exposed even to brilliant sunlight for long periods without bleaching. Kühne concluded that dry rhodopsin is not affected by light, and this has been the common understanding of workers in the field of vision ever since.

It occurred to us, however, that dry rhodopsin, like extremely cold rhodopsin, might undergo the light reaction, though with such small change in color as to have escaped notice. To test this possibility we prepared films of rhodopsin in gelatin, which could be dried thoroughly and were of a quality that permitted making accurate measurements of their color transmission throughout the spectrum.

We found that when dry gelatin films

of rhodopsin are exposed to light, the same change occurs as in very cold rhodopsin. The color is altered, but so slightly as easily to escape visual observation. In any case the change cannot be described as bleaching; if anything the color is a little intensified. Yet the light reaction is complete; if such exposed films are merely wetted with water, they bleach in the dark.

We have therefore two procedures—cooling to very low temperatures and removal of water—that clearly separate the light from the dark reactions in the bleaching of rhodopsin. Which of these reactions is responsible for stimulating rod vision? One cannot yet be certain, yet the response of the rods to light occurs so rapidly that only the light reaction seems fast enough to account for it.

What has been said, however, has a further consequence that brings it into direct relation with photography. Everyone knows that the photographic process also is divided into light and dark components. The result of exposing a film to light is usually invisible, a so-called "latent image." It is what later occurs in the darkroom, the dark reaction of development, that brings out the picture.

This now appears to be exactly what happens in vision. Here as in photography light produces an almost invisible result, a latent image, and this indeed is probably the process upon which retinal excitation depends. The visible loss of rhodopsin's color, its bleaching, is the result of subsequent dark reactions, of "development."

One can scarcely have notions like this without wanting to make a picture with a rhodopsin film; and we have been tempted into making one very crude rhodopsin photograph. Its subject is not exciting—only a row of black and white stripes—but we show it at the right for what interest it may have as the first such photograph. What is important is that it was made in typically photographic stages. The dry rhodopsin film was first exposed to light, producing a latent image. It was then developed in the dark by wetting. It then had to be fixed; and, though better ways are known, we fixed this photograph simply by redrying it. Since irradiated rhodopsin bleaches rather than blackens on development, the immediate result is a positive.

Photography with rhodopsin is only in its first crude stages, perhaps at the level that photography with silver bromide reached almost a century ago. I doubt that it has a future as a practical process. For us its primary interest is to pose certain problems in visual chemistry in a provocative form. It does, however, also add another chapter to the mingled histories of eye and camera.

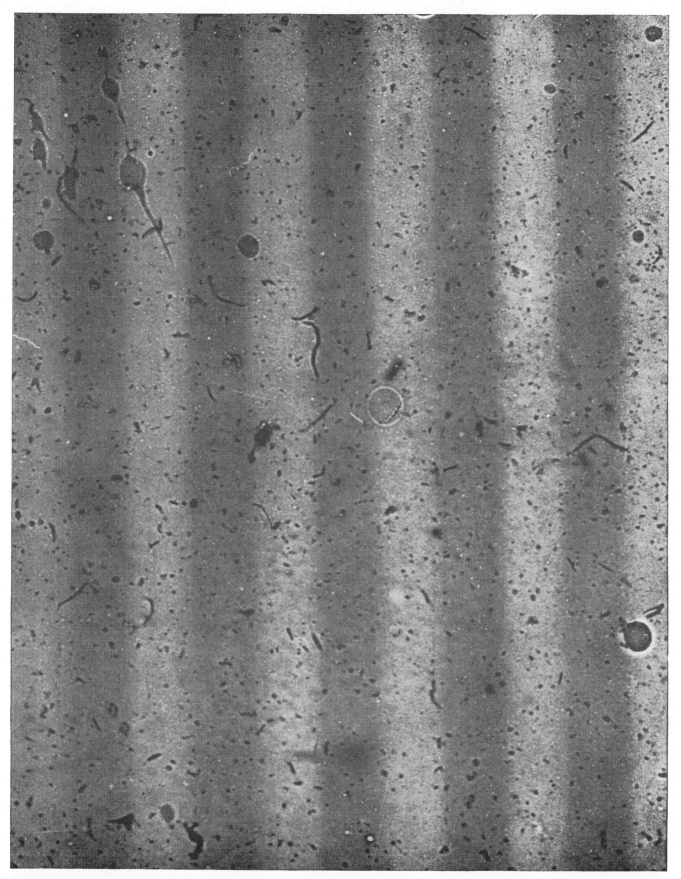

RHODOPSIN PHOTOGRAPH was made by the author and his associates Paul K. Brown and Oscar Starobin. Rhodopsin, the light-sensitive red pigment of rod vision, had been extracted from cattle retinas, mixed with gelatin and spread on celluloid. This was then dried and exposed to a pattern made up of black and white stripes. When the film was wetted in the dark with hydroxylamine, the rhodopsin bleached in the same pattern.

16

VISUAL PIGMENTS
IN MAN

W. A. H. RUSHTON   November 1962

Everyone knows that the eye is a camera—more properly a television camera—that not only forms a picture but also transmits it in code via the optic nerves to the brain. In this article I shall not discuss how the lens forms an image on the retina; it does so in virtually the same way that the lens of a photographic camera forms an image on a piece of film, and the process needs no explanation here. Nor shall I treat of the encoding of nerve messages in the eye, still less of their decoding in the brain, because on those topics reliable information remains extremely scanty. I shall deal rather with the light-sensitive constituents of the retina of the eye—the "silver bromide" of vision—and their relation to the perception of light and color.

It is no use taking a snapshot with color film if the illumination is poor; the only hope of getting a picture is to use sensitive black-and-white film. If the light signal is only sufficient to silhouette outlines, it cannot provide additional information for the discrimination of color. Thus for a camera to be well equipped to extract the maximum information from any kind of scene it must be provided with sensitive black-and-white film for twilight and color film for full daylight. The eye is furnished with a retina having precisely this dual purpose. The saying goes, "In the twilight all cats are gray," but by day some cats are tortoise-shell.

We cannot slip off our daylight retina and wind on the twilight roll; the two films must remain in place all the time. They are not situated one behind the other but are mixed together, the grains of the two "emulsions" lying side by side. The color grains are too insensitive to contribute to the twilight picture, which is therefore formed entirely by the black-and-white grains; these, on the other hand, give only a rather faint picture, which in daylight is quite overpowered by the color grains.

Of course the actual grains in the retina are not inorganic crystals such as silver bromide but are the specialized body cells known as rods and cones. The rods and cones do, however, contain a photosensitive pigment that is laid down in a molecular array so well ordered as to be quasi-crystalline. The rods are the grains responsible for twilight vision, and their photosensitive pigment is rhodopsin, often called visual purple. The cones are the grains of daylight vision, and the photosensitive pigments they contain will be one of the topics of this article.

It was first noticed almost a century ago that if a frog's eye was dissected in dim light and if the excised retina was then brought out into diffuse daylight, the initial rose-pink color of the retina would gradually fade and become almost transparent. The fading of the retina was the more rapid the stronger the light to which it was exposed; hence the term "bleaching" is used to describe the chemical change brought about when light falls on the photosensitive constituents of the rods and cones. If a microscope is employed to observe the retina as it bleaches, one can see that the pink color resides only in the rods. The cones appear to possess no colored pigment at all.

The presence of a photosensitive pigment in the rods does not prove that this is the chemical that catches the light with which we see; the pigment may be doing something quite different. There is one rather strict test that must be satisfied if rhodopsin, the pink pigment, is the starting point of vision.

Since the pigment looks pink by transmitted light, it obviously absorbs green and transmits red (and some blue). With a spectrophotometer it is quite easy to measure the absorption of a rhodopsin solution at various wavelengths. When this is done, one obtains a bell-shaped curve with a peak close to a wavelength of 500 millimicrons, in the blue-green region of the spectrum. If rhodopsin catches the light we see in twilight, we should see best precisely those wavelengths that are best caught. In other words, the spectral absorption curve of rhodopsin should coincide with the spectral sensitivity curve of human twilight vision. Actual measurements of the twilight sensitivity of the eye at various wavelengths leave no doubt that rhodopsin is indeed the pigment that enables us to see at night [*see illustration on page 156*].

The eye is able to discriminate differences in brightness efficiently over a range in which the brightest light is a billion times more intense than the dimmest. Any instrument that can do that must have a variable "gain," or sensitivity-multiplying factor, and some means of adjusting the gain to match the level of signal to be discriminated. It is common experience that the eye adjusts its gain so smoothly that when the sun goes behind a cloud, the details of the scene appear just as distinct as before, and indeed we have so little clue to the eye's automatic compensation that when (as in photography) we want an estimate of the light intensity, it is safer to use a photoelectric meter. The change in gain of the eye is called visual adaptation.

It is plain that visual adaptation adjusts itself automatically to the prevailing brightness. To explain how this could occur Selig Hecht of Columbia Uni-

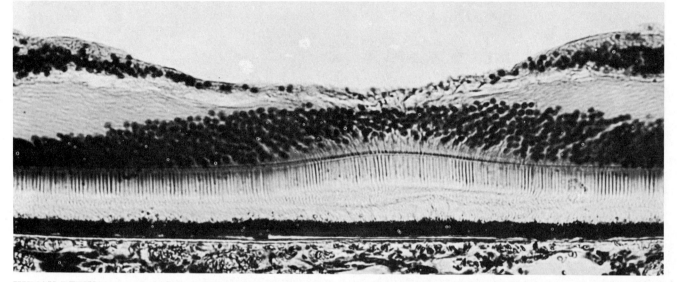

HUMAN RETINA, magnified about 370 diameters, is shown sectioned through the fovea, the tiny central region responsible for acute vision. The rods and cones, the photoreceptor cells containing the visual pigments, are the closely packed vertical stalks extending across the picture. Above the rods and cones are several layers composed chiefly of nerve cells that relay signals from the retina to the brain. At the fovea, which contains few if any rods, these layers are much thinned out to expose the light-sensitive part of the cones to incident light. This micrograph was made by C. M. H. Pedler of the Institute of Ophthalmology at University of London.

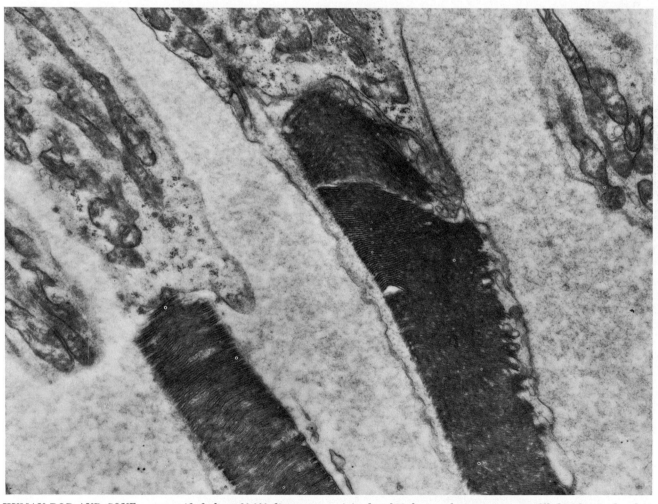

HUMAN ROD AND CONE are magnified about 20,000 diameters in this electron micrograph. The rod is on the left; the cone, on the right. The lamellated structures are the photoreceptor segments, believed to contain the visual pigments. These segments are joined at their base to the inner segments filled with mitochondria, which supply the cell with energy. The inner segments are positioned nearest the incoming light. The picture was made by Ben S. Fine of the Armed Forces Institute of Pathology in Washington.

versity 40 years ago drew attention to the visual pigments and suggested that their color intensity seems to vary with the level of light. He hypothesized that in bright light these pigments are somewhat bleached and that in the dark they are regenerated from precursors stored in the eye or conveyed by the blood. Under steady illumination a balance will be struck between these two processes, and the equilibrium level of rhodopsin will be lower the stronger the bleaching light is. Hecht suggested that the level of rod adaptation is controlled by the level of the rhodopsin in the rods.

One difficulty in accepting this rather plausible suggestion is that until one can measure the actual rhodopsin level in the eye and correlate it with the corresponding state of visual adaptation, the idea remains speculative and very insecure. This indeed was the situation for some 30 years, but now it is possible to measure rhodopsin and cone pigments in the normal human eye by a procedure requiring only about seven seconds. As a result one can now follow the time course of bleaching and regeneration and test Hecht's suggestion.

Most people have at one time or another seen the eyes of a cat in the

glare of an automobile headlight. The brilliant yellow-green eyes shining out of the darkness are a striking sight. The effect is caused simply by the reflection of light from the back of the cat's eye. What is important for our purpose is that these rays are reflected from behind the cat's retina and have therefore passed twice through the retina and the rhodopsin contained in the retina. This by itself would make the eye look pink, as it does in the case of the dissected frog retina. The cat, however, has a brilliant green backing to its retina and it is this backing that colors the returning light. To see the color of rhodopsin itself we need an animal whose retina has a white backing. If instead of a cat there were an alligator in the road, we should see the eye-shine colored pink by rhodopsin.

By using a photocell to analyze the returning light one can measure the rhodopsin no matter whether the eye is backed by green as it is in the cat, by white as in the alligator or even by black as in man. Regardless of its color, the reflectivity of the rear surface is unchanging, whereas the rhodopsin lying in front can be bleached away by strong light. It follows that if one measures not the color but the intensity of the returning light, one can find how much

of the light was absorbed by rhodopsin.

The illustration on this page shows schematically the instrument used to measure the bleaching of human eye pigments in my laboratory at the University of Cambridge. Light enters the eye through the upper half of the pupil, which has been dilated by a drug to allow more light to pass. It returns after reflection from the black rear surface, having twice traversed the retina. A small mirror intercepts the light from the lower half of the pupil and deflects it into a photomultiplier tube, which provides a measure of the light absorbed by the retinal pigments. If a powerful light is shined into the eye, the light bleaches away some of the pigment. This leaves less pigment to absorb the light traversing the retina; consequently the photocell output will be greater than before. The output can be returned to its former value by reduction of the measuring light. This is done by interposing a purple wedge in the beam of light entering the eye. The initial photocell output is restored when the amount of purple added by the wedge exactly matches the visual purple—the rhodopsin—removed by bleaching. The change in rhodopsin is thus measured simply by the change in wedge thickness that replaces it. The wedge scale is calibrated so that the reading is zero when all the rhodopsin is bleached away. Therefore the wedge setting for constant photocell output gives the rhodopsin density at that moment.

The intensity of the light reaching the photocell is only about a 20,000th of that falling on the eye, and the light striking the eye has to be so weak that it will not appreciably bleach the pigment it measures. Thus the equipment needs some rather careful compensations if measurements are to be reliable. We are not concerned here, however, with the technique of measurement but with the results in relation to the physiology of vision, and in particular with the question of the relation of rhodopsin level to visual adaptation.

The top illustration on page 157 shows the first measurements of this kind. They were made on my eye by F. W. Campbell at the University of Cambridge in 1955. The black dots show the wedge readings when a moderately bright bleaching light (one "bleaching unit") was applied to the dark-adapted eye. The pigment at first bleaches fast, then more slowly, and in five minutes it levels out, either because all the pigment is now bleached or because bleaching is just counterbalanced by the regeneration

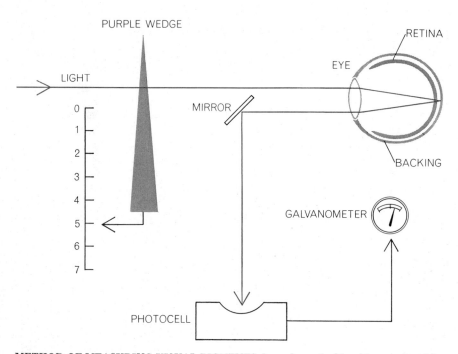

METHOD OF MEASURING VISUAL PIGMENTS depends on the bleaching produced by light. Light enters the eye through a purple wedge, and the amount reflected is measured by a photocell. When the pigment rhodopsin, or "visual purple," is bleached from the retina, an equivalent amount of wedge is inserted in the light beam to keep the electric output the same after bleaching as before. The change in pigment is measured by the wedge displacement; a change of one unit means reflectivity of the eye has changed by a factor of 10.

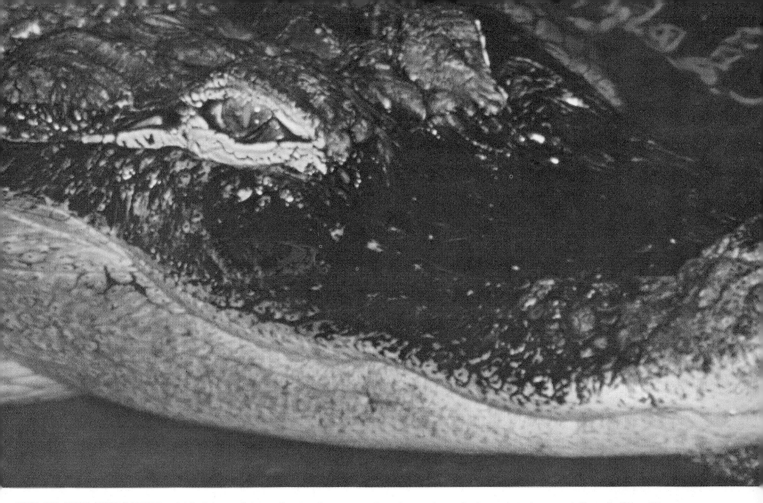

EYE OF THE ALLIGATOR, which has a white reflecting layer behind the retina, illustrates how rhodopsin bleaches in the light and regenerates in the dark. The eye of the alligator above is light-adapted; the light of a stroboscopic-flash lamp, reflected from the white layer through the retina, is essentially colorless. The eyes of the alligator below are dark-adapted; the light reflected is red. The photographs were made at the New York Zoological Park with the kind assistance of Herndon G. Dowling and Stephen Spencook.

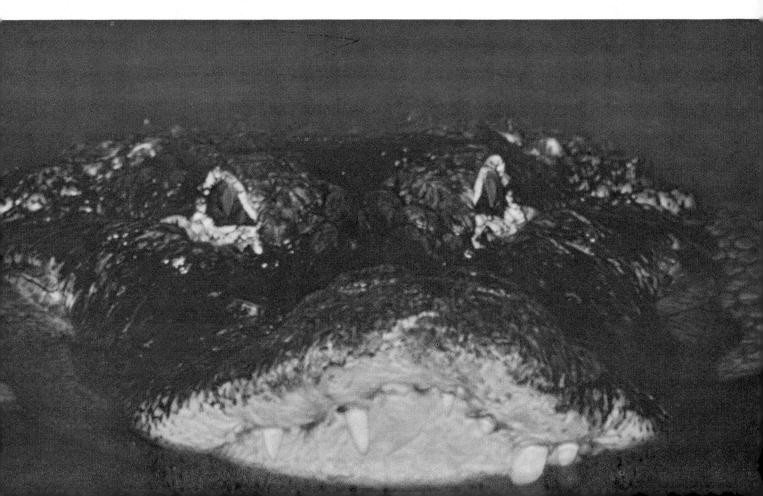

process. The latter is obviously the correct explanation, since by increasing the intensity of the bleaching light fivefold, further bleaching occurs and a lower level of equilibrium is achieved. In fact, a further increase of a hundredfold is needed to bleach the pigment entirely. The rate of pigment regeneration in the dark following total bleaching is plotted by the colored dots in the illustration. The regeneration follows an exponential curve and is about 90 per cent complete in 15 minutes.

Let us now examine Hecht's suggestion that it is the level of rhodopsin in the rods that defines the state of adaptation in twilight vision. But before doing so we must distinguish two quite different visual processes that are often designated by the word "adaptation." One process is exemplified by the quick changes in sensitivity that occur at night when the moon is fitfully obscured by passing clouds. This can be called field adaptation. When, on the other hand, we have got well adapted to bright light and then go into the dark—from sunlight into a theater, for instance—a different process occurs, which can be called adaptation of bleaching.

Now, field adaptation has nothing to do with the level of rhodopsin in the rods (or of visual pigments in the cones); the light intensity involved is only about a 100,000th of the bleaching unit referred to earlier, so that no appreciable bleaching can have occurred. Moreover, the time of adjustment to the new light level when the moon pops in and out of cloud is of the order of two seconds, rather than the 1,000 seconds required for the regeneration of rhodopsin. This rapid change of gain is in all likelihood produced entirely by the activity of nerve cells. Conceivably a feedback mechanism in the neural system maintains a constant signal strength by exchanging sensitivity for space-time discrimination. The adaptation of bleaching, on the other hand, turns out to be tightly linked to the level of rhodopsin in the rods.

The simplest way to examine this relation is to illuminate the eye with a powerful beam of light, a beam having an intensity of 100 bleaching units. After a minute or two all the rhodopsin will be bleached away and the course of pigment regeneration can be followed. The experiment is now repeated, but instead of measuring rhodopsin we determine the threshold of the eye by finding what is the weakest flash that can be detected at various intervals as the pigment regenerates. This is conveniently done by inserting a gray wedge to reduce the flash to threshold strength. The wedge displacement will now give the threshold directly on a logarithmic scale. A plot of this threshold yields the well-known dark-adaptation curve, shown in the bottom illustration on the next page.

As can be seen, the curve for the normal eye consists of two branches, the first of which corresponds to the log threshold of cones; the second, to the log threshold of rods. Only the rod threshold is related to rhodopsin, and it is a serious drawback that so much of this curve is hidden by the cone branch. Fortunately the complete rod curve can be obtained by using test subjects with a rare congenital abnormality in which rods are normal but cones entirely lack function. The dark-adaptation curve for such a subject is the black curve in the bottom illustration on the next page. It can be seen that the curve exactly follows the time course of the regeneration of rhodopsin, whether measured in the same subject or in a normal subject. It is therefore plain that the increase in light sensitivity of the rods waits precisely on the return of rhodopsin in the rods. What is far from plain, however, is what the increase in sensitivity waits for.

The change of sensitivity gain by nerve feedback in field adaptation is purposeful and efficient. The coupling of gain to the regeneration of rhodopsin in the adaptation of bleaching seems both pointless and clumsy. I have a far greater faith in nature, however, than in myself. I am sure that someone with deeper insight will eventually show that the deficiencies in dark adaptation, which to me seem unnecessary, are in fact inevitable.

The rapid and unconscious change of gain that makes absolute levels of light intensity hard to judge applies to cones as well as to rods, but in cones there is also the appreciation of color, which has its own adaptations. In judging brightness we estimate the brightness of parts with respect to the mean brightness of the whole. Thus the actual intensity of light reflected from black print in the noonday sun is far greater than that from white paper after sunset, yet the first looks black and the second white.

In color judgments wavelengths en-

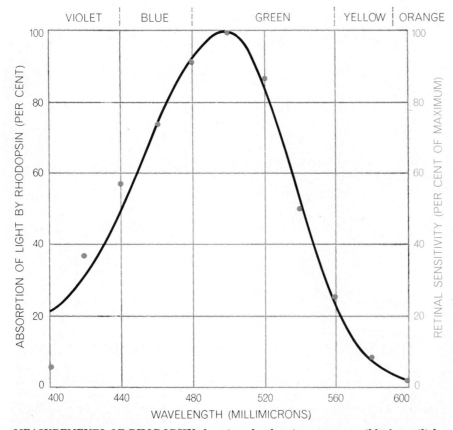

MEASUREMENTS OF RHODOPSIN show it to be the pigment responsible for twilight vision. The black curve indicates how a solution of rhodopsin, obtained from retinal rods, absorbs light of various wavelengths. Dots show sensitivity of the eye in twilight.

ter in, and we estimate the color of parts of a scene in relation to the mean wavelength of the whole. The fact that our perceptions of color can be independent of wavelength to a surprising degree has been brought into great prominence by the striking demonstrations of Edwin H. Land of the Polaroid Corporation [for further information, see "Experiments in Color Vision," by Edwin H. Land, Offprint #223]. Land has shown, for example, that two superimposed images of a scene, made on black-and-white film through different filters, will appear to contain a large range of color when one image is projected by red light and the other by white light. To say that the eye uses the average wavelength of such a red-and-white projection to judge the color of its parts is not meant to "explain" the Land phenomena, still less to suggest that no explanation is needed. It is merely a reminder that owing to some sort of adaptation—which Land has recently shown to be instantaneous—the eye is almost as bad at making absolute judgments of color as it is of brightness.

What the eye can do very well, however, is to make color *matches,* and these remain good even in the conditions of Land's projections. For instance, if monochromatic beams of red and green light are superimposed by projection on a screen, they can be made to match the yellow of a sodium lamp exactly, just by suitably adjusting the intensity of the red and of the green. If this red-green mixture is now substituted for the sodium yellow in one of Land's two-color projections, the colored picture resulting is exactly what it was before. Although many strange things appear in Land's pictures, one thing is clear: If red and green match yellow in one situation, they will match it in every other situation. Why, we may ask, are color matches stable under conditions where color appearance changes so greatly, and what colors can be matched by a mixture of others? A century ago James Clerk Maxwell showed that all colors could be matched by a suitable mixture of red, green and blue primaries, and indeed that any three colors could be chosen as primaries provided that no one of them could be matched by a mixture of the other two. The trichromaticity of color implies that the cones have three and only three ways of catching light. It seems reasonable, therefore, that there may be three and only three different cone pigments.

Since the rods have only one pigment,

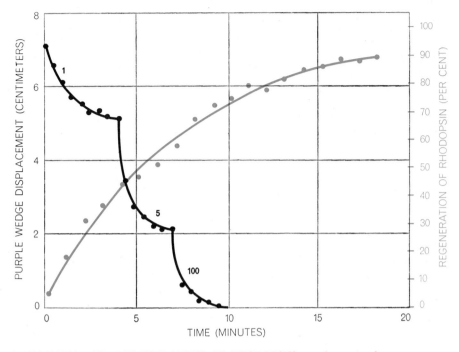

BLEACHING AND REGENERATION OF RHODOPSIN are shown in the two curves obtained by the method illustrated on page 154. The black curve records the time course of bleaching for a light of moderate intensity (1) and for lights five and 100 times brighter. In the dark, rhodopsin regenerates as shown by the colored curve. The measurements were made on the eye of the author by F. W. Campbell at the University of Cambridge.

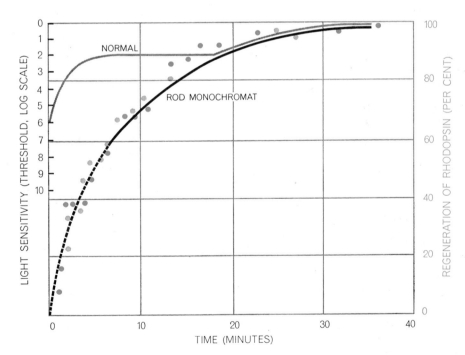

ROD AND CONE LIGHT SENSITIVITY can be distinguished by comparing a normal eye with that of a "rod monochromat," a person whose retinal cones do not function. The rhodopsin is fully bleached and the weakest detectable flash of light is measured. As the rhodopsin regenerates, the eye detects flashes that are weaker and weaker. The light sensitivity of the normal eye follows a discontinuous curve. The initial sensitivity increase is due to cones; the final increase is due to rods. In the rod monochromat the sensitivity rises more slowly but in a smooth curve. Independent measurements with the purple-wedge technique show that rhodopsin regeneration goes hand in hand with increased light sensitivity in the rod monochromat (*dark-colored dots*). In the normal eye, however, rhodopsin regeneration (*light-colored dots*) follows only the rod branch of the light-sensitivity curve.

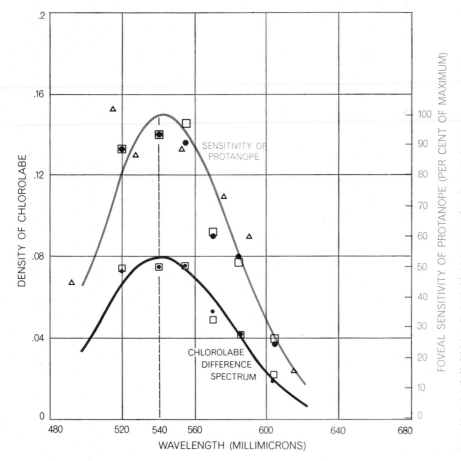

□ AFTER PARTIAL BLEACHING WITH
   BLUE-GREEN LIGHT

● AFTER PARTIAL BLEACHING WITH RED LIGHT

▣ AFTER BLEACHING WITH WHITE LIGHT

△ ACTION SPECTRUM

GREEN-CATCHING PIGMENT, called chlorolabe, can be measured in the eye of a protanope, the name given to a person who is red-blind. The pigment in the fovea of a protanope is partially bleached with red light and the change in reflectivity is measured at six wavelengths (*small dots*). The reflectivity change is then measured after partial bleaching with blue-green light (*small squares*). Since the protanope's fovea responds in the same way to both bleaches, it evidently contains only one pigment. The two sets of measurements define the difference spectrum of chlorolabe. Bleaching with white light, which shows total pigment present, shifts the foveal reflectivity upward at each wavelength (*larger squares and dots*). White-bleaching measurements coincide well with measurements of the protanope's sensitivity to white light (*colored curve*), made by F. H. G. Pitt of Imperial College. Still another way to measure bleaching, described in text, defines the "action spectrum" (*triangles*). It also supports the view that cones of the protanope contain one pigment.

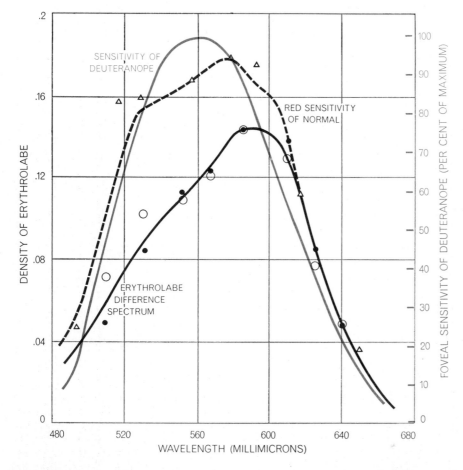

● AFTER BLEACHING WITH RED LIGHT

○ AFTER BLEACHING WITH BLUE-GREEN LIGHT

△ ACTION SPECTRUM

RED-CATCHING PIGMENT, erythrolabe, can be measured in the eye of a deuteranope, a person who is green-blind. The experiments are similar to those performed on the protanope. The black dots show the change in reflectivity of the fovea after partial bleaching with red light, the open circles after partial bleaching with blue-green light. The curve fitted to the two sets of circles is the difference spectrum of erythrolabe, the single visual pigment in the foveal cones of the deuteranope. The erythrolabe difference spectrum, however, does not coincide well with measurements by Pitt showing the deuteranope's sensitivity to white light (*colored curve*). This suggests that erythrolabe forms a colored photoproduct when bleached, which reduces foveal reflectivity below the values expected. The efficacy of bleaching as measured by the action spectrum comes closer to matching the deuteranope's visual sensitivity. It also agrees well with the sensitivity of the normal eye to red light alone (*broken curve*).

two lights of different wavelength composition will appear identical if they are scaled in intensity so that both are equally absorbed by rhodopsin. By the same token it should be possible to scale the intensity of two lights of different composition so that they will be absorbed equally by any one cone pigment. To that pigment the two lights would appear to have the same color. The scaling that will deceive the red pigment, however, will be detected by the green and blue pigments. It needs rather careful adjustment of two different color mixtures if they are to match; that is, if they are to deceive all three cone pigments at the same time. When this is achieved, the two inputs to the eye are in fact identical, and no one—not even Land—has the magic to show as different what all three cone pigments agree is the same.

Now we see why color matches are stable although color appearances change. Matches depend simply on the wavelength and intensity of light striking the three pigments and on the absorption spectra of these three chemicals. But appearances are subject to the whole complex of nervous interaction, not only between cone and cone in the retina but also between sensation and preconception in the mind. Let us therefore leave the rarefied atmosphere of color appearance and return to the solid ground of cone pigments.

If the cones contain three visual pigments, it should be possible to detect them and measure some of their properties by the method described for rhodopsin. To be sure, the human retina, like that of the frog, contains such a preponderance of rhodopsin that it is hard to measure anything else. Fortunately the fovea, that precious central square millimeter of the retina that we use for reading, contains no rods. It is also deficient in blue cones. Therefore if pigment-absorption measurements are confined to this tiny area, they should reveal the properties of just the red and green cones. One can simplify even further.

The common red-green color blindness is of two kinds: in one the color-blind individual is red-blind, in the other he is not. It turns out that the first individual lacks the red-sensitive pigment and that the second lacks the green-sensitive pigment. Therefore by measuring the fovea of the red-blind person, or protanope, we obtain information about the green-sensitive pigment only. The results of an analysis of this kind are

set forth in the top illustration on the opposite page.

It will be recollected that what we do is to adjust the wedge so that the output of the photocell is the same after bleaching as it was before. For the protanope experiment we use a gray wedge and express this displacement in terms of the corresponding change in optical density of the cone pigment. Since light passes through the pigment twice, once on entering and once on returning, measurements indicate a "double density" of pigment. Such measurements, made in lights of six wavelengths, are shown by the squares and dots in the illustration. The change in the reflectivity of the fovea, caused by bleaching, is maximal when measured with light that has a wavelength of 540 millimicrons and diminishes on each side. The small squares represent change in the reflectivity after bleaching with blue-green light; the small dots, after bleaching with very bright red light. These changes define

a curve that we call a difference spectrum. The fact that both curves coincide means that there is only one pigment present. If there had been a mixture, the more red-sensitive of the two would have shown a greater change after bleaching with red light; the other, after bleaching with blue-green light. A second series of measurements made after bleaching with a bright white light shows the total pigment present.

To discover whether or not this photosensitive pigment is indeed the basis of cone vision in the protanope we apply the test discussed earlier for rhodopsin. We simply ask: Does the spectral absorption coincide with the spectral sensitivity? The colored curve in the top illustration on the facing page shows how the cone sensitivity of the protanope does in fact correspond to the absorption measurements. We may conclude, therefore, that the protanope in daylight sees by this pigment, which is called chlorolabe, after the Greek words for "green-catching."

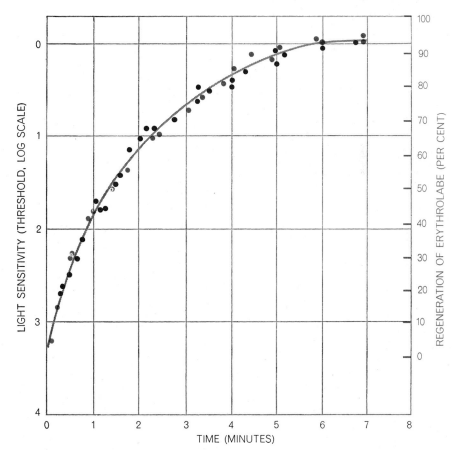

LIGHT SENSITIVITY AND REGENERATION OF ERYTHROLABE follow the same time course. The black dots show how the deuteranope's fovea becomes increasingly sensitive to brief flashes of light as the eye becomes dark-adapted. The colored dots are separate measurements made after the erythrolabe has been fully bleached. They show that the foveal pigment regenerates in seven minutes. The coincidence of the two sets of measurements implies that cones, like rods, have dark adaptation linked to pigment regeneration.

The other type of color-blind person, lacking chlorolabe, is known as a deuteranope. That he also has a single foveal pigment is established by the similar change in foveal reflectivity produced by either deep red light or blue-green light, as shown in the bottom illustration on page 158. It is plain that this pigment, which is called erythrolabe ("red-catching"), has a difference spectrum extending much further into the red than chlorolabe. If erythrolabe is the pigment that catches the light by which the deuteranope sees, he ought to be able to see further into the red end of the spectrum than the protanope can. This indeed is the case, but it is also apparent that the difference spectrum of chlorolabe does not coincide too well with the spectral sensitivity of the deuteranope, shown by the colored curve. Therefore the matter needs to be studied further.

If erythrolabe is the cone pigment of the deuteranope, lights of various wavelengths adjusted in intensity so that each appears equally bright to the deuteranope ought also to prove equivalent in the rate at which they bleach erythrolabe. Measurements of bleaching efficacy for lights of various wavelengths produce an "action spectrum," shown by triangles in the two illustrations on page 158. It can be seen in the bottom illustration that the action spectrum coincides reasonably well with the sensitivity of the deuteranope and also with the sensitivity of the red mechanism in the normal eye, shown by the broken curve. Thus there is fair agreement between sensitivity and bleaching power, and erythrolabe has a strong claim as the visual pigment of the deuteranope and of the normal red color mechanism.

It is also possible to measure the time required for the erythrolabe in the deuteranope's fovea to regenerate after bleaching. The curve in the illustration at the left resembles that for rhodopsin but rises about four times faster. It can be seen that the light sensitivity of the deuteranope, also plotted, increases precisely in step with the return of erythrolabe. So we are reasonably confident that erythrolabe is the pigment with which the deuteranope catches light.

Now we are in a position to prove that the normal fovea contains both green-sensitive chlorolabe and red-sensitive erythrolabe. The pertinent measurements are shown in the illustration below. The black dots show the bleaching produced by deep red light, and it is evident that they define a curve identical to the difference spectrum of erythrolabe, as measured in the deuteranope.

If in the deuteranope we changed the bleaching light from red to blue-green, no alteration would occur, since both lights bleach the deuteranope's single pigment equally. But when blue-green light is used to bleach the normal eye, one discovers that additional bleaching takes place, which cannot be attributed to erythrolabe. This additional bleaching is shown by the open circles in the illustration. Since no change in erythrolabe can contribute to this increment, it must represent the pure change in a second pigment in the normal eye. To see if this pigment is chlorolabe we draw on the same chart the difference spectrum of chlorolabe, as measured in the protanope, and we find that it closely follows the open circles. Thus the normal fovea is seen to contain both erythrolabe and chlorolabe.

A person with normal color vision can distinguish colors in the red-orange-yellow-green range of the spectrum because all of these colors affect the pigments erythrolabe and chlorolabe in different proportions. In this range protanopes and deuteranopes have only the one, or only the other, of these pigments; hence they have no more means of distinguishing these colors by day than a person with normal vision has by night. They can see only one color because they have only one pigment.

The reader will ask: What about the blues? Is there a "blue-catcher"—a cyanolabe—to complete the triad of cone pigments? I think there is, but it is much harder than the others to measure and there is not much at present to be said about it.

Practically all the ideas in this article have been entertained long ago by acute investigators; they have also often been disputed. What the measurement of pigments in man has done is to bring some degree of exactness and security to ideas that were enticing but speculative. The precision of measurement, however, lies not in the investigator who turns the knobs but in the subjects who sit with clamped head and fixed eye gazing steadfastly 20 minutes at a time through flashing and gloom. These are my students, some normal, some color-blind—volunteers from the classes in physiology in the University of Cambridge.

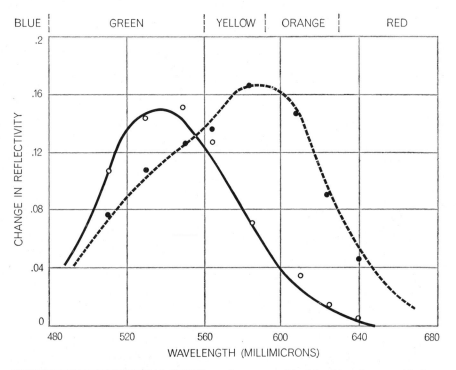

TWO PIGMENTS IN NORMAL CONES are demonstrated by bleaching the eye with deep red light, then with blue-green light and recording the change in reflectivity of the fovea at eight wavelengths. Bleaching with red light gives the results shown by black dots and coincides with the erythrolabe difference spectrum (*broken curve*) found in the deuteranope (*see bottom illustration on page 158*). When the bleaching light is blue-green, the reflectivity of the fovea increases beyond that observed when the bleach is red. The additional reflectivity is shown by open circles and conforms to the difference spectrum of chlorolabe (*solid curve*), as measured in the protanope (*see top illustration on page 158*).

Part V

NERVOUS
INTEGRATION

V

Nervous Integration

INTRODUCTION Most biologists would probably agree that the central nervous systems of animals represent the highest achievement of multicellular organization. The brains of some mammals contain more than a billion cells, linked to one another in an astonishingly intricate and precise way. Each cell is capable of translating input information, which arrives in the form of chemical transmitter substance from one or many nerve cells connecting with it, into a patterned sequence of electrical impulses. These propagate without signal loss to another region, often many centimeters away, and there release the chemical that will excite another nerve cell, a muscle fiber, or perhaps a gland.

The physiological performance of this remarkable system has intrigued biologists since the days of Galvani and Volta, when knowledge of the nature of "animal electricity" played an important role in the understanding of purely physical systems for generating currents and voltages. Contemporary investigations have sought an understanding of nervous function at two levels; and the aims and strategies of these two approaches are so different that they are really almost separate disciplines.

The first approach is fundamentally cellular. It is the attempt to comprehend the way in which impulses are carried along the long fibrous processes, the axons, of nerve cells. Such investigations have made impressive progress, especially in the past fifteen years. We now know that the propagating impulse results from sudden and selective changes in the permeability of the nerve membrane, first to sodium ions and then to potassium. Because these two ions are quite different from one another in their internal and external concentrations, the first of these changes in permeability produces an inward current across the membrane, temporarily reversing the normal outside-positive resting potential. The second change allows potassium to flow outward and re-establish the original resting condition. Other studies have provided a description of the events that ensue when this traveling sequence of potential changes reaches the end of the nerve fiber and causes the release of a small amount of transmitter chemical. Such analyses are being pursued further—toward a thorough understanding of those structural changes in the membrane fabric that allow sodium or potassium to pass with such sudden freedom and that encourage release of a certain amount of transmitter chemical at the nerve terminal.

The second approach is focused on the behavior of aggregates of nerve cells, in an attempt to explain how the properties of the cells, and of the junctions between them, are employed to convey information. Investigations of this kind may be concerned with the connections in the spinal cord that underlie a particular reflex act, or with the effect of a particular pattern of impulses in a set of sensory nerve fibers upon a higher-order cell, or with the behavioral results of stimulating a certain cluster of cells in the brain. These experiments depend, of course, upon accurate information about the properties of the cellular elements composing the nervous system; but this need be only a complete description of the properties and behavior, and not an explanation of the mechanism responsible for them. The position is analogous to that of a traffic engineer who wishes to understand the rules governing the movement of vehicles through a complex system of freeways. He requires information about the size, range, speed, and behavior of the cars, but he need not know the principles of the internal combustion engine. Indeed, for his purposes they could all run on steam.

The articles in this section all relate to the second approach, because they deal with ensembles of cellular elements. Their inclusion is appropriate, since the theme of this book is the transition from cellular to supracellular properties. In no area are the problems and the challenges of this transition more clearly exemplified than in the central nervous system, where remarkable perceptual and behavioral feats emerge from these large aggregates of basically similar, "standard" units. The nerve cells of the invertebrates and those of the higher vertebrates appear to be surprisingly similar: they conduct impulses in much the same fashion, they have junctions or synapses with nearly identical properties, and they both assume complex patterns of branching. It is equally clear, however, that larger brains show extraordinary other capacities that can be explained only as the result of the additional opportunity for permutation of connection offered by increased numbers of cells. Such problems are never easy to solve, but some promising beginnings are discussed in the four articles that follow.

The first, "The Synapse" by Sir John Eccles, provides a useful transition from cellular to multi-unit performance. By focusing on the characteristics of transmission between nerve cells in the mammalian spinal cord, Eccles has disclosed some of the basic properties of synapses and clarified the organization of a reflex pathway mediating one of the simplest forms of mammalian behavior. The organization of the reflex pathway is only indirectly alluded to in the article: the motoneurons are excited by sensory fibers which respond to stretching of the muscle served by those same motoneurons; they are inhibited by other fibers that respond to the stretching of muscles having the opposite action at the same joint. Eccles' careful analysis of the mechanisms of synaptic excitation and inhibition has thus provided a refined understanding of the operation of a simple reflex, whereby a muscle tends to maintain its length against an applied stretch and at the same time prevents contraction in its antagonists.

In "The Visual Cortex of the Brain," David H. Hubel describes some remarkable findings on the organization of a nervous center that processes information from a complex sense organ. The analysis has proceeded step by step from the neurons in the retina of the eye to those in the highest receiving area in the visual cortex. Each set of cells receives connections from another set nearer to the receptors, and in turn passes excitation on to a still higher level. At each successive stage, more complex stimuli are required to actuate the cells; at each level of synaptic integration, a highly specific set of connections appears to be made by the preceding layer of neurons. Thus optic nerve fibers will discharge when simple spots of light are shone onto the region of retina from which they come, whereas simple cortical cells require elongate linear stimuli with a specific angular orientation. Complex cortical cells may have responses that are similar but distributed over a wider area. Hubel shows that the preferences of cortical cells for a particular linear orientation have an anatomical correlate; cells with similar requirements are aligned in vertical columns. The specificity of these cells seems to be a direct outcome of their receiving connections from a particular set of cells from the level below—a situation for which the structure is obviously appropriate. Such connections, or the developmental machinery producing them, must be the result of a long natural selection process that has suited the nature of the extracted information to the needs of the animal. The cortical connections ensure that certain properties of a stimulus will be attended to and many others disregarded; this finding suggests that studies on the central organization of sensory systems will provide a portrait of the subject's own perceptual world. This exciting prospect can be reversed: an animal's behavior may predict the behavior of his sensory cortex.

A major emphasis of Hubel's studies is clearly upon the developmental processes that give rise to these crucially important connections between one level and the next higher one. It is not surprising that Hubel and his colleague Torsten Wiesel are now investigating the consequences of altered development upon cortical organization. They have recently found, for example, that newborn kittens, before they have opened their eyes, possess a "ready-made" pattern of connections resembling that in adult cats. If, however, one eye is occluded, or made to diverge from the other in its gaze by an operation, the input to cortical cells that had previously been binocular is soon restricted to one eye alone. Continued traffic is apparently necessary for the maintenance of connections that are initially established by a genetically determined mechanism.

The nature of the establishment of nervous connections during development is discussed by R. W. Sperry in "The Growth of Nerve Circuits." Sperry's own experiments, and those of others, have eclipsed the theory that the central nervous system is an essentially "plastic" tissue, in which new connections can always be grown to circumvent derangements and in which one group of cells can assume functions once handled by others. On the contrary, the evidence suggests that extremely precise and specific connections must be maintained or re-established if behavioral order is to be preserved. Along with this conclusion, Sperry cites evidence suggesting that the environment of the peripheral terminations of a nerve cell determines what central connections the cell will make. If this is indeed the usual case, then we must explain the nature of the biochemical influence that passes up the nerve fiber and learn how it determines the acceptability of different targets for central connection.

Each of the three articles discussed above concentrates upon a certain part of the vertebrate central nervous system. The fourth, "Learning in the Octopus" by Brian B. Boycott, is a useful reminder that brains of invertebrates, even though they have a different construction and are produced by an entirely independent line of evolution, can achieve high levels of complexity. Octopuses are excellent subjects for learning experiments, and Boycott, J. Z. Young, and their colleagues have located the parts of the brain concerned with various sorts of memory (visual and tactile centers, for example, are separate). These critical areas are usually occupied by large numbers of small cells; this supports the suggestion that complex functions develop when the opportunity for permutation is great.

17

THE SYNAPSE

SIR JOHN ECCLES   January 1965

The human brain is the most highly organized form of matter known, and in complexity the brains of the other higher animals are not greatly inferior. For certain purposes it is expedient to regard the brain as being analogous to a machine. Even if it is so regarded, however, it is a machine of a totally different kind from those made by man. In trying to understand the workings of his own brain man meets his highest challenge. Nothing is given; there are no operating diagrams, no maker's instructions.

The first step in trying to understand the brain is to examine its structure in order to discover the components from which it is built and how they are related to one another. After that one can attempt to understand the mode of operation of the simplest components. These two modes of investigation—the morphological and the physiological—have now become complementary. In studying the nervous system with today's sensitive electrical devices, however, it is all too easy to find physiological events that cannot be correlated with any known anatomical structure. Con-

versely, the electron microscope reveals many structural details whose physiological significance is obscure or unknown.

At the close of the past century the Spanish anatomist Santiago Ramón y Cajal showed how all parts of the nervous system are built up of individual nerve cells of many different shapes and sizes. Like other cells, each nerve cell has a nucleus and a surrounding cytoplasm. Its outer surface consists of numerous fine branches—the dendrites—that receive nerve impulses from other nerve cells, and one relatively long branch—the axon—that transmits nerve impulses. Near its end the axon divides into branches that terminate at the dendrites or bodies of other nerve cells. The axon can be as short as a fraction of a millimeter or as long as a meter, depending on its place and function. It has many of the properties of an electric cable and is uniquely specialized to conduct the brief electrical waves called nerve impulses [for further information, see "how Cells Communicate," by Bernhard Katz, Offprint #98]. In very thin axons these impulses travel at less

than one meter per second; in others, for example in the large axons of the nerve cells that activate muscles, they travel as fast as 100 meters per second.

The electrical impulse that travels along the axon ceases abruptly when it comes to the point where the axon's terminal fibers make contact with another nerve cell. These junction points were given the name "synapses" by Sir Charles Sherrington, who laid the foundations of what is sometimes called synaptology. If the nerve impulse is to continue beyond the synapse, it must be regenerated afresh on the other side. As recently as 15 years ago some physiologists held that transmission at the synapse was predominantly, if not exclusively, an electrical phenomenon. Now, however, there is abundant evidence that transmission is effectuated by the release of specific chemical substances that trigger a regeneration of the impulse. In fact, the first strong evidence showing that a transmitter substance acts across the synapse was provided more than 40 years ago by Sir Henry Dale and Otto Loewi.

It has been estimated that the hu-

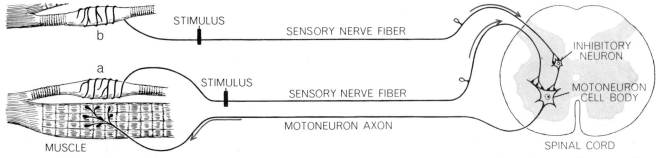

REFLEX ARCS provide simple pathways for studying the transmission of nerve impulses from one nerve cell to another. This transmission is effectuated at the junction points called synapses. In the illustration the sensory fiber from one muscle stretch receptor (a) makes direct synaptic contact with a motoneuron in the spinal cord. Nerve impulses generated by the moto- neuron activate the muscle to which the stretch receptor is attached. Stretch receptor b responds to the tension in a neighboring antagonistic muscle and sends impulses to a nerve cell that can inhibit the firing of the motoneuron. By electrically stimulating the appropriate stretch-receptor fibers one can study the effect of excitatory and inhibitory impulses on motoneurons.

man central nervous system, which of course includes the spinal cord as well as the brain itself, consists of about 10 billion (10^{10}) nerve cells. With rare exceptions each nerve cell receives information directly in the form of impulses from many other nerve cells—often hundreds—and transmits information to a like number. Depending on its threshold of response, a given nerve cell may fire an impulse when stimulated by only a few incoming fibers or it may not fire until stimulated by many incoming fibers. It has long been known that this threshold can be raised or lowered by various factors. Moreover, it was conjectured some 60 years ago that some of the incoming fibers must inhibit the firing of the receiving cell rather than excite it [*see illustration at right*]. The conjecture was subsequently confirmed, and the mechanism of the inhibitory effect has now been clarified. This mechanism and its equally fundamental counterpart—nerve-cell excitation—are the subject of this article.

Probing the Nerve Cell

At the level of anatomy there are some clues to indicate how the fine axon terminals impinging on a nerve cell can make the cell regenerate a nerve impulse of its own. The top illustration on the next page shows how a nerve cell and its dendrites are covered by fine branches of nerve fibers that terminate in knoblike structures. These structures are the synapses.

The electron microscope has revealed structural details of synapses that fit in nicely with the view that a chemical transmitter is involved in nerve transmission [*see lower two illustrations on next page*]. Enclosed in the synaptic knob are many vesicles, or tiny sacs, which appear to contain the transmitter substances that induce synaptic transmission. Between the synaptic knob and the synaptic membrane of the adjoining nerve cell is a remarkably uniform space of about 20 millimicrons that is termed the synaptic cleft. Many of the synaptic vesicles are concentrated adjacent to this cleft; it seems plausible that the transmitter substance is discharged from the nearest vesicles into the cleft, where it can act on the adjacent cell membrane. This hypothesis is supported by the discovery that the transmitter is released in packets of a few thousand molecules.

The study of synaptic transmission was revolutionized in 1951 by the introduction of delicate techniques for recording electrically from the interior

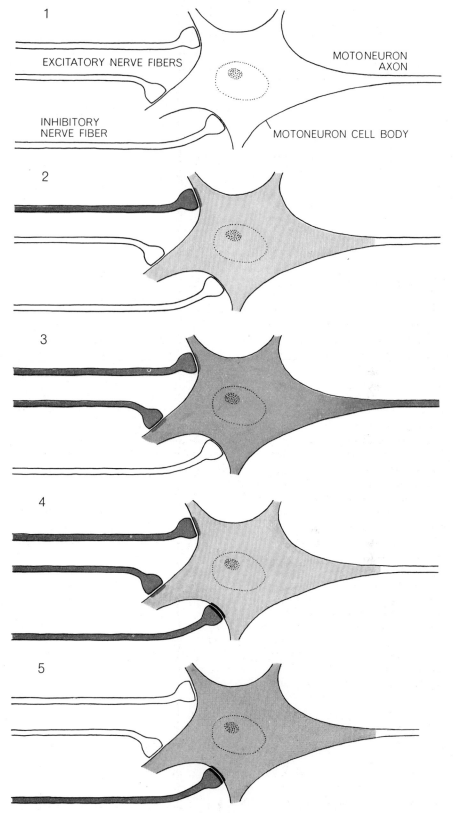

EXCITATION AND INHIBITION of a nerve cell are accomplished by the nerve fibers that form synapses on its surface. Diagram *1* shows a motoneuron in the resting state. In *2* impulses received from one excitatory fiber are inadequate to cause the motoneuron to fire. In *3* impulses from a second excitatory fiber raise the motoneuron to firing threshold. In *4* impulses carried by an inhibitory fiber restore the subthreshold condition. In *5* the inhibitory fiber alone is carrying impulses. There is no difference in the electrical impulses carried by excitatory and inhibitory nerve fibers. They achieve opposite effects because they release different chemical transmitter substances at their synaptic endings.

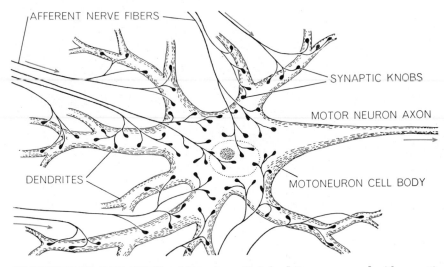

MOTONEURON CELL BODY and branches called dendrites are covered with synaptic knobs, which represent the terminals of axons, or impulse-carrying fibers, from other nerve cells. The axon of each motoneuron, in turn, terminates at a muscle fiber.

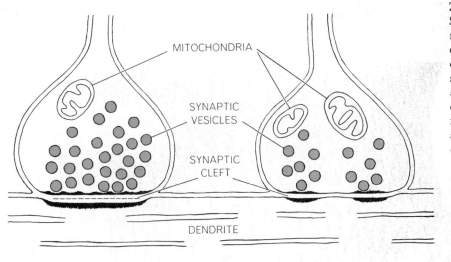

SYNAPTIC KNOBS are designed to deliver short bursts of a chemical transmitter substance into the synaptic cleft, where it can act on the surface of the nerve-cell membrane below. Before release, molecules of the chemical transmitter are stored in numerous vesicles, or sacs. Mitochondria are specialized structures that help to supply the cell with energy.

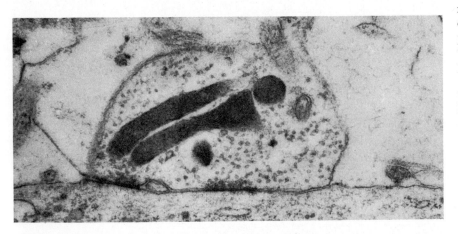

ASSUMED INHIBITORY SYNAPSE on a nerve cell is magnified 28,000 diameters in this electron micrograph by the late L. H. Hamlyn of University College London. Synaptic vesicles, believed to contain the transmitter substance, are bunched in two regions along the synaptic cleft. The darkening of the cleft in these regions is so far unexplained.

of single nerve cells. This is done by inserting into the nerve cell an extremely fine glass pipette with a diameter of .5 micron—about a fifty-thousandth of an inch. The pipette is filled with an electrically conducting salt solution such as concentrated potassium chloride. If the pipette is carefully inserted and held rigidly in place, the cell membrane appears to seal quickly around the glass, thus preventing the flow of a short-circuiting current through the puncture in the cell membrane. Impaled in this fashion, nerve cells can function normally for hours. Although there is no way of observing the cells during the insertion of the pipette, the insertion can be guided by using as clues the electric signals that the pipette picks up when close to active nerve cells.

When my colleagues and I in New Zealand and later at the John Curtin School of Medical Research in Canberra first employed this technique, we chose to study the large nerve cells called motoneurons, which lie in the spinal cord and whose function is to activate muscles. This was a fortunate choice: intracellular investigations with motoneurons have proved to be easier and more rewarding than those with any other kind of mammalian nerve cell.

We soon found that when the nerve cell responds to the chemical synaptic transmitter, the response depends in part on characteristic features of ionic composition that are also concerned with the transmission of impulses in the cell and along its axon. When the nerve cell is at rest, its physiological makeup resembles that of most other cells in that the water solution inside the cell is quite different in composition from the solution in which the cell is bathed. The nerve cell is able to exploit this difference between external and internal composition and use it in quite different ways for generating an electrical impulse and for synaptic transmission.

The composition of the external solution is well established because the solution is essentially the same as blood from which cells and proteins have been removed. The composition of the internal solution is known only approximately. Indirect evidence indicates that the concentrations of sodium and chloride ions outside the cell are respectively some 10 and 14 times higher than the concentrations inside the cell. In contrast, the concentration of potassium ions inside the cell is about 30 times higher than the concentration outside.

How can one account for this remarkable state of affairs? Part of the explanation is that the inside of the cell is negatively charged with respect to the outside of the cell by about 70 millivolts. Since like charges repel each other, this internal negative charge tends to drive chloride ions (Cl⁻) outward through the cell membrane and, at the same time, to impede their inward movement. In fact, a potential difference of 70 millivolts is just sufficient to maintain the observed disparity in the concentration of chloride ions inside the cell and outside it; chloride ions diffuse inward and outward at equal rates. A drop of 70 millivolts across the membrane therefore defines the "equilibrium potential" for chloride ions.

To obtain a concentration of potassium ions (K⁺) that is 30 times higher inside the cell than outside would require that the interior of the cell membrane be about 90 millivolts negative with respect to the exterior. Since the

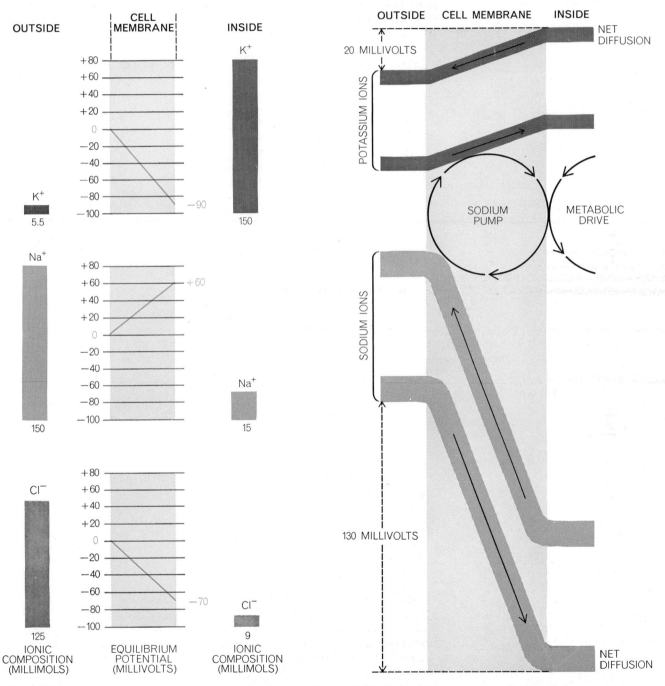

IONIC COMPOSITION outside and inside the nerve cell is markedly different. The "equilibrium potential" is the voltage drop that would have to exist across the membrane of the nerve cell to produce the observed difference in concentration for each type of ion. The actual voltage drop is about 70 millivolts, with the inside being negative. Given this drop, chloride ions diffuse inward and outward at equal rates, but the concentration of sodium and potassium must be maintained by some auxiliary mechanism (*right*).

METABOLIC PUMP must be postulated to account for the observed concentrations of potassium and sodium ions on opposite sides of the nerve-cell membrane. The negative potential inside is 20 millivolts short of the equilibrium potential for potassium ions. Thus there is a net outward diffusion of potassium ions that must be balanced by the pump. For sodium ions the potential across the membrane is 130 millivolts in the wrong direction, so very energetic pumping is needed. Chloride ions are in equilibrium.

actual interior is only 70 millivolts negative, it falls short of the equilibrium potential for potassium ions by 20 millivolts. Evidently the thirtyfold concentration can be achieved and maintained only if there is some auxiliary mechanism for "pumping" potassium ions into the cell at a rate equal to their spontaneous net outward diffusion.

The pumping mechanism has the still more difficult task of pumping sodium ions (Na^+) out of the cell against a potential gradient of 130 millivolts. This figure is obtained by adding the 70 millivolts of internal negative charge to the equilibrium potential for sodium ions, which is 60 millivolts of internal *positive* charge [*see illustrations on preceding page*]. If it were not for this postulated pump, the concentration of sodium ions inside and outside the cell would be almost the reverse of what is observed.

In their classic studies of nerve-impulse transmission in the giant axon of the squid, A. L. Hodgkin, A. F. Huxley and Bernhard Katz of Britain demonstrated that the propagation of the impulse coincides with abrupt changes in the permeability of the axon membrane. When a nerve impulse has been triggered in some way, what can be described as a gate opens and lets sodium ions pour into the axon during the advance of the impulse, making the interior of the axon locally positive. The process is self-reinforcing in that the flow of some sodium ions through the membrane opens the gate further and makes it easier for others to follow. The sharp reversal of the internal polarity of the membrane constitutes the nerve impulse, which moves like a wave until it has traveled the length of the axon. In the wake of the impulse the sodium gate closes and a potassium gate opens, thereby restoring the normal polarity of the membrane within a millisecond or less.

With this understanding of the nerve impulse in hand, one is ready to follow the electrical events at the excitatory synapse. One might guess that if the nerve impulse results from an abrupt inflow of sodium ions and a rapid change in the electrical polarity of the axon's interior, something similar must happen at the body and dendrites of the nerve cell in order to generate the impulse in the first place. Indeed, the function of the excitatory synaptic terminals on the cell body and its dendrites is to depolarize the interior of the cell membrane essentially by permitting an inflow of sodium ions. When the depolarization reaches a threshold value, a nerve impulse is triggered.

As a simple instance of this phenomenon we have recorded the depolarization that occurs in a single motoneuron activated directly by the large nerve fibers that enter the spinal cord from special stretch-receptors known as annulospiral endings. These receptors in turn are located in the same muscle that

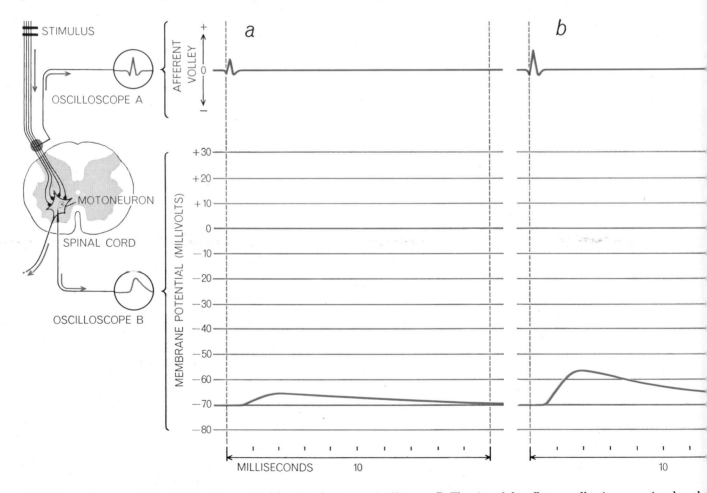

EXCITATION OF A MOTONEURON is studied by stimulating the sensory fibers that send impulses to it. The size of the "afferent volleys" reaching the motoneuron is displayed on oscilloscope *A*. A microelectrode implanted in the motoneuron measures the changes in the cell's internal electric potential. These changes, called excitatory postsynaptic potentials (EPSP's), appear on oscilloscope *B*. The size of the afferent volley is proportional to the number of fibers stimulated to fire. It is assumed here that one to four fibers can be activated. When only one fiber is activated (*a*) the potential inside the motoneuron shifts only slightly. When two fibers are activated (*b*), the shift is somewhat greater. When three fibers are activated (*c*), the potential reaches the threshold

is activated by the motoneuron under study. Thus the whole system forms a typical reflex arc, such as the arc responsible for the patellar reflex, or "knee jerk" [see illustration on page 166].

To conduct the experiment we anesthetize an animal (most often a cat) and free by dissection a muscle nerve that contains these large nerve fibers. By applying a mild electric shock to the exposed nerve one can produce a single impulse in each of the fibers; since the impulses travel to the spinal cord almost synchronously they are referred to collectively as a volley. The number of impulses contained in the volley can be reduced by reducing the stimulation applied to the nerve. The volley strength is measured at a point just outside the spinal cord and is displayed on an oscilloscope. About half a millisecond after detection of a volley there is a wavelike change in the voltage inside the motoneuron that has received the volley. The change is detected by a microelectrode

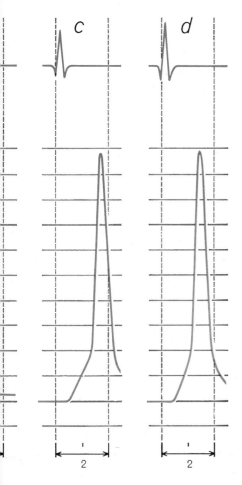

inserted in the motoneuron and is displayed on another oscilloscope.

What we find is that the negative voltage inside the cell becomes progressively less negative as more of the fibers impinging on the cell are stimulated to fire. This observed depolarization is in fact a simple summation of the depolarizations produced by each individual synapse. When the depolarization of the interior of the motoneuron reaches a critical point, a "spike" suddenly appears on the second oscilloscope, showing that a nerve impulse has been generated. During the spike the voltage inside the cell changes from about 70 millivolts negative to as much as 30 millivolts positive. The spike regularly appears when the depolarization, or reduction of membrane potential, reaches a critical level, which is usually between 10 and 18 millivolts. The only effect of a further strengthening of the synaptic stimulus is to shorten the time needed for the motoneuron to reach the firing threshold [see illustration at left]. The depolarizing potentials produced in the cell membrane by excitatory synapses are called excitatory postsynaptic potentials, or EPSP's.

Through one barrel of a double-barreled microelectrode one can apply a background current to change the resting potential of the interior of the cell membrane, either increasing it or decreasing it. When the potential is made more negative, the EPSP rises more steeply to an earlier peak. When the potential is made less negative, the EPSP rises more slowly to a lower peak. Finally, when the charge inside the cell is reversed so as to be positive with respect to the exterior, the excitatory synapses give rise to an EPSP that is actually the reverse of the normal one [see illustration at right].

These observations support the hypothesis that excitatory synapses produce what amounts virtually to a short circuit in the synaptic membrane potential. When this occurs, the membrane no longer acts as a barrier to the passage of ions but lets them flow through in response to the differing electric potential on the two sides of the membrane. In other words, the ions are momentarily allowed to travel freely down their electrochemical gradients, which means that sodium ions flow into the cell and, to a lesser degree, potassium ions flow out. It is this net flow of positive ions that creates the excitatory postsynaptic potential. The flow of negative ions, such as the chloride ion, is apparently not involved. By artificially

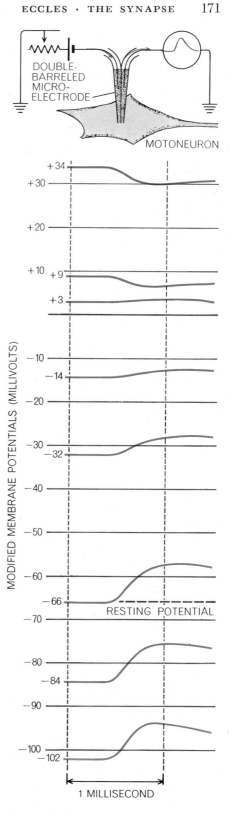

MANIPULATION of the resting potential of a motoneuron clarifies the nature of the EPSP. A steady background current applied through the left barrel of a microelectrode (top) shifts the membrane potential away from its normal resting level (minus 66 millivolts in this particular cell). The other barrel records the EPSP. The equilibrium potential, the potential at which the EPSP reverses direction, is about zero millivolts.

at which depolarization proceeds swiftly and a spike appears on oscilloscope B. The spike signifies that the motoneuron has generated a nerve impulse of its own. When four or more fibers are activated (d), the motoneuron reaches the threshold more quickly.

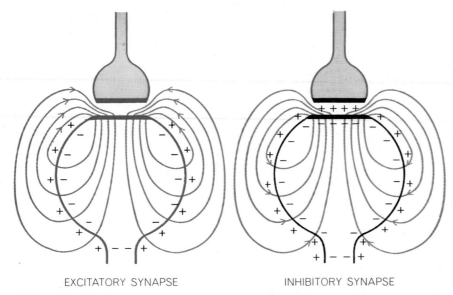

EXCITATORY SYNAPSE   INHIBITORY SYNAPSE

CURRENT FLOWS induced by excitatory and inhibitory synapses are respectively shown
at left and right. When the nerve cell is at rest, the interior of the cell membrane is uniformly negative with respect to the exterior. The excitatory synapse releases a chemical
substance that depolarizes the cell membrane below the synaptic cleft, thus letting current flow into the cell at that point. At an inhibitory synapse the current flow is reversed.

altering the potential inside the cell one
can establish that there is no flow of
ions, and therefore no EPSP, when the
voltage drop across the membrane is
zero.

How is the synaptic membrane converted from a strong ionic barrier into
an ion-permeable state? It is currently
accepted that the agency of conversion
is the chemical transmitter substance
contained in the vesicles inside the syn-

aptic knob. When a nerve impulse
reaches the synaptic knob, some of the
vesicles are caused to eject the transmitter substance into the synaptic cleft
[see illustration below]. The molecules
of the substance would take only a few
microseconds to diffuse across the cleft
and become attached to specific receptor sites on the surface membrane of the
adjacent nerve cell.

Presumably the receptor sites are as-

sociated with fine channels in the membrane that are opened in some way by
the attachment of the transmitter-sub-
stance molecules to the receptor sites.
With the channels thus opened, sodium
and potassium ions flow through the
membrane thousands of times more
readily than they normally do, thereby
producing the intense ionic flux that depolarizes the cell membrane and produces the EPSP. In many synapses the
current flows strongly for only about a
millisecond before the transmitter substance is eliminated from the synaptic
cleft, either by diffusion into the surrounding regions or as a result of being
destroyed by enzymes. The latter process is known to occur when the transmitter substance is acetylcholine, which
is destroyed by the enzyme acetylcho-
linesterase.

The substantiation of this general picture of synaptic transmission requires
the solution of many fundamental problems. Since we do not know the specific
transmitter substance for the vast majority of synapses in the nervous system
we do not know if there are many different substances or only a few. The
only one identified with reasonable cer-
tainty in the mammalian central nervous
system is acetylcholine. We know practically nothing about the mechanism
by which a presynaptic nerve impulse
causes the transmitter substance to be
injected into the synaptic cleft. Nor do
we know how the synaptic vesicles not
immediately adjacent to the synaptic
cleft are moved up to the firing line to
replace the emptied vesicles. It is conjectured that the vesicles contain the
enzyme systems needed to recharge
themselves. The entire process must be
swift and efficient: the total amount of
transmitter substance in synaptic terminals is enough for only a few minutes of
synaptic activity at normal operating
rates. There are also knotty problems
to be solved on the other side of the
synaptic cleft. What, for example, is the
nature of the receptor sites? How are
the ionic channels in the membrane
opened up?

The Inhibitory Synapse

Let us turn now to the second type
of synapse that has been identified in
the nervous system. These are the synapses that can inhibit the firing of a
nerve cell even though it may be receiving a volley of excitatory impulses.
When inhibitory synapses are examined
in the electron microscope, they look
very much like excitatory synapses.

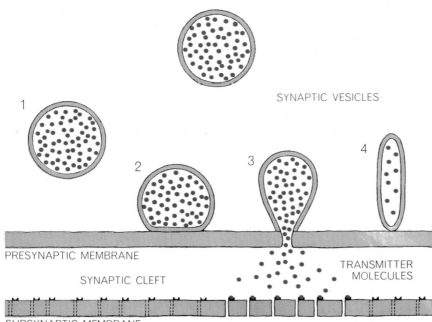

SYNAPTIC VESICLES

1

2   3   4

PRESYNAPTIC MEMBRANE

SYNAPTIC CLEFT   TRANSMITTER
                                        MOLECULES

SUBSYNAPTIC MEMBRANE

SYNAPTIC VESICLES containing a chemical transmitter are distributed throughout the
synaptic knob. They are arranged here in a probable sequence, showing how they move
up to the synaptic cleft, discharge their contents and return to the interior for recharging.

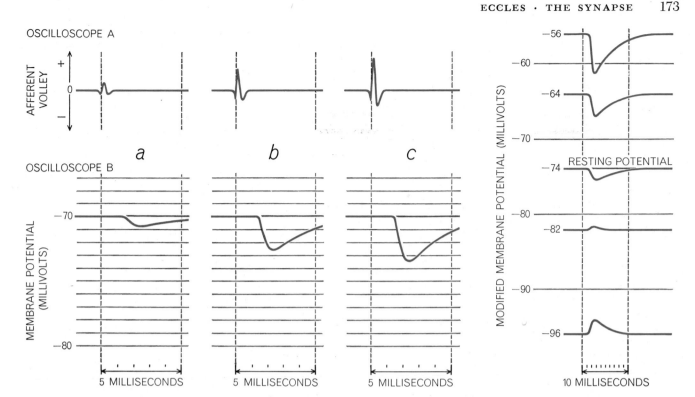

INHIBITION OF A MOTONEURON is investigated by methods like those used for studying the EPSP. The inhibitory counterpart of the EPSP is the IPSP: the inhibitory postsynaptic potential. Oscilloscope *A* records an afferent volley that travels to a number of inhibitory nerve cells whose axons form synapses on a nearby motoneuron (*see illustration on page 166*). A microelec-trode in the motoneuron is connected to oscilloscope *B*. The sequence *a*, *b* and *c* shows how successively larger afferent volleys produce successively deeper IPSP's. Curves at right show how the IPSP is modified when a background current is used to change the motoneuron's resting potential. The equilibrium potential where the IPSP reverses direction is about minus 80 millivolts.

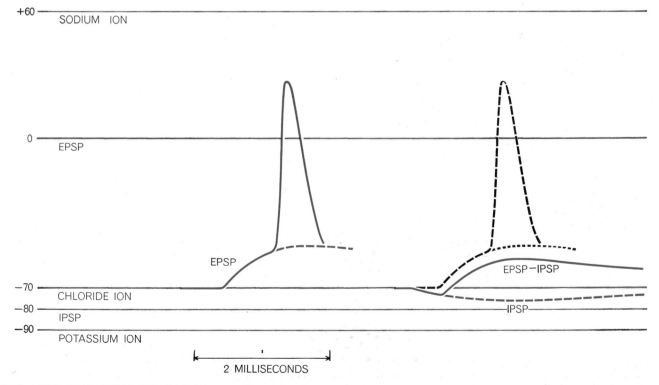

INHIBITION OF A SPIKE DISCHARGE is an electrical subtraction process. When a normal EPSP reaches a threshold (*left*), it will ordinarily produce a spike. An IPSP widens the gap between the cell's internal potential and the firing threshold. Thus if a cell is simultaneously subjected to both excitatory and inhibitory stimulation, the IPSP is subtracted from the EPSP (*right*) and no spike occurs. The five horizontal lines show equilibrium potentials for the three principal ions as well as for the EPSP and IPSP.

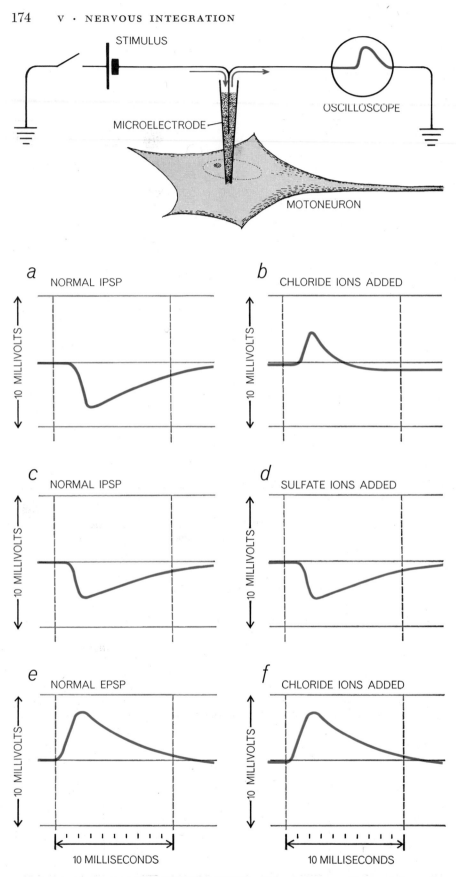

MODIFICATION OF ION CONCENTRATION within the nerve cell gives information about the permeability of the cell membrane. The internal ionic composition is altered by injecting selected ions through a microelectrode a minute or so before applying an afferent volley and recording the EPSP or IPSP. In the first experiment a normal IPSP (*a*) is changed to a pseudo-EPSP (*b*) by an injection of chloride ions. When sulfate ions are similarly injected, the IPSP is practically unchanged (*b, c*). The third experiment shows that an injection of chloride ions has no significant effect on the EPSP (*e, f*).

(There are probably some subtle differences, but they need not concern us here.) Microelectrode recordings of the activity of single motoneurons and other nerve cells have now shown that the inhibitory postsynaptic potential (IPSP) is virtually a mirror image of the EPSP [*see top illustration on preceding page*]. Moreover, individual inhibitory synapses, like excitatory synapses, have a cumulative effect. The chief difference is simply that the IPSP makes the cell's internal voltage more negative than it is normally, which is in a direction opposite to that needed for generating a spike discharge.

By driving the internal voltage of a nerve cell in the negative direction inhibitory synapses oppose the action of excitatory synapses, which of course drive it in the positive direction. Hence if the potential inside a resting cell is 70 millivolts negative, a strong volley of inhibitory impulses can drive the potential to 75 or 80 millivolts negative. One can easily see that if the potential is made more negative in this way the excitatory synapses find it more difficult to raise the internal voltage to the threshold point for the generation of a spike. Thus the nerve cell responds to the algebraic sum of the internal voltage changes produced by excitatory and inhibitory synapses [*see bottom illustration on preceding page*].

If, as in the experiment described earlier, the internal membrane potential is altered by the flow of an electric current through one barrel of a double-barreled microelectrode, one can observe the effect of such changes on the inhibitory postsynaptic potential. When the internal potential is made less negative, the inhibitory postsynaptic potential is deepened. Conversely, when the potential is made more negative, the IPSP diminishes; it finally reverses when the internal potential is driven below minus 80 millivolts.

One can therefore conclude that inhibitory synapses share with excitatory synapses the ability to change the ionic permeability of the synaptic membrane. The difference is that inhibitory synapses enable ions to flow freely down an electrochemical gradient that has an equilibrium point at minus 80 millivolts rather than at zero, as is the case for excitatory synapses. This effect could be achieved by the outward flow of positively charged ions such as potassium or the inward flow of negatively charged ions such as chloride, or by a combination of negative and positive ionic flows such that the interior reaches equilibrium at minus 80 millivolts.

In an effort to discover the permeability changes associated with the inhibitory potential my colleagues and I have altered the concentration of ions normally found in motoneurons and have introduced a variety of other ions that are not normally present. This can be done by impaling nerve cells with micropipettes that are filled with a salt solution containing the ion to be injected. The actual injection is achieved by passing a brief current through the micropipette.

If the concentration of chloride ions within the cell is in this way increased as much as three times, the inhibitory postsynaptic potential reverses and acts as a depolarizing current; that is, it resembles an excitatory potential. On the other hand, if the cell is heavily injected with sulfate ions, which are also negatively charged, there is no such reversal [*see illustration on opposite page*]. This simple test shows that under the influence of the inhibitory transmitter substance, which is still unidentified, the subsynaptic membrane becomes permeable momentarily to chloride ions but not to sulfate ions. During the generation of the IPSP the outflow of chloride ions is so rapid that it more than outweighs the flow of other ions that generate the normal inhibitory potential.

My colleagues have now tested the effect of injecting motoneurons with more than 30 kinds of negatively charged ion. With one exception the hydrated ions (ions bound to water) to which the cell membrane is permeable under the influence of the inhibitory transmitter substance are smaller than the hydrated ions to which the membrane is impermeable. The exception is the formate ion (HCO_2^-), which may have an ellipsoidal shape and so be able to pass through membrane pores that block smaller spherical ions.

Apart from the formate ion all the ions to which the membrane is permeable have a diameter not greater than 1.14 times the diameter of the potassium ion; that is, they are less than 2.9 angstrom units in diameter. Comparable investigations in other laboratories have found the same permeability effects, including the exceptional behavior of the formate ion, in fishes, toads and snails. It may well be that the ionic mechanism responsible for synaptic inhibition is the same throughout the animal kingdom.

The significance of these and other studies is that they strongly indicate that the inhibitory transmitter substance opens the membrane to the flow of potassium ions but not to sodium ions. It

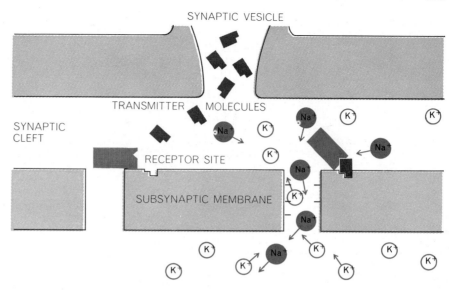

EXCITATORY SYNAPSE may employ transmitter molecules that open large channels in the nerve-cell membrane. This would permit sodium ions, which are plentiful outside the cell, to pour through the membrane freely. The outward flow of potassium ions, driven by a smaller potential gradient, would be at a much slower rate. Chloride ions (*not shown*) may be prevented from flowing by negative charges on the channel walls.

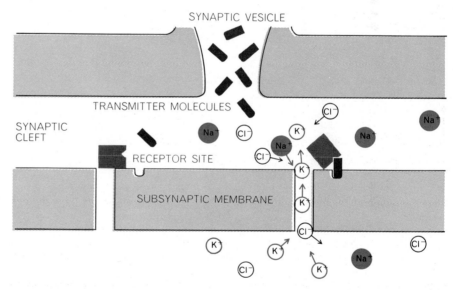

INHIBITORY SYNAPSE may employ another type of transmitter molecule that opens channels too small to pass sodium ions. The net outflow of potassium ions and inflow of chloride ions would account for the hyperpolarization that is observed as an IPSP.

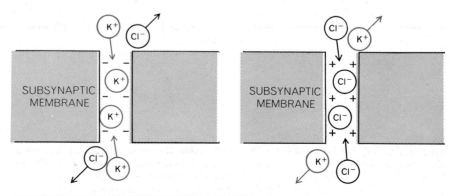

MODIFICATIONS OF INHIBITORY SYNAPSE may involve channels that carry either negative or positive charges on their walls. Negative charges (*left*) would permit only potassium ions to pass. Positive charges (*right*) would permit only chloride ions to pass.

is known that the sodium ion is somewhat larger than any of the negatively charged ions, including the formate ion, that are able to pass through the membrane during synaptic inhibition. It is not possible, however, to test the effectiveness of potassium ions by injecting excess amounts into the cell because the excess is immediately diluted by an osmotic flow of water into the cell.

As I have indicated, the concentration of potassium ions inside the nerve cell is about 30 times greater than the concentration outside, and to maintain this large difference in concentration without the help of a metabolic pump the inside of the membrane would have to be charged 90 millivolts negative with respect to the exterior. This implies that if the membrane were suddenly made porous to potassium ions, the resulting outflow of ions would make the inside potential of the membrane even more negative than it is in the resting state, and that is just what happens during synaptic inhibition. The membrane must not simultaneously become porous to sodium ions, because they exist in much higher concentration outside the cell than inside and their rapid inflow would more than compensate for the potassium outflow. In fact, the fundamental difference between synaptic excitation and synaptic inhibition is that the membrane freely passes sodium ions in response to the former and largely excludes the passage of sodium ions in response to the latter.

Channels in the Membrane

This fine discrimination between ions that are not very different in size must be explained by any hypothesis of synaptic action. It is most unlikely that the channels through the membrane are created afresh and accurately maintained for a thousandth of a second every time a burst of transmitter substance is released into the synaptic cleft. It is more likely that channels of at least two different sizes are built directly into the membrane structure. In some way the excitatory transmitter substance would selectively unplug the larger channels and permit the free inflow of sodium ions. Potassium ions would simultaneously flow out and thus would tend to counteract the large potential change that would be produced by the massive sodium inflow. The inhibitory transmitter substance would selectively unplug the smaller channels that are large enough to pass potassium and chloride ions but not sodium ions [see upper two illustrations on previous page].

To explain certain types of inhibition other features must be added to this hypothesis of synaptic transmission. In the simple hypothesis chloride and potassium ions can flow freely through pores of all inhibitory synapses. It has been shown, however, that the inhibition of the contraction of heart muscle by the vagus nerve is due almost exclusively to potassium-ion flow. On the other hand, in the muscles of crustaceans and in nerve cells in the snail's brain synaptic inhibition is due largely to the flow of chloride ions. This selective permeability could be explained if there were fixed charges along the walls of the channels. If such charges were negative, they would repel negatively charged ions and prevent their passage; if they were positive, they would similarly prevent the passage of positively charged ions. One can now suggest that the channels opened by the excitatory transmitter are negatively charged and so do not permit the passage of the negatively charged chloride ion, even though it is small enough to move through the channel freely.

One might wonder if a given nerve cell can have excitatory synaptic action at some of its axon terminals and inhibitory action at others. The answer is no. Two different kinds of nerve cell are needed, one for each type of transmission and synaptic transmitter substance. This can readily be demonstrated by the effect of strychnine and tetanus toxin in the spinal cord; they specifically prevent inhibitory synaptic action and leave excitatory action unaltered. As a result the synaptic excitation of nerve cells is uncontrolled and convulsions result. The special types of cell responsible for inhibitory synaptic action are now being recognized in many parts of the central nervous system.

This account of communication between nerve cells is necessarily oversimplified, yet it shows that some significant advances are being made at the level of individual components of the nervous system. By selecting the most favorable situations we have been able to throw light on some details of nerve-cell behavior. We can be encouraged by these limited successes. But the task of understanding in a comprehensive way how the human brain operates staggers its own imagination.

18

THE VISUAL CORTEX
OF THE BRAIN

DAVID H. HUBEL   November 1963

An image of the outside world striking the retina of the eye activates a most intricate process that results in vision: the transformation of the retinal image into a perception. The transformation occurs partly in the retina but mostly in the brain, and it is, as one can recognize instantly by considering how modest in comparison is the achievement of a camera, a task of impressive magnitude.

The process begins with the responses of some 130 million light-sensitive receptor cells in each retina. From these cells messages are transmitted to other retinal cells and then sent on to the brain, where they must be analyzed and interpreted. To get an idea of the magnitude of the task, think what is involved in watching a moving animal, such as a horse. At a glance one takes in its size, form, color and rate of movement. From tiny differences in the two retinal images there results a three-dimensional picture. Somehow the brain manages to compare this picture with previous impressions; recognition occurs and then any appropriate action can be taken.

The organization of the visual system —a large, intricately connected population of nerve cells in the retina and brain —is still poorly understood. In recent years, however, various studies have begun to reveal something of the arrangement and function of these cells. A decade ago Stephen W. Kuffler, working with cats at the Johns Hopkins Hospital, discovered that some analysis of visual patterns takes place outside the brain, in the nerve cells of the retina. My colleague Torsten N. Wiesel and I at the Harvard Medical School, exploring the first stages of the processing that occurs in the brain of the cat, have mapped the visual pathway a little further: to what appears to be the sixth step from the retina to the cortex of the cerebrum. This kind of

work falls far short of providing a full understanding of vision, but it does convey some idea of the mechanisms and circuitry of the visual system.

In broad outline the visual pathway is clearly defined [*see bottom illustration on following page*]. From the retina of each eye visual messages travel along the optic nerve, which consists of about a million nerve fibers. At the junction known as the chiasm about half of the nerves cross over into opposite hemispheres of the brain, the other nerves remaining on the same side. The optic nerve fibers lead to the first way stations in the brain: a pair of cell clusters called the lateral geniculate bodies. From here new fibers course back through the brain to the visual area of the cerebral cortex. It is convenient, although admittedly a gross oversimplification, to think of the pathway from retina to cortex as consisting of six types of nerve cells, of which three are in the retina, one is in the geniculate body and two are in the cortex.

Nerve cells, or neurons, transmit messages in the form of brief electrochemical impulses. These travel along the outer membrane of the cell, notably along the membrane of its long principal fiber, the axon. It is possible to obtain an electrical record of impulses of a single nerve cell by placing a fine electrode near the cell body or one of its fibers. Such measurements have shown that impulses travel along the nerves at velocities of between half a meter and 100 meters per second. The impulses in a given fiber all have about the same amplitude; the strength of the stimuli that give rise to them is reflected not in amplitude but in frequency.

At its terminus the fiber of a nerve cell makes contact with another nerve cell (or with a muscle cell or gland

cell), forming the junction called the synapse. At most synapses an impulse on reaching the end of a fiber causes the release of a small amount of a specific substance, which diffuses outward to the membrane of the next cell. There the substance either excites the cell or inhibits it. In excitation the substance acts to bring the cell into a state in which it is more likely to "fire"; in inhibition the substance acts to prevent firing. For most synapses the substances that act as transmitters are unknown. Moreover, there is no sure way to determine from microscopic appearances alone whether a synapse is excitatory or inhibitory.

It is at the synapses that the modification and analysis of nerve messages take place. The kind of analysis depends partly on the nature of the synapse: on how many nerve fibers converge on a single cell and on how the excitatory and inhibitory endings distribute themselves. In most parts of the nervous system the anatomy is too intricate to reveal much about function. One way to circumvent this difficulty is to record impulses with microelectrodes in anesthetized animals, first from the fibers coming into a structure of neurons and then from the neurons themselves, or from the fibers they send onward. Comparison of the behavior of incoming and outgoing fibers provides a basis for learning what the structure does. Through such exploration of the different parts of the brain concerned with vision one can hope to build up some idea of how the entire visual system works.

That is what Wiesel and I have undertaken, mainly through studies of the visual system of the cat. In our experiments the anesthetized animal faces a wide screen 1.5 meters away, and we shine various patterns of white light on the screen with a projector. Simultaneously we penetrate the visual portion of the cortex with microelectrodes. In that way we can record the responses of individual cells to the light patterns. Sometimes it takes many hours to find the region of the retina with which a particular visual cell is linked and to work out the optimum stimuli for that cell. The reader should bear in mind the relation between each visual cell—no matter how far along the visual pathway it may be—and the retina. It requires an image on the retina to evoke a meaningful response in any visual cell, however indirect and complex the linkage may be.

The retina is a complicated structure, in both its anatomy and its physiology, and the description I shall give is highly simplified. Light coming through the lens of the eye falls on the mosaic of receptor cells in the retina. The receptor cells do not send impulses directly through the optic nerve but instead connect with a set of retinal cells called bipolar cells. These in turn connect with retinal ganglion cells, and it is the latter set of cells, the third in the visual pathway, that sends its fibers—the optic nerve fibers—to the brain.

This series of cells and synapses is no simple bucket brigade for impulses: a receptor may send nerve endings to more than one bipolar cell, and several receptors may converge on one bipolar cell. The same holds for the synapses between the bipolar cells and the retinal ganglion cells. Stimulating a single receptor by light might therefore be expected to have an influence on many bipolar or ganglion cells; conversely, it should be possible to influence one bipolar or retinal ganglion cell from a number of receptors and hence from a substantial area of the retina.

The area of receptor mosaic in the retina feeding into a single visual cell is called the receptive field of the cell. This term is applied to any cell in the visual system to refer to the area of retina with which the cell is connected—the retinal area that on stimulation produces a response from the cell.

Any of the synapses with a particular cell may be excitatory or inhibitory, so that stimulation of a particular point on the retina may either increase or decrease the cell's firing rate. Moreover, a single cell may receive several excitatory and inhibitory impulses at once, with the result that it will respond according to the net effect of these inputs. In considering the behavior of a single cell an observer should remember that it is just one of a huge popu-

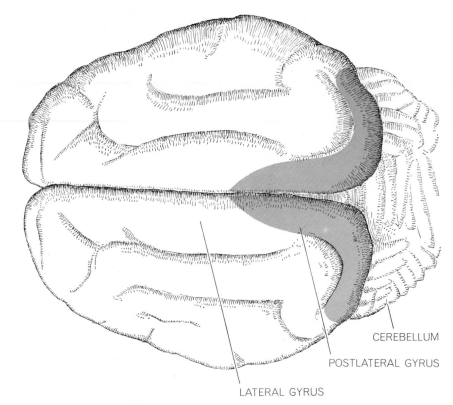

CEREBELLUM

POSTLATERAL GYRUS

LATERAL GYRUS

CORTEX OF CAT'S BRAIN is depicted as it would be seen from the top. The colored region indicates the cortical area that deals at least in a preliminary way with vision.

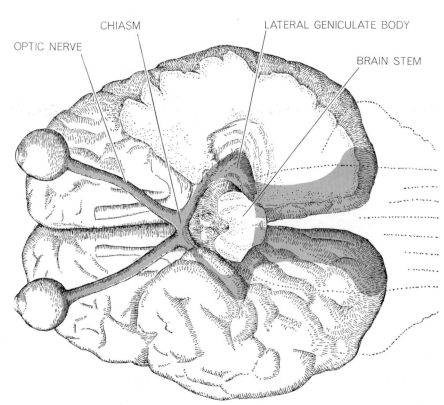

CHIASM   LATERAL GENICULATE BODY

OPTIC NERVE   BRAIN STEM

VISUAL SYSTEM appears in this representation of the human brain as viewed below. Visual pathway from retinas to cortex via the lateral geniculate body is shown in color.

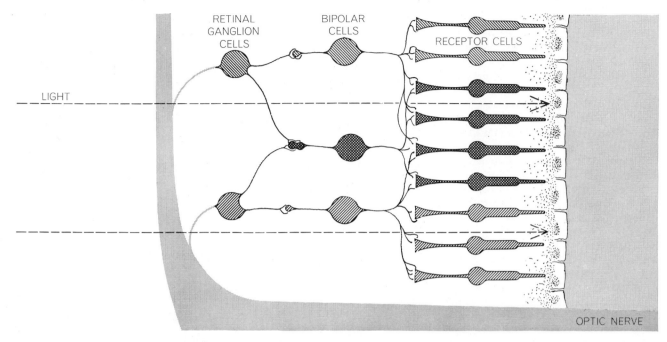

STRUCTURE OF RETINA is depicted schematically. Images fall on the receptor cells, of which there are about 130 million in each retina. Some analysis of an image occurs as the receptors transmit messages to the retinal ganglion cells via the bipolar cells. A group of receptors funnels into a particular ganglion cell, as indicated by the shading; that group forms the ganglion cell's receptive field. Inasmuch as the fields of several ganglion cells overlap, one receptor may send messages to several ganglion cells.

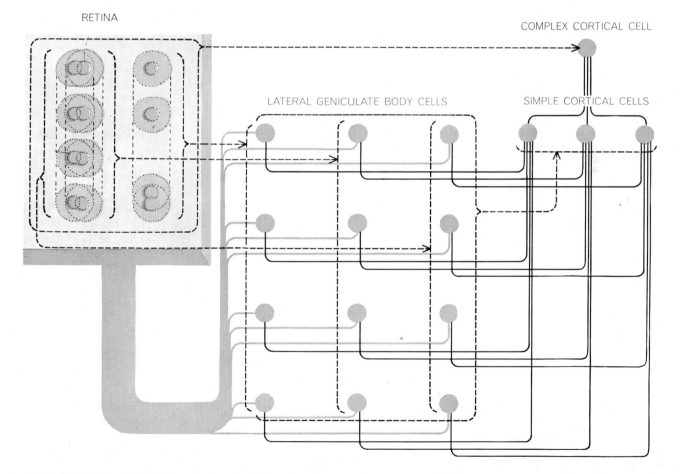

VISUAL PROCESSING BY BRAIN begins in the lateral geniculate body, which continues the analysis made by retinal cells. In the cortex "simple" cells respond strongly to line stimuli, provided that the position and orientation of the line are suitable for a particular cell. "Complex" cells respond well to line stimuli, but the position of the line is not critical and the cell continues to respond even if a properly oriented stimulus is moved, as long as it remains in the cell's receptive field. Broken lines indicate how receptive fields of all these cells overlap on the retina; solid lines, how several cells at one stage affect a single cell at the next stage.

lation of cells: a stimulus that excites one cell will undoubtedly excite many others, meanwhile inhibiting yet another array of cells and leaving others entirely unaffected.

For many years it has been known that retinal ganglion cells fire at a fairly steady rate even in the absence of any stimulation. Kuffler was the first to observe how the retinal ganglion cells of mammals are influenced by small spots of light. He found that the resting discharges of a cell were intensified or diminished by light in a small and more or less circular region of the retina. That region was of course the cell's receptive field. Depending on where in the field a spot of light fell, either of two responses could be produced. One was an "on" response, in which the cell's firing rate increased under the stimulus of light. The other was an "off" response, in which the stimulus of light decreased the cell's firing rate. Moreover, turning the light off usually evoked a burst of impulses from the cell. Kuffler called the retinal regions from which these responses could be evoked "on" regions and "off" regions.

On mapping the receptive fields of a large number of retinal ganglion cells into "on" and "off" regions, Kuffler discovered that there were two distinct cell types. In one the receptive field consisted of a small circular "on" area and a surrounding zone that gave "off" responses. Kuffler termed this an "on"-center cell. The second type, which he called "off"-center, had just the reverse form of field—an "off" center and an "on" periphery [*see top illustration on this page*]. For a given cell the effects of light varied markedly according to the place in which the light struck the receptive field. Two spots of light shone on separate parts of an "on" area produced a more vigorous "on" response than either spot alone, whereas if one spot was shone on an "on" area and the other on an "off" area, the two effects tended to neutralize each other, resulting in a very weak "on" or "off" response. In an "on"-center cell, illuminating the entire central "on" region evoked a maximum response; a smaller or larger spot of light was less effective.

Lighting up the whole retina diffusely, even though it may affect every receptor in the retina, does not affect a retinal ganglion cell nearly so strongly as a small circular spot of exactly the right size placed so as to cover precisely the receptive-field center. The main concern of these cells seems to be the contrast in illumination between one retinal region and surrounding regions.

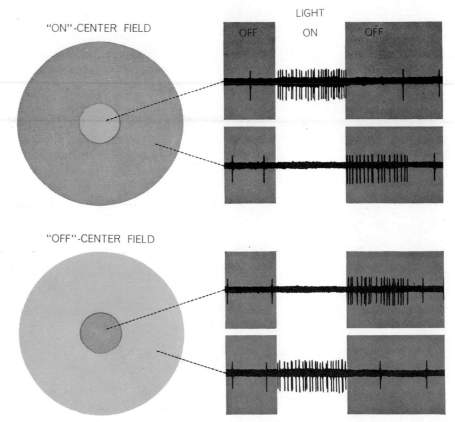

CONCENTRIC FIELDS are characteristic of retinal ganglion cells and of geniculate cells. At top an oscilloscope recording shows strong firing by an "on"-center type of cell when a spot of light strikes the field center; if the spot hits an "off" area, the firing is suppressed until the light goes off. At bottom are responses of another cell of the "off"-center type.

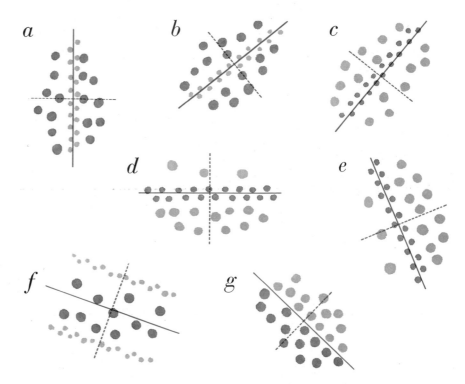

SIMPLE CORTICAL CELLS have receptive fields of various types. In all of them the "on" and "off" areas, represented by colored and gray dots respectively, are separated by straight boundaries. Orientations of fields vary, as indicated particularly at *a* and *b*. In the cat's visual system such fields are generally one millimeter or less in diameter.

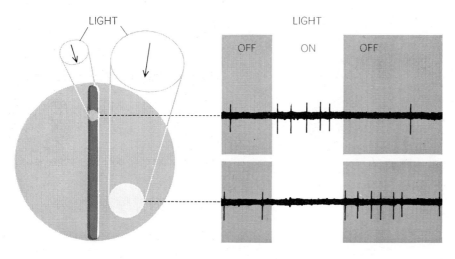

RESPONSE IS WEAK when a circular spot of light is shone on receptive field of a simple cortical cell. Such spots get a vigorous response from retinal and geniculate cells. This cell has a receptive field of type shown at *a* in bottom illustration on preceding page.

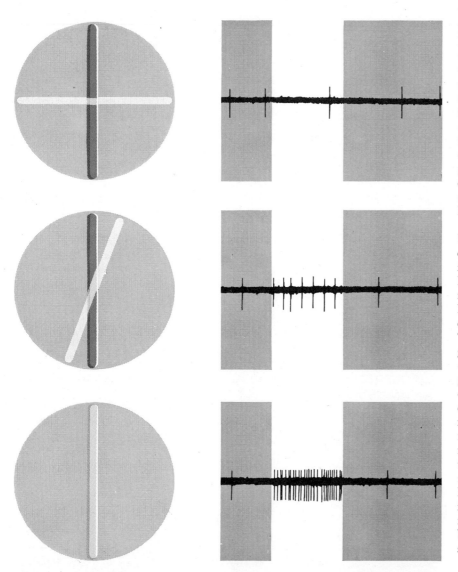

IMPORTANCE OF ORIENTATION to simple cortical cells is indicated by varying responses to a slit of light from a cell preferring a vertical orientation. Horizontal slit *(top)* produces no response, slight tilt a weak response, vertical slit a strong response.

Retinal ganglion cells differ greatly in the size of their receptive-field centers. Cells near the fovea (the part of the retina serving the center of gaze) are specialized for precise discrimination; in the monkey the field centers of these cells may be about the same size as a single cone—an area subtending a few minutes of arc at the cornea. On the other hand, some cells far out in the retinal periphery have field centers up to a millimeter or so in diameter. (In man one millimeter of retina corresponds to an arc of about three degrees in the 180-degree visual field.) Cells with such large receptive-field centers are probably specialized for work in very dim light, since they can sum up messages from a large number of receptors.

Given this knowledge of the kind of visual information brought to the brain by the optic nerve, our first problem was to learn how the messages were handled at the first central way station, the lateral geniculate body. Compared with the retina, the geniculate body is a relatively simple structure. In a sense there is only one synapse involved, since the incoming optic nerve fibers end in cells that send their fibers directly to the visual cortex. Yet in the cat many optic nerve fibers converge on each geniculate cell, and it is reasonable to expect some change in the visual messages from the optic nerve to the geniculate cells.

When we came to study the geniculate body, we found that the cells have many of the characteristics Kuffler described for retinal ganglion cells. Each geniculate cell is driven from a circumscribed retinal region (the receptive field) and has either an "on" center or an "off" center, with an opposing periphery. There are, however, differences between geniculate cells and retinal ganglion cells, the most important of which is the greatly enhanced capacity of the periphery of a geniculate cell's receptive field to cancel the effects of the center. This means that the lateral geniculate cells must be even more specialized than retinal ganglion cells in responding to spatial differences in retinal illumination rather than to the illumination itself. The lateral geniculate body, in short, has the function of increasing the disparity—already present in retinal ganglion cells—between responses to a small, centered spot and to diffuse light.

In contrast to the comparatively simple lateral geniculate body, the cerebral cortex is a structure of stupendous complexity. The cells of this great plate of

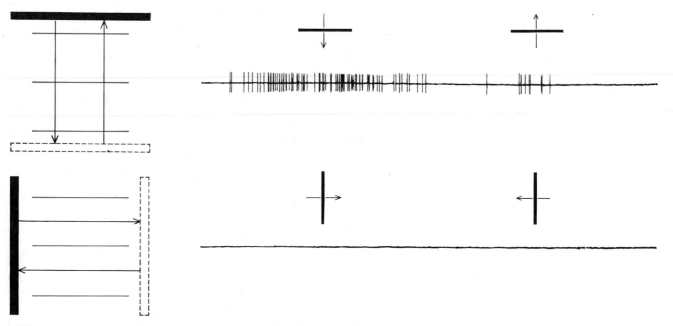

COMPLEX CORTICAL CELL responded vigorously to slow downward movement of a dark, horizontal bar. Upward movement of bar produced a weak response and horizontal movement of a vertical bar produced no response. For other shapes, orientations and movements there are other complex cells showing maximum response. Such cells may figure in perception of form and movement.

gray matter—a structure that would be about 20 square feet in area and a tenth of an inch thick if flattened out—are arranged in a number of more or less distinct layers. The millions of fibers that come in from the lateral geniculate body connect with cortical cells in the layer that is fourth from the top. From here the information is sooner or later disseminated to all layers of the cortex by rich interconnections between them. Many of the cells, particularly those of the third and fifth layers, send their fibers out of the cortex, projecting to centers deep in the brain or passing over to nearby cortical areas for further processing of the visual messages. Our problem was to learn how the information the visual cortex sends out differs from what it takes in.

Most connections between cortical cells are in a direction perpendicular to the surface; side-to-side connections are generally quite short. One might therefore predict that impulses arriving at a particular area of the cortex would exert their effects quite locally. Moreover, the retinas project to the visual cortex (via the lateral geniculate body) in a systematic topologic manner; that is, a given area of cortex gets its input ultimately from a circumscribed area of retina. These two observations suggest that a given cortical cell should have a small receptive field; it should be influenced from a circumscribed retinal region only, just as a geniculate or retinal ganglion cell is. Beyond this the anatomy provides no hint of what the cortex does

with the information it receives about an image on the retina.

In the face of the anatomical complexity of the cortex, it would have been surprising if the cells had proved to have the concentric receptive fields characteristic of cells in the retina and the lateral geniculate body. Indeed, in the cat we have observed no cortical cells with concentric receptive fields; instead there are many different cell types, with fields markedly different from anything seen in the retinal and geniculate cells.

The many varieties of cortical cells may, however, be classified by function into two large groups. One we have called "simple"; the function of these cells is to respond to line stimuli—such shapes as slits, which we define as light lines on a dark background; dark bars (dark lines on a light background), and edges (straight-line boundaries between light and dark regions). Whether or not a given cell responds depends on the orientation of the shape and its position on the cell's receptive field. A bar shone vertically on the screen may activate a given cell, whereas the same cell will fail to respond (but others will respond) if the bar is displaced to one side or moved appreciably out of the vertical. The second group of cortical cells we have called "complex"; they too respond best to bars, slits or edges, provided that, as with simple cells, the shape is suitably oriented for the particular cell under observation. Complex cells, how-

ever, are not so discriminating as to the exact position of the stimulus, provided that it is properly oriented. Moreover, unlike simple cells, they respond with sustained firing to moving lines.

From the preference of simple and complex cells for specific orientation of light stimuli, it follows that there must be a multiplicity of cell types to handle the great number of possible positions and orientations. Wiesel and I have found a large variety of cortical cell responses, even though the number of individual cells we have studied runs only into the hundreds compared with the millions that exist. Among simple cells, the retinal region over which a cell can be influenced—the receptive field—is, like the fields of retinal and geniculate cells, divided into "on" and "off" areas. In simple cells, however, these areas are far from being circularly symmetrical. In a typical example the receptive field consists of a very long and narrow "on" area, which is adjoined on each side by larger "off" regions. The magnitude of an "on" response depends, as with retinal and geniculate cells, on how much either type of region is covered by the stimulating light. A long, narrow slit that just fills the elongated "on" region produces a powerful "on" response. Stimulation with the slit in a different orientation produces a much weaker effect, because the slit is now no longer illuminating all the "on" region but instead includes some of the antagonistic "off" region. A slit at right angles to the optimum orientation for a

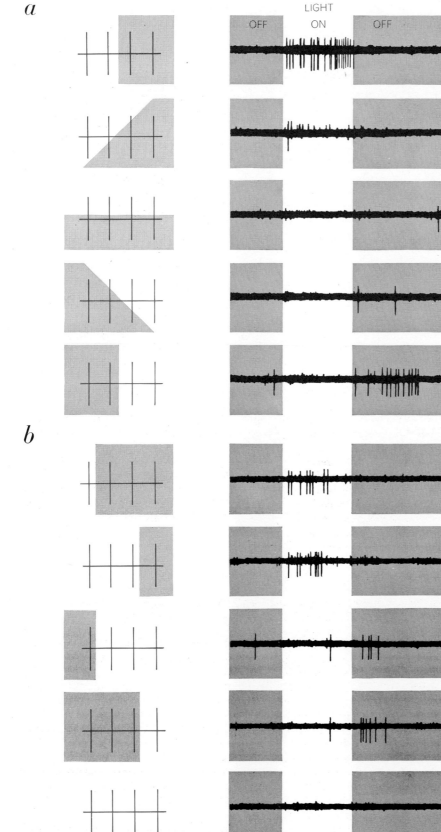

a

b

LIGHT

OFF   ON   OFF

SINGLE COMPLEX CELL showed varying responses to an edge projected on the cell's receptive field in the retina. In group *a* the stimulus was presented in differing orientations. In group *b* all the edges were vertical and all but the last evoked responses regardless of where in the receptive field the light struck. When a large rectangle of light covered entire receptive field, however, as shown at bottom, cell failed to respond.

cell of this type is usually completely ineffective.

In the simple cortical cells the process of pitting these two antagonistic parts of a receptive field against each other is carried still further than it is in the lateral geniculate body. As a rule a large spot of light—or what amounts to the same thing, diffuse light covering the whole retina—evokes no response at all in simple cortical cells. Here the "on" and "off" effects apparently balance out with great precision.

Some other common types of simple receptive fields include an "on" center with a large "off" area to one side and a small one to the other; an "on" and an "off" area side by side; a narrow "off" center with "on" sides; a wide "on" center with narrow "off" sides. All these fields have in common that the border or borders separating "on" and "off" regions are straight and parallel rather than circular [*see bottom illustration on page 180*]. The most efficient stimuli—slits, edges or dark bars—all involve straight lines. Each cell responds best to a particular orientation of line; other orientations produce less vigorous responses, and usually the orientation perpendicular to the optimum evokes no response at all. A particular cell's optimum, which we term the receptive-field orientation, is thus a property built into the cell by its connections. In general the receptive-field orientation differs from one cell to the next, and it may be vertical, horizontal or oblique. We have no evidence that any one orientation, such as vertical or horizontal, is more common than any other.

How can one explain this specificity of simple cortical cells? We are inclined to think they receive their input directly from the incoming lateral geniculate fibers. We suppose a typical simple cell has for its input a large number of lateral geniculate cells whose "on" centers are arranged along a straight line; a spot of light shone anywhere along that line will activate some of the geniculate cells and lead to activation of the cortical cell. A light shone over the entire area will activate all the geniculate cells and have a tremendous final impact on the cortical cell [*see bottom illustration on page 179*].

One can now begin to grasp the significance of the great number of cells in the visual cortex. Each cell seems to have its own specific duties; it takes care of one restricted part of the retina, responds best to one particular shape of stimulus and to one particular orientation. To look at the problem from the opposite direction, for each stimulus—each area of the retina stimulated, each type of line (edge, slit or bar) and each orientation of stimulus—there is a particular set of simple cortical cells that will respond; changing any of the stimulus arrangements will cause a whole new population of cells to respond. The number of populations responding successively as the eye watches a slowly rotating propeller is scarcely imaginable.

Such a profound rearrangement and analysis of the incoming messages might seem enough of a task for a single structure, but it turns out to be only part of what happens in the cortex. The next major transformation involves the cortical cells that occupy what is probably the sixth step in the visual pathway: the complex cells, which are also present in this cortical region and to some extent intermixed with the simple cells.

Complex cells are like simple ones in several ways. A cell responds to a stimulus only within a restricted region of retina: the receptive field. It responds best to the line stimuli (slits, edges or dark bars) and the stimulus must be oriented to suit the cell. But complex fields, unlike the simple ones, cannot be mapped into antagonistic "on" and "off" regions.

A typical complex cell we studied happened to fire to a vertical edge, and it gave "on" or "off" responses depending on whether light was to the left or to the right. Other orientations were almost completely without effect [*see illustration on following page*]. These re-

sponses are just what could be expected from a simple cell with a receptive field consisting of an excitatory area separated from an inhibitory one by a vertical boundary. In this case, however, the cell had an additional property that could not be explained by such an arrangement. A vertical edge evoked responses anywhere within the receptive field, "on" responses with light to the left, "off" responses with light to the right. Such behavior cannot be understood in terms of antagonistic "on" and "off" subdivisions of the receptive field, and when we explored the field with small spots we found no such regions. Instead the spot either produced responses at both "on" and "off" or evoked no responses at all.

Complex cells, then, respond like simple cells to one particular aspect of the stimulus, namely its orientation. But when the stimulus is moved, without changing the orientation, a complex cell differs from its simple counterpart chiefly in responding with sustained firing. The firing continues as the stimulus is moved over a substantial retinal area, usually the entire receptive field of the cell, whereas a simple cell will respond to movement only as the stimulus crosses a very narrow boundary separating "on" and "off" regions.

It is difficult to explain this behavior by any scheme in which geniculate cells project directly to complex cells. On the other hand, the findings can be explained fairly well by the supposition

that a complex cell receives its input from a large number of simple cells. This supposition requires only that the simple cells have the same field orientation and be all of the same general type. A complex cell responding to vertical edges, for example, would thus receive fibers from simple cells that have vertically oriented receptive fields. All such a scheme needs to have added is the requirement that the retinal positions of these simple fields be arranged throughout the area occupied by the complex field.

The main difficulty with such a scheme is that it presupposes an enormous degree of cortical organization. What a vast network of connections must be needed if a single complex cell is to receive fibers from just the right simple cells, all with the appropriate field arrangements, tilts and positions! Yet there is unexpected and compelling evidence that such a system of connections exists. It comes from a study of what can be called the functional architecture of the cortex. By penetrating with a microelectrode through the cortex in many directions, perhaps many times in a single tiny region of the brain, we learned that the cells are arranged not in a haphazard manner but with a high degree of order. The physiological results show that functionally the cortex is subdivided like a beehive into tiny columns, or segments [*see illustration on next page*], each of which extends from the surface to the white matter lower in the brain. A column is de-

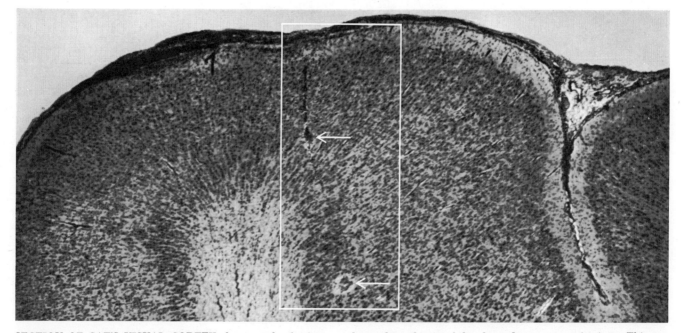

SECTION OF CAT'S VISUAL CORTEX shows track of microelectrode penetration and, at arrows, two points along the track where lesions were made so that it would be possible to ascertain later where the tip of the electrode was at certain times. This section of cortex is from a single gyrus, or fold of the brain; it was six millimeters wide and is shown here enlarged 30 diameters.

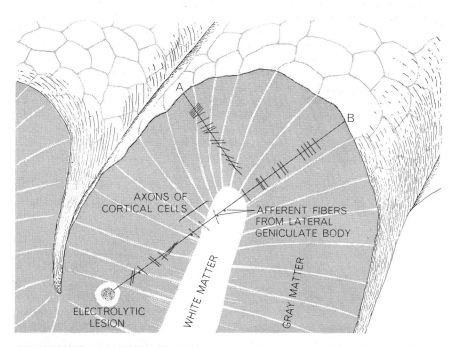

FUNCTIONAL ARRANGEMENT of cells in visual cortex resembled columns, although columnar structure is not apparent under a microscope. Lines *A* and *B* show paths of two microelectrode penetrations; colored lines show receptive-field orientations encountered. Cells in a single column had same orientation; change of orientation showed new column.

fined not by any anatomically obvious wall—no columns are visible under the microscope—but by the fact that the thousands of cells it contains all have the same receptive-field orientation. The evidence for this is that in a typical microelectrode penetration through the cortex the cells—recorded in sequence as the electrode is pushed ahead—all have the same field orientation, provided that the penetration is made in a direction perpendicular to the surface of the cortical segment. If the penetration is oblique, as we pass from column to column we record several cells with one field orientation, then a new sequence of cells with a new orientation, and then still another.

The columns are irregular in cross-sectional shape, and on the average they are about half a millimeter across. In respects other than receptive-field orientation the cells in a particular column tend to differ; some are simple, others complex; some respond to slits, others prefer dark bars or edges.

Returning to the proposed scheme for explaining the properties of complex cells, one sees that gathered together in a single column are the very cells one should expect to be interconnected: cells whose fields have the same orientation and the same general retinal position, although not the same position. Furthermore, it is known from

the anatomy that there are rich interconnections between neighboring cells, and the preponderance of these connections in a vertical direction fits well with the long, narrow, more or less cylindrical shape of the columns. This means that a column may be looked on as an independent functional unit of cortex, in which simple cells receive connections from lateral geniculate cells and send projections to complex cells.

It is possible to get an inkling of the part these different cell types play in vision by considering what must be happening in the brain when one looks at a form, such as, to take a relatively simple example, a black square on a white background. Suppose the eyes fix on some arbitrary point to the left of the square. On the reasonably safe assumption that the human visual cortex works something like the cat's and the monkey's, it can be predicted that the near edge of the square will activate a particular group of simple cells, namely cells that prefer edges with light to the left and dark to the right and whose fields are oriented vertically and are so placed on the retina that the boundary between "on" and "off" regions falls exactly along the image of the near edge of the square. Other populations of cells will obviously be called into action by the other three edges of the square. All the cell populations will change if the eye strays from the point fixed on, or if

the square is moved while the eye remains stationary, or if the square is rotated.

In the same way each edge will activate a population of complex cells, again cells that prefer edges in a specific orientation. But a given complex cell, unlike a simple cell, will continue to be activated when the eye moves or when the form moves, if the movement is not so large that the edge passes entirely outside the receptive field of the cell, and if there is no rotation. This means that the populations of complex cells affected by the whole square will be to some extent independent of the exact position of the image of the square on the retina.

Each of the cortical columns contains thousands of cells, some with simple fields and some with complex. Evidently the visual cortex analyzes an enormous amount of information, with each small region of visual field represented over and over again in column after column, first for one receptive-field orientation and then for another.

In sum, the visual cortex appears to have a rich assortment of functions. It rearranges the input from the lateral geniculate body in a way that makes lines and contours the most important stimuli. What appears to be a first step in perceptual generalization results from the response of cortical cells to the orientation of a stimulus, apart from its exact retinal position. Movement is also an important stimulus factor; its rate and direction must both be specified if a cell is to be effectively driven.

One cannot expect to "explain" vision, however, from a knowledge of the behavior of a single set of cells, geniculate or cortical, any more than one could understand a wood-pulp mill from an examination of the machine that cuts the logs into chips. We are now studying how still "higher" structures build on the information they receive from these cortical cells, rearranging it to produce an even greater complexity of response.

In all of this work we have been particularly encouraged to find that the areas we study can be understood in terms of comparatively simple concepts such as the nerve impulse, convergence of many nerves on a single cell, excitation and inhibition. Moreover, if the connections suggested by these studies are remotely close to reality, one can conclude that at least some parts of the brain can be followed relatively easily, without necessarily requiring higher mathematics, computers or a knowledge of network theories.

19

THE GROWTH OF NERVE CIRCUITS

R. W. SPERRY   November 1959

Severe damage to the principal motor nerve of the face may leave a person afflicted with a condition known as "crocodile tears." As the injured nerve regenerates, fibers that originally activated a salivary gland can go astray and connect themselves to the lachrymal gland of one eye. Thereafter every situation calling for salivation induces weeping from that eye. Often the regenerating salivary fibers invade sweat glands and related organs in the skin, causing profuse sweating and flushing in areas of the face and temple. The random shuffling of motor-nerve connections to the muscles of the face characteristically deranges facial expression, causing a grimace-like contraction of the affected side. Sometimes, to prevent atrophy of the facial muscles when the injured facial nerve fails to regenerate, surgeons will connect the denervated facial muscles to a nearby healthy nerve: the motor nerve of the tongue or the motor nerve of the shoulder muscle. The restored facial movements still lack meaningful expression and tend to be associated with the chewing movements of the tongue or the action of the shoulder muscles.

Naturally the primary concern of the patient in such cases is whether or not normal function can be restored. If the symptoms do not clear up spontaneously, can they be corrected by training and re-education? By faithful practice in front of a mirror, for example, can a patient learn to inhibit the crocodile tears and regain control of facial expression?

Not so long ago the reply to such questions was a confident "Yes." For most of the present century investigators and physicians were agreed that the central nervous system is plastic enough so that any muscle nerve might be reconnected to any other muscle with good function-

al success. Sensory-nerve fibers were thought to be equally interchangeable within a given sensory system. It was believed that the central pathways in the nervous system were first laid down in the embryo in randomized equipotential networks. By use and learning these pathways became channelized; connections that proved adaptive in function were reinforced, while the nonadaptive ones underwent "disuse atrophy." Learning thus determined not only the function but also the structure of the nervous system. This theoretical picture was sustained by experiments on animals in which, according to the literature, the crossing of major nerve-trunks was followed by full restoration of function. In the prevailing mood of optimism physicians were able to report encouraging progress by their patients.

During the past 15 years, however, scientific and medical opinion has undergone a major shift, amounting to an almost complete about-face. No longer do physicians encourage the patient with a regenerated facial nerve to try to regain control of facial expression by training; their advice today is to inhibit all expression, to practice a "poker face" in order to make the two sides of the face match in appearance. The outlook is equally dim for restoration of coordination in cases of severe nerve injury in other parts of the body.

This changed viewpoint reflects a revision in the picture of the entire nervous system. According to the new picture, the connections necessary for normal coordination arise in embryonic development according to a biochemically determined plan that precisely connects the various nerve endings in the body to their corresponding points in the nerve centers of the brain and spinal cord. Although the higher cen-

ters in the brain are capable of extensive learning, the lower centers in the brain stem and spinal cord are quite implastic. Because their function is dictated by their structure, it cannot be significantly modified by use or learning. Nor can the disordered connections set up by the random regeneration of injured nerves be corrected by re-education.

The evidence for this view, which comes from new experiments and from exacting clinical observations, is so persuasive that it is difficult to understand how the opposite view could have prevailed so long. It appears that most of the earlier reports of the high functional plasticity of the nervous system will go down in the record as unfortunate examples of how an erroneous medical or scientific opinion, once implanted, can snowball until it biases experimental observations and crushes dissenting interpretations.

Hundreds of experiments seemed to support the now-discounted opinion. One of the experiments most frequently cited was first reported in 1912 and was repeated with concurring interpretation as recently as 1941. In several monkeys opposing pairs of eye muscles were interchanged to reverse the movement of one eye. Upon recovery from surgery, the movements of the abnormally connected eye were said to coordinate with those of the normal one. In another oft-repeated experiment the nerves that control the lifting of the foot were crossed. Instead of a reversal of foot action, the animals showed recovery of muscle coordination hardly distinguishable from that on the normal side. Even when nerves from the forelimb were cross-connected to nerves of the hindlimb, the animals appeared to make complete

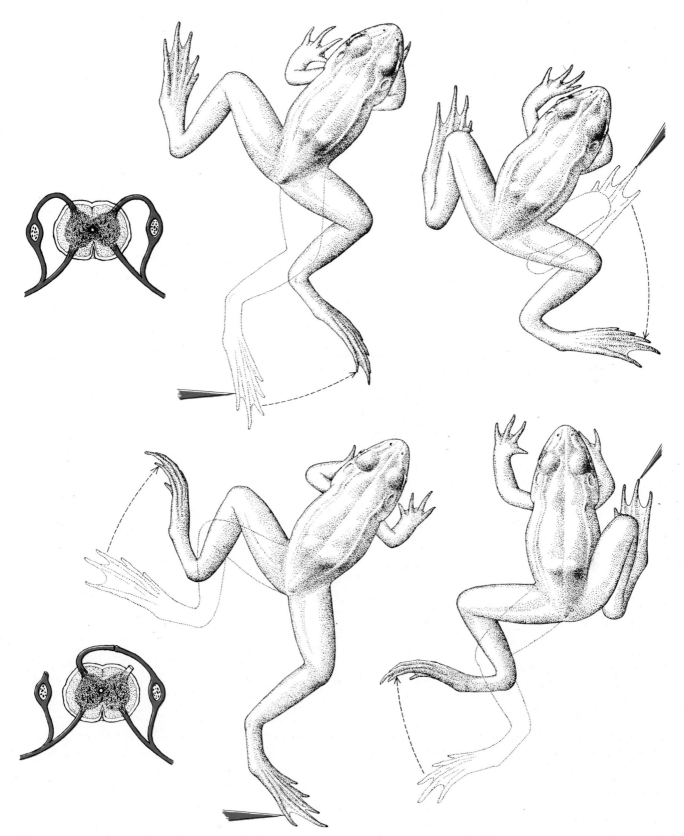

CROSSED SENSORY NERVES produce incorrect postural reflexes in the frog. The normal connections shown in the cross section of the spinal cord at top left cause the frog to withdraw an extended leg (*top center*) and to extend a flexed leg (*top right*) when a stimulus (*colored pointer*) is applied. If the sensory root entering the right side of the spinal cord is cut and surgically attached to the left side, as shown in the cross section at bottom left, the reactions of the left leg will be determined by the posture of the right.

functional readjustment. With such results in the literature there seemed to be few limits to the restorative possibilities of peripheral nerve surgery.

The doctrine of the functional plasticity of the nervous system was sharply challenged in 1938 by Frank R. Ford and Barnes Woodhall of the Johns Hopkins School of Medicine. In an account of their clinical experience with functional disorders following the regeneration of nerves, they declared that these disorders persisted stubbornly in many of their patients for years without improvement. Their report cast serious doubt upon the accepted methods of therapy and the theory that rationalized them.

That same year I began a series of experiments in the laboratory of Paul Weiss at the University of Chicago. The initial aim of this investigation was to find out if functional plasticity was a property of the higher brain-centers only or whether it extended to the lowest levels of the spinal cord as alleged in some earlier reports. To explore the question I started to experiment on the simple reflexes involved in coordinating the foot movements of the rat. These reflexes depend upon a relatively simple circuit called a reflex arc. The fibers that activate the muscle connect to an association neuron in the spinal-reflex center; the association neuron is connected in turn to a particular type of sensory cell, the proprioceptive neuron, the terminal fibers of which are embedded in the muscle. This circuit, with the sensory nerve indicating the state of contraction or relaxation of the muscle and its orientation in space, provides the feedback necessary for proper muscle timing and coordination.

I switched the nerve connections between opposing muscles in the hindlimb of rats in such a way as to reverse the movement at the ankle joint. The nerves were cut, crossed and reunited end-to-end within tubes of dissected rat artery. Now whenever a nerve is severed, all fibers beyond the break degenerate and are absorbed. Even when they are united with the mechanical aid of an arterial tube, cut nerves do not heal together directly. New fibers sprout from the end of the central stump and grow into the muscles within the degenerate framework of the old nerve. After the surgery, I assumed, new coordinating circuits would be established and functional adjustment would follow quickly.

Much to my surprise the anticipated adjustment never occurred. The rats seemed unable to correct the reversals of motor coordination produced by the operation. When they tried to lift the

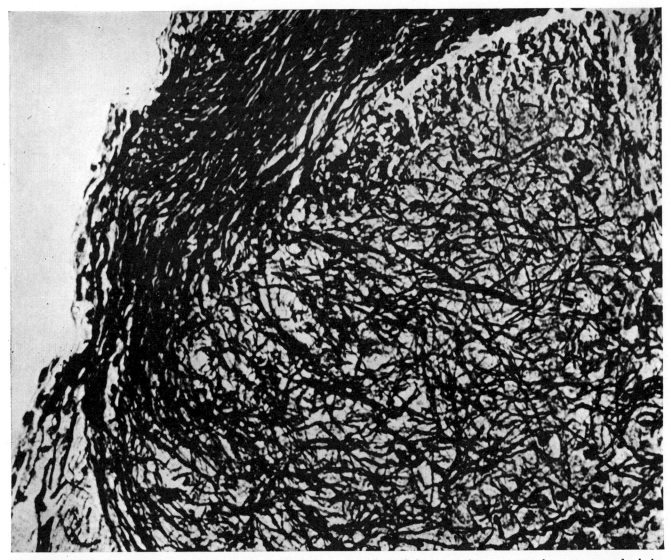

POST-MORTEM VIEW of the frog's crossed sensory root (magnified 500 diameters) shows that the cut nerve-fibers have regenerated into the spinal cord. Despite their tangled appearance, the fibers have formed the connections necessary for proper muscle timing and coordination. Because the nerves have been transposed, however, frogs display abnormal reflexes shown on preceding page.

affected foot, it pulled downward; when they tried to rise on the ball of the foot, their toes swung up and they fell back on their heels [*see illustration on next page*].

In a parallel experiment I presented the rats with a simpler readjustment problem by switching muscles instead of nerves. The muscles involved were transposed by cutting and crossing their tendons. Strong muscle-action was restored within two or three weeks. Yet the rats were still unable to correct the reversal of ankle movement. To check the experiment I tried crossing both muscles and nerves in a control group. Here the two reversals mutually canceled their effects, and the rat was able to raise and lower the foot in proper timing.

The rats with reversed foot-movements were put on a program of special training: they were forced to climb ladders and stretch upward on their hindlegs many times a day to get food pellets from automatic feeders. Yet the affected feet continued to work backward in machine-like fashion. We carried out a similar series of experiments on the forelimb, in which voluntary movements are under better control. But the rats still could not adapt to the rearrangement of the nerve connections.

When it became clear that re-education had little or no effect in the rat's motor system, we turned to the sensory system. In the laboratory of Karl S. Lashley at Harvard University I transposed the nerves connecting to the skin of the left and right hindfeet [*see illustration at right*]. As the crossed hindlimb nerves regenerated, the rats began to exhibit false reference of sensations. In response to a mild electric shock to the sole of the right foot, the animals withdrew their left foot. This movement shifted their weight to the right foot, thereby increasing its contact with the offending electrode.

During the course of these experiments several rats developed a persistent sore on the sole of the right foot. Until the sores responded to medication, the animals hopped about on three feet with the fourth raised protectively—but it was always the wrong, uninjured foot that they raised. When they were prompted to lick the injury, they repeatedly licked the uninjured foot. Although this accidental soreness in the re-innervated foot presented the best kind of training situation, the rats still were unable to readapt.

Having concluded that it would be

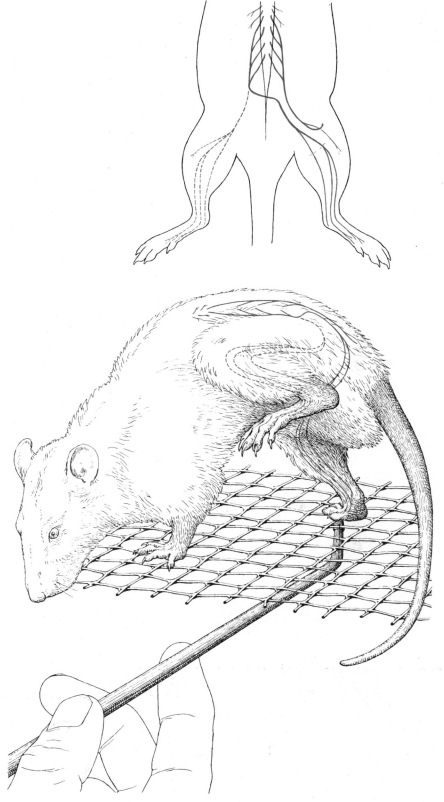

INCORRECT WITHDRAWAL REFLEX and referred sensations appear in the rat after two sensory nerves are crossed. Here the main trunk-nerve to the left foot (*broken line*) has been crossed and connected to the corresponding nerve on the right side (*colored solid line*). Afterward, when sole of right foot is stimulated electrically, rat withdraws its left foot; if the shock is strong enough to produce soreness, rat licks its uninjured foot.

CROSSED MUSCLES produce the same effects as crossed nerves. In a normal rat (*top right*) the leg muscles are connected as shown at top left. When these connections are transposed (*bottom left*), the posture and movements of the ankle joint are reversed (*bottom right*). When the rat attempts to lift the affected foot, it pulls downward; when the animal tries to rise on the ball of the foot, the toes swing up and it falls back on its heels. Even with prolonged training, the rat can never learn to correct these movements.

difficult or impossible to demonstrate any sort of plasticity in the rat, we began a similar experiment on monkeys at the Yerkes Laboratories of Primate Biology in Orange Park, Fla. At first the monkeys seemed to make more progress than the rats: After we had transposed the nerves of the biceps and triceps muscles of the upper arm, they were quick to notice and to halt the reversed arm-movements that began to appear as the nerves regenerated. Thereafter the mon-

keys did attain a minor degree of readjustment in arm movement under the most simplified training routine. But after three years of testing and observation, they too failed to achieve any generalized positive correction in the action of the cross-innervated muscles.

The marked conflict between our results and those previously reported prompted us to reproduce the procedures of the earlier studies more closely.

Instead of crossing isolated branch-nerves to single muscles I now crossed large trunk-nerves carrying motor impulses to many different muscles. All the muscles in the region were left intact and the nerves were permitted to regenerate into their respective areas at random.

In the hindlimb of the rat this operation produced neither a reversal of movement nor good functional readjustment; instead it caused a spastic contraction of all the muscles of the lower leg. Because of the greater strength of the postural or antigravity muscles, the contraction produced a stiltlike stiffening of the ankle joint in the extended position. This result was highly illuminating. Although the extended leg-posture was clearly abnormal in the rat, it could very easily be mistaken for a return of normal coordination in an animal that walks on its toes, like a dog or a cat. The fact that dogs and cats had been used in most of the earlier investigations made it apparent that nearly all of the hundreds of earlier reports of good functional recovery were subject to reinterpretation.

We know that central nervous system plasticity can be substantiated to some extent in cases where muscles have been transplanted in human patients. But even here we must make a distinction between the degree of plasticity demonstrated after muscle transposition and that shown after nerve regeneration. When a muscle is transplanted with its nerves intact, the motor cells that activate it continue to work together as a unit. On the other hand, in human beings and higher animals the fibers of a regenerating motor-nerve become haphazardly redistributed among the muscles it previously supplied. To restore these muscles to their previous control and coordination the reflex connections in the spinal cord would have to be reestablished down to the level of individual nerve cells. Even man's superb nervous system does not possess this degree of plasticity.

In most cases humans can learn to control transplanted muscles only in simple, slow, voluntary movements. The control of complex, rapid and reflex movements is limited at best, and is subject to relapse under conditions of fatigue, shock or surprise. Humans seem to have a much greater capacity for adjustment than the subhuman primates. Such re-education as does occur must therefore be due to the greater development of higher learning-centers in the human brain. Contrary to earlier supposition it does not reflect an intrinsic plasticity of nerve networks in general.

The nervous systems of reptiles, birds, and mammals other than primates show even less functional plasticity than those of primates. Farther down the evolutionary scale in the lower vertebrates, however, we find an entirely different type of neural plasticity: a structural plasticity not possessed by higher animals. Fishes, frogs and salamanders can regenerate any part of the central nervous system—even the tissue of the brain itself. Furthermore, if one of the large motor nerves of a salamander or a fish is cut, the animals re-establish normal reflex-arcs and recover coordination. This occurs even if several nerve stumps in a limb or fin are deliberately crossconnected to produce gross abnormalities in the distribution of the regenerating fibers.

In his pioneering investigations during the 1920's Paul Weiss was able to rule out the possibility that these spectacular recoveries could be based on learning or any other sort of functional plasticity. He transposed the developing forelimb buds of salamander embryos and reimplanted them with their frontto-back axes reversed. When function later appeared, the motor coordination in the transplanted limbs was perfectly normal, indicating that normal reflexarcs had been established. Because the forelimbs were reversed in orientation, however, they pushed the animal backward when it tried to go forward, and vice versa. The perfectly coordinated reversed action persisted indefinitely without correction.

I was able to confirm the absence of any appreciable functional plasticity in amphibians in a set of experiments on the visual system of frogs and salamanders. In some I inverted the eyeballs surgically, producing upside-down vision; in others I cross-connected the eyes to the wrong sides of the brain, producing vision that was reversed from side to side. The animals never learned to correct the erroneous responses caused by this surgical rearrangement of their eyes [see "The Eye and the Brain," by R. W. Sperry; SCIENTIFIC AMERICAN, May, 1956]. Even when the eyes of frog and salamander embryos were rotated prior to the onset of vision (in later experiments by L. S. Stone at Yale University and George Szekely in Hungary), the same visual disorientation developed and persisted throughout life.

The evidence at present thus indicates that the structural plasticity observed in lower vertebrates is inherent in the growth process and is quite independent of function. It is as if the forces of embryonic development that laid down the circuits in the beginning continue to operate in regeneration. We have as yet only preliminary insight into the nature of these forces.

In studies now in progress at the California Institute of Technology Harbans Arora has made an interesting observation on the regeneration of the nerve controlling the eye muscle in fishes. His findings suggest that fibers directed by chance to their own muscles make connections more readily than foreign fibers reaching the same muscle. As a result the fibers that originally con-

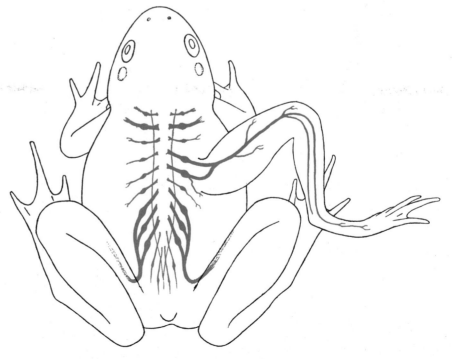

ABNORMAL REFLEX ARCS develop when a hindlimb bud is grafted onto the back of a tadpole. As the tadpole grows, sensory nerves destined for the skin of the back, flank and belly invade the nerveless extra limb and form spinal connections appropriate for limb reflexes. When grafted limb is stimulated in adult frog, muscles of the right hindlimb respond.

trolled the muscle tend to recapture control in regeneration. Such selective reaffiliation of nerve and muscle indicates that some chemical specificity must match one to the other.

Selective outgrowth of regenerating nerves to their proper end-organs seems not to be the rule, however, even in lower forms. Among mammals it has not been found at all, except on the much more gross scale that differentiates sensory from motor endings, smooth muscle from striated muscle, muscle from gland, and so on. Nor does simple selective outgrowth account for the restoration of function in salamanders. The early studies by Weiss showed that fiber outgrowth and muscle re-innervation generally proceed in these animals in a random, nonselective manner, comparable to that in mammals. Upon re-innervation, however, salamander muscles regain their former coordination and timing, even when their function is disoriented by nerve-crossing.

These observations suggest that the rearrangement of connections in the periphery of the salamander nervous system has chemical repercussions that result in a compensatory shift of reflex relations at the centers. It is postulated that the motor-nerve cells regenerating into new muscles take on a new chemical flavor, as it were. Thereupon their old central associations dissolve, and new ones form to match the new terminals in the periphery. The reflex circuit would thus be restored to its original state, with the peripheral and central terminals linked by a new pathway. Higher animals, lacking this embryonic type of structural plasticity, show no restoration of function.

This explanation at first seemed rather far-fetched, especially from the standpoint of electrophysiology, which offers no evidence for such qualitative specificity among nerve fibers. However, the underlying idea is well supported by recent experiments on the regeneration of sensory nerves.

At the University of Chicago Nancy M. Miner, one of my former associates, is responsible for a significant series of experiments indicating the role of some sort of chemical specificity in the hookup of the nervous system. She grafted extra hindlimb buds onto the backs of tadpoles; the buds became connected to the sensory fibers that would normally innervate the skin of the belly, flank and back [see illustration on preceding page]. The grafted leg served only as a sensory field for the nearby sensory nerves because there are no nearby limb nerves to invade it. When a stimulus was applied to the grafted limb in the mature frog, the animal moved the normal hindlimb on the same side, just as it would if the normal limb had received the stimulus. The belly and trunk nerves connected to the grafted limb had evidently taken on a hindlimb "flavor" and then formed the appropriate reflex connections in the central nervous system. In another experiment Miner removed a strip of skin from the trunk of a tadpole, cut its nerves and replaced it so that the skin of the back now covered the belly, and vice versa [see illustration below]. When the grown frogs were stimulated in the grafted area of the back, they responded by wiping at the belly with the forelimb; when they were stimulated in the grafted area on the belly, they wiped at the back with the hindlimb.

To account for these experimental findings it is necessary to conclude that the sensory fibers that made connections to the grafted tissues must have been modified by the character of these tissues. It is therefore unnecessary to postulate that each nerve fiber in embryonic development makes some predestined contact with a particular terminal point in the skin. Growing freely into the nearest area not yet innervated, the fibers establish their peripheral terminals at random. Thereafter they must proceed to form central hookups appropriate for the particular kind of skin to which they have become attached. It seems clearly to be some quality in the skin at the outer end of the circuit that determines the pattern of the reflex connections established at the center.

No attraction from a distance need be invoked in this selective patterning of the central hookup. The multiple branches of each nerve fiber undergo extensive ramification among the central nerve cells, with the tips of the branches making numerous contacts with all the cells in the vicinity. Presumably most of the contacts do not affect the growing fiber-tips. It is only when contact is made with central nerve cells which have the appropriate chemical specificity that the growing fiber adheres and forms the specialized synaptic ending capable of transmitting the nerve impulse.

In man these observations and interpretations provide the basis for the new

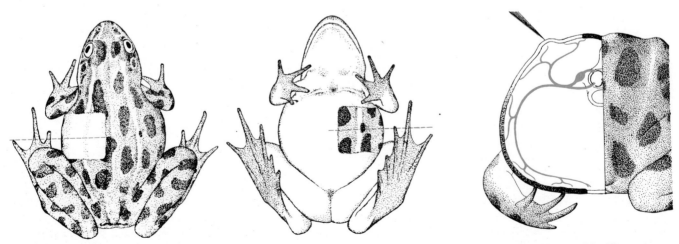

ROTATED PATCH OF SKIN demonstrates how embryonic nerves respond to the biochemical "flavor" of the tissues they innervate. Here a strip of skin was removed from a tadpole, cut free of all connections, and replaced so that the skin of the back now covered the belly and vice versa. In the grown frog the skin retained its original color and flavor despite its location, as shown in the dorsal and ventral views (left and center, respectively). The cutaway view at right (made along broken lines) shows how new nerves have invaded the graft and formed spinal reflex-arcs appropriate to the skin's flavor rather than to its location. Thus when the belly skin on the back is stimulated, the frog wipes at its belly; when the back graft on the belly is stimulated, the frog wipes at its back.

view of the nervous system which holds that its networks are determined by biochemical processes in the course of embryonic growth. Let us consider this scheme in connection with the extreme localization of skin sensations that makes it possible to locate a pinprick, for example, anywhere on the body surface. This "local sign" quality depends upon the precise matching between the central and peripheral connection of each one of thousands upon thousands of cutaneous nerve fibers connecting the skin surfaces to the spinal cord and brain. During embryonic growth we may assume that the skin undergoes a highly refined differentiation until each spot on the skin acquires a unique chemical make-up. A mosaic is not envisaged here but rather smooth gradients of differentiation extending from front to back and from top to bottom, with local elaborations of these basic gradients in the regions of the limbs. Each skin locus becomes distinguished by a given latitude and longitude, so to speak, expressed in the tissues as a combination of biochemical properties. The cutaneous nerves, as they grow out from their central ganglia, may terminate largely at random in their respective local areas. Through intimate terminal contacts the specific local flavor is imparted to each nerve fiber. This specificity is then transmitted along the fiber to all parts of the nerve cell including its ramification within the central nervous system. In this way the local-sign properties of the skin become stamped secondarily upon the cutaneous nerves and are carried into the sensory centers of the brain and spinal cord. Precise localization is further enhanced by the overlapping of the terminal connections formed by the fibers in the skin [*see illustration at right*].

Implicit in this theory is the assumption that in the embryo the cerebral cortex and the lower relay-centers also undergo a differentiation that parallels in miniature that of the body surface. In other words, just as from the skin to the first central connection point in the spinal cord, so from relay to relay and finally to the cortex the central linkages arise on the basis of selective chemical affinities. At each of its ascending levels the nervous system forms a maplike projection of the body surface.

This mechanism presumably operates not only in the sensory system but in the nervous system in general. Since the organization of the lower nerve centers and the peripheral nerve circuits in higher animals seems to take place only in the

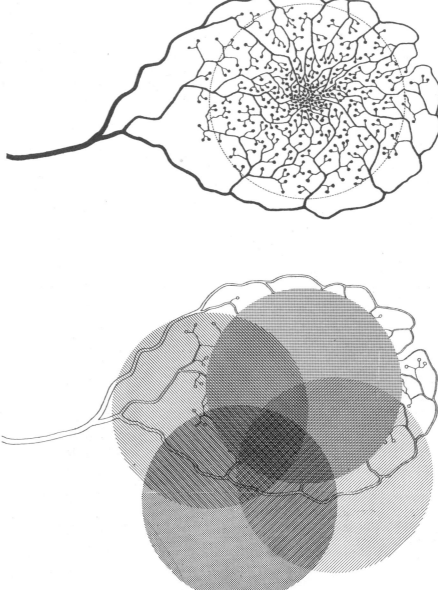

OVERLAP OF SENSORY FIBERS permits a subject to localize a pinprick accurately. The schematic diagram at top shows the terminal branches of a single cutaneous fiber; the branches are most abundant at the center of the area they innervate (**broken circle**). Bottom diagram shows how this area overlaps those of three other fibers. As shading indicates, each of the areas transmits a recognizably different signal to the central nervous system.

early plastic stages of growth, it is clear that injury to them later on cannot be repaired by any amount of re-education and training. In the structural plasticity of the system in lower animals, however, we are able to observe the processes by which our own nervous systems develop.

The new approach provides a sound biological basis for the explanation of built-in behavior mechanisms generally, from the simplest reflexes to the most complicated patterns of inherited behavior. It brings the study of behavior into the realm of experimental embryology on the same basis as other organs and organ systems. Although learning must now yield its former monopolistic status, we must not infer that it has no role in brain development. Particularly in man, whose brain grows and matures for many years, learning is a powerful method of imposing additional organization on the higher levels of the nervous system. Until the neural basis of learning is discovered, however, we cannot say whether it produces this added organization by changing the actual layout and hookups of cerebral networks or simply by increasing the conductance of certain pre-established pathways.

20

LEARNING
IN THE OCTOPUS

BRIAN B. BOYCOTT   March 1965

In recent years a number of British students of animal behavior, of whom I am one, have done much of their experimental work at the Stazione Zoologica in Naples. The reason why these investigations have been pursued in Naples rather than in Britain is that our chosen experimental animal—*Octopus vulgaris*, or the common European octopus—is found in considerable numbers along the shores of the Mediterranean. *Octopus vulgaris* is a cooperative experimental subject. If it is provided with a shelter of bricks at one end of a tank of running seawater, it takes up residence in the shelter. When a crab or some other food object is placed at the other end of the tank, the octopus swims or walks the length of the tank, catches the prey with its arms and carries it home to be poisoned and eaten. Since it responds so consistently to the presence of prey, the animal is readily trained. It is also tolerant of surgery and survives the removal of the greater part of its brain. This makes the octopus an ideal animal with which to test directly the relation between the various parts of the brain and the various kinds of perception and learning.

There are many unanswered questions about such relations. We now know a great deal about conduction in nerve fibers, transmission from nerve fiber to nerve fiber at the synapses and the integrative action of nerve fibers in such aggregations of nerve cells as the spinal cord; we are almost wholly ignorant, however, of the levels of neural integration involved in such long-term activities as memory. We can still quote with sympathy the remark of the late Karl S. Lashley of Harvard University: "I sometimes feel, in reviewing the evidence on the localization of the memory trace, that the necessary conclusion is that learning is just not possible!"

It was J. Z. Young, then at the University of Oxford, who first began to exploit the possibility of using for memory studies various marine mollusks of the class Cephalopoda. Shortly before World War II he undertook to work with the cuttlefish *Sepia officinalis*. In a simple experiment he and F. K. Sanders removed from a cuttlefish that part of the brain known as the vertical lobe. They found that a cuttlefish so deprived would respond normally—that is, attack—when it was shown a prawn. If the prawn was pulled out of sight around a corner after the attack began, however, the cuttlefish could not pursue it. The animal might advance to where the prawn had first been presented, but it was apparently unable to make whatever associations were necessary to follow the prawn around the corner. One might say it could not remember to hunt when the prey was no longer in sight. Young and Sanders found that surgical lesions in certain other parts of the cuttlefish's brain did not affect this hunting behavior.

In 1947 I had the privilege of joining Young in his studies. Financed by the Nuffield Foundation, we began

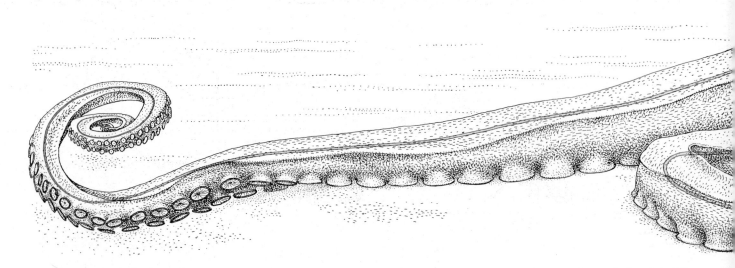

COMMON EUROPEAN OCTOPUS (*Octopus vulgaris*) is the experimental animal the author and his fellow-workers in Naples use for their investigations of perception and learning. The animal's brain (*in color between the eyes*) is about two cubic

work at the Stazione Zoologica, where both seawater aquariums and *Octopus vulgaris* were in abundant supply. The octopus was chosen in preference to the other common laboratory cephalopods—cuttlefishes and squids—because they do not survive so well in tanks and are less tolerant of surgery. At Naples today, in addition to Young's associates from University College London, there are investigators from the University of Oxford led by Stuart Sutherland and from the University of Cambridge led by Martin J. Wells, all going their various ways toward using the brains of octopuses for the analysis of perception and memory. At present most of the work is financed by the Office of Aerospace Research of the U.S. Air Force.

In our early experiments we attempted to train octopuses to do a variety of things, such as taking crabs out of one kind of pot but not out of another, to run a maze and so on. Our most successful experiment was to put a crab in the tank together with some kind of geo-metric figure—say a Plexiglas square five centimeters on a side—and give the octopus an electric shock when it made the normal attacking response. With this simple method we found that octopuses could learn not to attack a crab shown with a square but to go on attacking a crab shown without one [*see bottom illustrations on pages 198 and 199*]. Or we could train the animals to stop taking crabs but to go on eating sardines or vice versa. The purpose of these experiments was to elucidate the anatomy and connections of the animal's brain and relate them to its learning behavior.

Like the brains of most other invertebrates, the brain of the octopus surrounds its esophagus [*see illustrations on next page*]. The lobes of the brain under the esophagus contain nerve fibers that stimulate peripheral nerve centers, for example the ganglia in the arms and the mantle. These peripheral ganglia contain the nerve cells whose fibers in turn stimulate the muscles and other effectors of the body; through

them local reflexes can occur. When all of the brain except the lobes under the esophagus is removed, the octopus remains alive but lies at the bottom of the tank; it breathes regularly but maintains no definite posture. If it is sufficiently stimulated, it responds with stereotyped behavior.

A greater variety of behavior can be obtained if some of the brain lobes above the esophagus are left intact. For instance, the upper brain's median basal lobe and anterior basal lobe send their fibers down to the lower lobes and through them evoke the patterns of nerve activity involved in walking and swimming. Above these two lobes are the vertical lobe, the superior frontal lobe and the inferior frontal lobe; their surgical removal does not result in any defects of behavior that are immediately obvious.

It is with these three lobes and the two optic lobes—which lie on each side of the central mass of the brain—that this article is mostly concerned. Using the electric-shock method of training

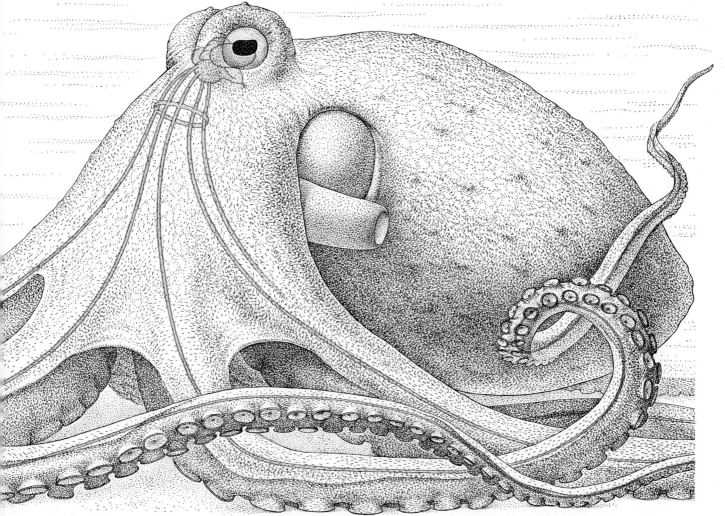

centimeters in size; the basket-like structure below it is composed of the eight major nerves of the arms, some of which are also outlined in color. The octopus adapts readily to life in a tank of seawater and can be trained easily through reward and punishment.

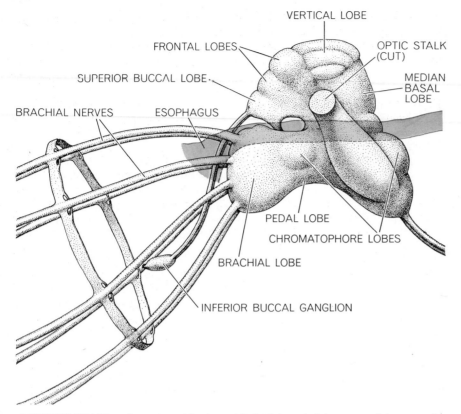

VERTICAL LOBE

FRONTAL LOBES

OPTIC STALK (CUT)

SUPERIOR BUCCAL LOBE

MEDIAN BASAL LOBE

BRACHIAL NERVES   ESOPHAGUS

PEDAL LOBE

CHROMATOPHORE LOBES

BRACHIAL LOBE

INFERIOR BUCCAL GANGLION

OCTOPUS BRAIN is shown in a side view with the left optic lobe removed (*see top view of brain below*). The labels identify external anatomical features of the brain and its nerve connections. As is the case with many other invertebrates, the brain of the octopus completely surrounds the animal's esophagus. Excision of the entire upper part of the brain is not fatal, but the octopus's behavior then exhibits neither learning nor memory.

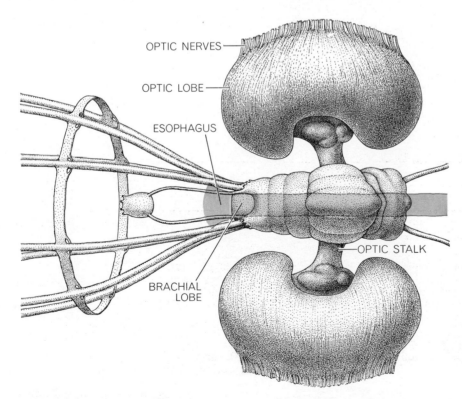

OPTIC NERVES

OPTIC LOBE

ESOPHAGUS

OPTIC STALK

BRACHIAL LOBE

TOP VIEW OF BRAIN relates the two large optic lobes and their stalks to the central brain structure situated above and below the octopus's esophagus (*color*). Combined, the mass of the two optic lobes roughly equals that of the brain's central structure; the fringe of nerves at each lobe's outer edge connects to the retinal structures of the octopus's eyes.

we soon found that, as far as visual learning goes, removing either the vertical lobe, the superior frontal lobe or both, or cutting the nerve tracts between these two lobes, left the octopus unable to learn the required discriminations (or, if they had already been learned, unable to retain them). Since operations on other parts of the brain—performed on control animals—had no effect either on learning or on previously learned behavior, we seemed to have demonstrated that the vertical lobe and superior frontal lobe of the octopus brain are memory centers. In a sense they are, but this is an unduly simple view; in a recent summary of findings Young has listed no fewer than six different effects caused by the removal of or damage to the vertical-lobe system alone.

Karl Lashley, who studied the cerebral cortex of mammals, concluded that, in the organization of a memory, the involvement of specific groups of nerve cells is not as important as the total number of nerve cells available for organization. A similar situation appears to hold true in the functioning of the vertical lobe of the octopus brain; there is a definite relation between the amount of vertical lobe left intact and the accuracy with which a learned response is performed [*see top illustration on page 200*]. This seems to suggest that, at least in the octopus's vertical lobe and the mammalian cerebral cortex, memory is both everywhere and nowhere in particular.

Some of the difficulties such a conclusion presents may be due to a failure to distinguish experimentally between the two constituents of a memory. Whatever its nature, a memory must consist not only of a representation, in neural terms, of the learned situation but also of a mechanism that enables that representation to persist. A distinction must be made between the topology of what persists (the coding and spatial relations involved in the memory of a particular animal) and the mechanisms of persistence (the neural change that is presumably the same in the memory of any animal). Indeed, it may be that some of the theoretical confusion in the study of memory arises from the fact that experiments showing a quantitative relation between memory and nerve tissue tell us something about how the neural representation of memory is organized but nothing about how the representation is kept going.

In our experiments demonstrating that an octopus deprived of its vertical lobe could not be trained to discriminate between a crab alone (that is, reward) and a crab accompanied by a geometric figure and a shock (that is, punishment) our groups of trials were separated by intervals of approximately two hours. When we spaced the trials so that they were only five minutes apart, however, we found that such animals were capable of learning [*see bottom illustration on page 200*]. Using the number of trials required as a criterion of learning, we found that these animals attained a level of performance as good as that of normal animals trained with longer time intervals between trials.

One significant difference remained: a normal octopus has a learning-retention period of two weeks or longer, but animals without a vertical lobe had retention periods of only 30 minutes to two hours. These observations suggest that the establishment of a memory involves two mechanisms. There is first a short-term, or transitory, memory that, by its continuing activity between intervals of training, leads to a long-term change in the brain. If there were no reinforcement, the short-term memory

would wane; with reinforcement it keeps going and so induces the long-term—and by implication slower—changes that enable a brain to retain memories for long periods.

In 1957 Eliot Stellar of the University of Pennsylvania School of Medicine pointed out the parallels between our results with invertebrates and the unexpected discovery of a similar effect in man by Wilder Penfield, Brenda Milner and W. B. Scoville of the Montreal Neurological Institute. Epileptic patients who have been treated by surgical removal of the temporal lobes of the brain score as well in I.Q. tests after the operation as they do in tests before the onset of epilepsy. They remember their past, their profession and their relatives. They cannot, however, retain new information for more than short periods. Articles can be read and understood, but they are not remembered once they are finished and another topic is taken up. A relative may die but his death goes unremembered after an hour or so. This surgery involves the hippocampal system of the

human brain; its effects seem to suggest that, although man's cerebral cortex incorporates a long-term memory system, the hippocampal system is essential to the establishment of new long-term memories.

Today a considerable body of behavioral and psychological evidence favors the separation of memory into short-term and long-term systems. At the neurological level this distinction has brought about a reaffirmation of the role in memory of what are called self-reexciting chains. A few years ago the concept of such chains had gone out of fashion because it had been found that neither convulsive shocks nor cooling the brain to a temperature so low that all activity ceased would abolish learned responses. It is now known that if such treatments are given during the early stages of learning—that is, before a memory is fully established—they have an effect; supposedly this is because they have interfered with the more active part of the process. As the surgical operations for epilepsy indicate, a long-term memory system is intact after removal of the temporal lobes. A short-

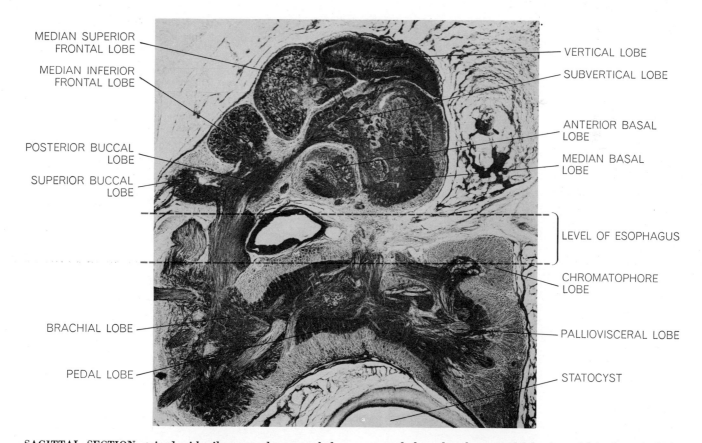

MEDIAN SUPERIOR FRONTAL LOBE

MEDIAN INFERIOR FRONTAL LOBE

POSTERIOR BUCCAL LOBE

SUPERIOR BUCCAL LOBE

BRACHIAL LOBE

PEDAL LOBE

VERTICAL LOBE

SUBVERTICAL LOBE

ANTERIOR BASAL LOBE

MEDIAN BASAL LOBE

LEVEL OF ESOPHAGUS

CHROMATOPHORE LOBE

PALLIOVISCERAL LOBE

STATOCYST

SAGITTAL SECTION stained with silver reveals some of the structures in the octopus brain. Broken lines (*color*) show the route of the esophagus, the boundary between the upper and lower parts of the brain. Labels identify eight lobes in the upper brain and four in the lower; experiments before and after surgical removal show that the vertical lobe (*top right*) plays a role in visual learning and that the inferior frontal lobe (*top left*) is one of two involved in tactile learning. The statocyst (*bottom right*) is not a part of the brain; it is one of the twin organs responsible for the octopus's sense of balance. Magnification is 15 diameters.

UNSCHOOLED OCTOPUS leaves the shelter at one end of its tank (*first photograph*) and walks toward the bait at the opposite end. The advancing animal uses only one of its eyes to guide it. When the bait, a crab, is in range, the octopus throws its leading

term memory system must also remain, however, because the patients can remember new information for short periods, particularly when they use mnemonic devices. On the basis of this interpretation it would appear that the hippocampal system may have the role of linking the two memory mechanisms—whatever that may mean.

For octopuses in our training situation it seems at first that when the vertical lobe of the brain is removed, the long-term memory system of the animal is completely abolished, leaving only the short-term system. We obtain a different result, however, if instead of showing such an animal a crab with or without a geometric figure we present it with figures only, rewarding it with a crab for an attack on one figure and punishing it with a shock for an attack on another. Under such conditions an octopus without its vertical lobe can learn the required discriminations and retain them. At least two conclusions can be drawn from this kind of result. The first is that the vertical lobe is essential to the memory system if the learned response involves a change in what might be termed innate behavior toward an object as familiar to an octopus as a crab. The second is that a long-term memory system for some responses

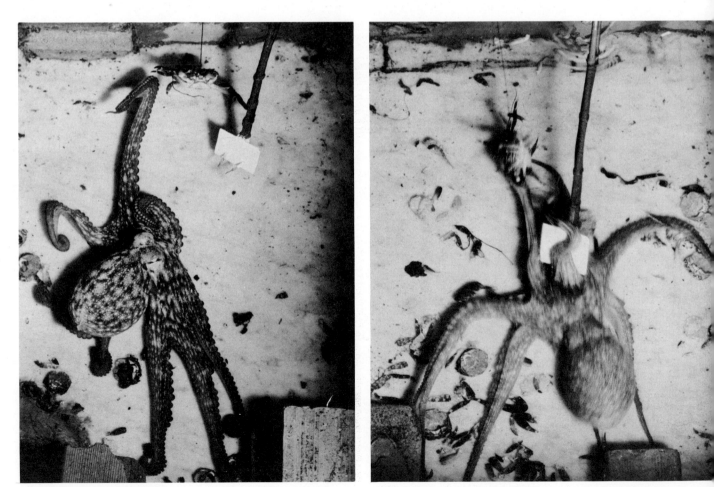

TRAINED OCTOPUS is cautious in its approach when a crab and a geometric figure are presented together (*first photograph*). If the animal seizes the crab, it receives an electric shock (*note darkened region at the base of the arm in second photograph*). As

arms forward to seize it (*second photograph*). Next it tucks the crab up toward its mouth (*third photograph*). The octopus then returns to its shelter (*fourth photograph*), where it kills the crab with a poisonous secretion from its salivary glands and eats it.

can be maintained in the absence of the vertical lobe.

Since we do not know (and probably never will know because it is so difficult to rear *Octopus vulgaris* from its larval stage) whether the octopus's response to a crab is learned or innate, our studies over the past eight years have involved experiments in which reward or punishment is given only after the animal has responded to an artificial situation, that is, the presentation of a figure of a given size, shape or color. It has been shown that animals without a vertical lobe can learn to attack unfamiliar figures for a reward, although they do so more slowly than normal animals. Once they have learned to attack such figures these octopuses retain their response for as long a time as normal animals do. If octopuses without a vertical lobe are required to reverse a learned visual response, however, they find it particularly difficult. When a shock is received for attacking a figure that formerly brought a reward, the animals can still learn to discriminate, but they make between four and five times as many mistakes as normal animals; moreover, their period of retention is shorter.

In addition to its large visual system

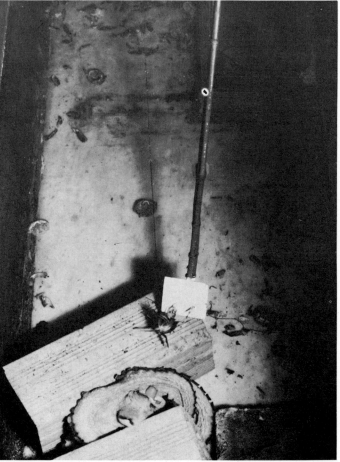

training continues, the octopus will often not even leave its shelter when crab and figure are presented (*third photograph*). If crab and figure are brought near a fully trained animal, it pales and pumps a jet of water at them (*fourth photograph*).

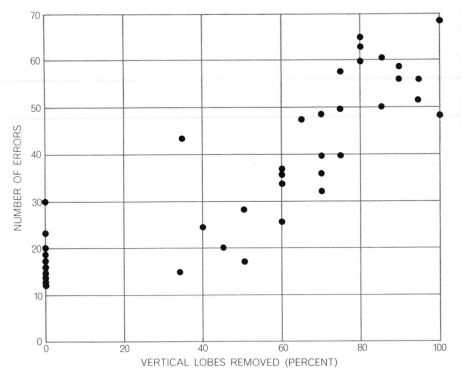

CONTRAST IN PERFORMANCE of normal (*far left*) and surgically altered octopuses shows that the number of errors increased more than threefold as larger and larger portions of the brain's vertical lobe were excised. This finding supports the conclusion that the organization of memory depends primarily on the number of brain cells available.

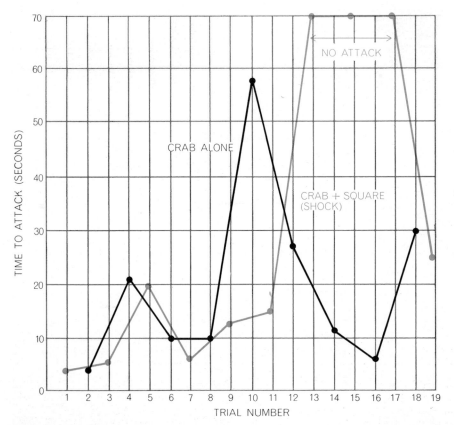

ABILITY TO LEARN can be demonstrated by an octopus deprived of the vertical lobes of its brain, provided that the trials are only a few minutes apart. In the example illustrated, little learning was apparent during 12 alternating exposures to negative and positive stimuli. Thereafter three successive negative stimuli were avoided by the octopus.

the octopus has a complex chemo-tactile sensory system. Most of the investigation of this system has been done by Martin Wells and his wife Joyce. By applying methods similar to those used for training the animals to make visual discriminations, they have been able to show that tactile learning in the octopus is about as rapid as visual learning. Octopuses have been trained to discriminate between a live bivalve and a counterfeit one consisting of shells of the same species that have been cleaned and filled with wax. They can discriminate between a bivalve with a ribbed shell and another species of comparable size but with a smooth shell. Just recently Wells has found that octopuses can detect hydrochloric acid, sucrose or quinine dissolved in sea-water at concentrations 100 times less than those the human tongue can detect in distilled water. Presented with artificial objects, they can distinguish between grooved cylinders and smooth ones, although they cannot distinguish between two grooved objects that differ only in the direction in which the grooves run [*see illustration on opposite page*]. After intensive training they can discriminate a cube from a sphere about 75 percent of the time.

Through each arm of an octopus, which is studded with two rows of suckers, runs a cord of nerve fibers and ganglia. In these ganglia occur local reflexes along the arm and between the rows of suckers. It is supposed that, when the octopus makes a tactile discrimination, the state of excitation in the ganglia above each sucker is determined by the proportion of sense organs excited, and the degree to which these sense organs are stimulated determines the frequency with which nerve impulses are discharged in the fibers running from the ganglia to the brain. Learning in the isolated arm ganglia is probably not possible. Wells has found that for tactile learning to occur the upper brain's median inferior frontal lobe and subfrontal lobe are necessary. Damage to these regions of the brain does not affect visual learning, and for that reason the two lobes have often been used as the sites for control lesions in the investigation of visual learning.

The role of the median inferior frontal lobe seems to be to interrelate the information received from each of the octopus's eight arms; if the lobe is removed and one arm is trained to reject an object, then the other arms continue to accept the object. Without the subfrontal lobe the animals cannot even learn to reject objects by touch. As in the case of the vertical lobe in visual learning, the retention of small portions of the subfrontal lobe allows adequate learned performance. Wells believes that as few as 13,000 of the five million subfrontal-lobe cells may be sufficient for some learning to occur. The subfrontal lobe is structurally very similar to the vertical lobe; it must be considered the vertical lobe's counterpart in the chemotactile system. Removal of the vertical lobe nonetheless has an effect on chemotactile discrimination, mainly in the direction of slowing the rate at which learning occurs.

This account has discussed the main lines of work on memory systems that have been carried out with octopuses as experimental animals, together with some comparisons with human memory. Recently Young has summarized all the work on the cephalopod brain of the past 17 years and has devised a scheme of how such brains may work in the formation, storage and translation of memory into effective action.

Young proposes that in the course of evolution chemotactile and visual centers developed out of a primitive taste-and-bite reflex mechanism. As these "distance receptor" systems evolved, providing information as to where food might be obtained other than that received from direct contact, there came to be a more indirect relation between a change in the environment and the responses that such a change produced in the animal. As this happened, signal systems of greater duration than are provided by simple reflex mechanisms also had to evolve; learning had to become possible so that the animal could assess the significance of each distant environmental change.

Suppose, for example, a crab appears at a distance in the visual field of an octopus; as a result of what can be called "cue signals" there arises in the octopus brain a system for producing "graduated commands to attack." This command system will be weak at first but will grow stronger with reinforcement. The actual strengthening process will vary according to the reward or punishment met at each attack, because the outcome of each attack gives rise to a "result" signal. Such signals condition the distance-receptor systems that initially cued the attack—in the present example, the visual-receptor system. These result signals

become distributed throughout the nervous tissue that carries a record of a particular event.

There is, of course, a delay between the moment the cue signals are received in the brain and the moment the result signals arrive. If the result signals are to produce the appropriate conditioning of memory elements, the address of these elements, so to speak, has to be held to allow correct delivery of the information of, say, taste or pain. In the brain of the octopus each optic lobe contains "classifying" cells, among them vertically and horizontally oriented sets of nerve fibers that are presumably related to the vertical and horizontal arrangement of elements in the retina of the octopus's eye. These classifying cells form synapses with "memory" cells in the optic lobes that in their turn activate the cells that signal either attack or retreat. According to Young's hypothesis, each of the memory cells at first has a pair of alternative pathways; the actual neural change during learning consists in closing one of the two pathways. This closing may be accomplished by small cells that are

abundant in these learning centers and that can perhaps be switched on so as to produce a substance that inhibits transmission.

Suppose an attack has been evoked by means of this system; the memory cells activate not only an attack circuit but also a circuit reaching the vertical lobe of the upper brain. The signals indicating the results of the attack, such as taste or pain, arrive back and further reinforce the memory cells in the optic lobes, which have been under the influence of the appropriate pathways set up in the vertical lobe during the time interval between the cue signal and the result signal [see illustration on next page].

The hypothesis that the actual change represented by memory is produced by the small cells agrees with the fact that these cells are also present in the part of the brain that was shown by the Wellses to be the minimum necessary for tactile memory. Young suggests that the small cells were originally part of the primitive taste-and-bite reflex system, serving the function of

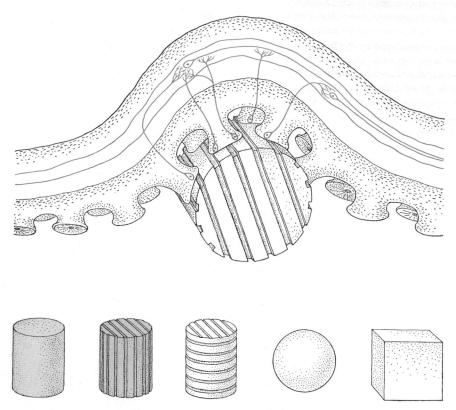

LEARNING BY TOUCH in the octopus was investigated by presenting objects with a variety of shapes and textures. In the case of a grooved cylinder (top) only the sense organs in contact with the surface are excited; those resting over the grooves remain inactive. Thus the octopus can learn to discriminate between a smooth cylinder (gray) and a grooved one (color), and even between a cube and a sphere; it cannot, however, discriminate between two cylinders that differ only in the orientation of the grooves.

temporary inhibition. The evolution of the memory consisted in making the inhibition last longer. The sets of auxiliary lobes associated with the memory system arose to allow for various combinations of inputs to be set up, to be combined with the signals that report the results of actions and finally to be "delivered to the correct address" in the memory.

There is much that is speculative about this description, but the fact remains that both the visual and the tactile memory systems of the octopus embrace sets of brain lobes arranged in similar circuits. This organization provides opportunities for study of the memory process that are made more challenging by Young's conviction that comparable circuits exist in the brains of mammals, including man.

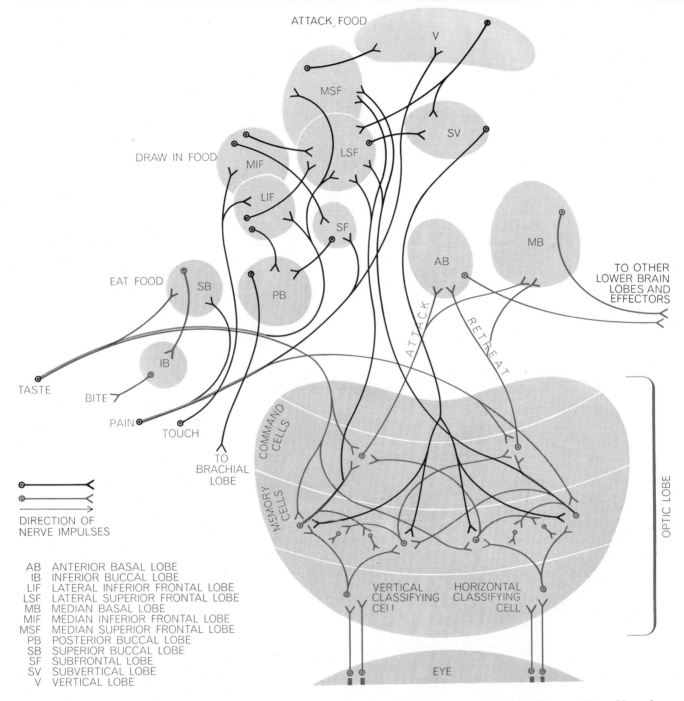

DUAL NATURE OF MEMORY can be traced out on an exploded view of the octopus brain. Circuits leading from the optic lobe (*color*) are the first to be activated on receipt of a visual cue by the lobe's classifying cells. The cue is then recorded in the memory cells and relayed to the command cells; the latter induce the octopus to attack or retreat. If an attack is rewarded, the returning "result" signal will reinforce a memory that registers the initiating cue favorably. If, instead, the attack brings pain, the reinforced memory will register the cue unfavorably and any similar cues encountered in the future will be channeled to the circuits governing retreat rather than attack. Additional circuits (*black*) connect the memory and command regions of the optic lobe to various lobes of the upper brain; thus each event and its outcome are also recorded and reinforced in these nervous tissues. In due course what appear to be the long-term components of the memory system become localized in individual upper brain lobes.

Part VI

BEHAVIOR

VI
Behavior

INTRODUCTION The result of most complex nervous and endocrine activity is an overt response on the part of an animal. Some responses are clearly related to a stimulus; some appear to be caused by an internal state. Behavior takes the form of a coordinated set of movements that change the animal's relation to its environment; it may range from reflexes as simple as limb withdrawal, through such cyclically repeated locomotor patterns as the flight of an insect, to highly complex ritualized activities like some of those in courtship. Much behavior appears to be entirely innate, that is, dependent upon a set of genetically determined central nervous connections; other actions are obviously learned, and many others contain elements of both types.

The study of behavior is comprehensive, drawing upon many disciplines other than biology. Since this book is concerned with the elaboration of cellular elements into interacting systems, the emphasis of the following four selections is upon the physiological mechanisms underlying behavior. Analysis of these mechanisms is extremely difficult, because in most situations so many underlying elements are involved that a meaningful cause-and-effect relationship is hard to extract. Fortunately, some systems exhibit either an extreme specialization or a simplification of cellular elements that makes them more promising subjects for study.

The first two articles in this section are based on such systems. In "Electric Location by Fishes," H. W. Lissmann describes a kind of navigation employed by several groups of fish that generate weak repetitive electric discharges and respond to external alterations in the field thus produced. This system shares with the "sonar" mechanism of guidance in bats a unique property: the organism orients itself by generating its own signals and measuring distortion in those signals returned from the environment. This circumstance affords a special advantage to the student of behavior, who can inspect the way the animal controls the output (in itself a complex behavioral act) as well as the way it interprets the returned signal. Lissmann's studies, which demonstrate clearly the utility of electrical navigation, have opened up investigations of the properties of the sensory system which receives the signals. The receptors are located, Lissmann suggests, along the lateral line. Electrical recording from single fibers in the lateral line nerve show that the sensory cells respond with a single impulse or a burst of impulses to each outgoing discharge from the electric organ. Changes in the number of impulses per burst, or in the total frequency, are produced when conductors or dielectrics are placed in the vicinity of the receptor. More recently, recordings have been made from single nerve cells at a higher level in the receiving system, the cerebellum. Many of these nerve cells collect input from a number of receptors along the lateral line, and, like some of the units in the visual cortex described in the preceding section by Hubel, they may respond selectively to *moving* objects that sequentially distort the field created by the fish's electrical discharge. With these findings, analysis of the central nervous machinery that underlies the behavioral performance has begun.

In the second article, "Moths and Ultrasound," Kenneth D. Roeder discusses another navigation system; this one is composed of just four sensory cells, and is employed in evading a predator. The predator is a bat, which emits ultrasonic cries and uses the echoes of these cries to determine the location of insect prey. Some of the prey species, however, have evolved acoustical receptors sensitive to the same frequency range employed by the bats. The sensory cells are capable of some localization, and the complex interaction between this ability of the sensory cells and the variable position of the wings during the normal flight cycle has been analyzed in detail. Especially interesting is the fact that different frequencies of sensory discharge, which depend upon differences in sound intensity, may produce alternate forms of behavior. Faint sounds cause the insect to turn away from the sound source, whereas loud sounds evoke erratic, evasive flight. In most sensory systems, changes in stimulus intensity merely change the intensity of the response; but in this system, a curious switch, which changes the *character* of the response instead, is apparently present.

In the third and fourth articles, the emphasis shifts from the influence of the nervous system upon behavior to the influence of the endocrine system. It is well known that the hormonal state of an animal significantly influences its behavioral performance, but to date we know very little about the specific relationships. In "The Reproductive Behavior of Ring Doves," Daniel S. Lehrman combines an analysis of the hormonal state of birds during the reproductive cycle with careful observations of their mating, nest-building, and incubation behavior. The results clearly demonstrate that the dove's reactions (presumably "innate" reactions) to environmental stimuli, including the mate, affect the behavior in a way that depends on the bird's endocrine balance at the time. The willingness of birds to incubate, for example, can be changed both by the injection of estrogen and by manipulation of stimuli.

In "Pheromones," Edward O. Wilson discusses an entirely different use of chemical communication. Unlike hormones, which transmit signals from one tissue to another in the same organism, pheromones provide an avenue of chemical information transfer *between individuals*. This communication may be part of courtship and mating behavior, as it is in those female moths that secrete a volatile attractant detected by chemoreceptors on the male's antennae; or it may enable one individual to control the development of another, as in the termite nest where the "king" and "queen" secrete substances that inhibit the sexual maturation of other members of the colony. Pheromones may even provide orientation and guidance, as do the ant "trail substances" described in detail by Wilson.

21

ELECTRIC LOCATION BY FISHES

H. W. LISSMANN   March 1963

Study of the ingenious adaptations displayed in the anatomy, physiology and behavior of animals leads to the familiar conclusion that each has evolved to suit life in its particular corner of the world. It is well to bear in mind, however, that each animal also inhabits a private subjective world that is not accessible to direct observation. This world is made up of information communicated to the creature from the outside in the form of messages picked up by its sense organs. No adaptation is more crucial to survival; the environment changes from place to place and from moment to moment, and the animal must respond appropriately in every place and at every moment. The sense organs transform energy of various kinds—heat and light, mechanical energy and chemical energy—into nerve impulses. Because the human organism is sensitive to the same kinds of energy, man can to some extent visualize the world as it appears to other living things. It helps in considering the behavior of a dog, for example, to realize that it can see less well than a man but can hear and smell better. There are limits to this procedure; ultimately the dog's sensory messages are projected onto its brain and are there evaluated differently.

Some animals present more serious obstacles to understanding. As I sit writing at my desk I face a large aquarium that contains an elegant fish about 20 inches long. It has no popular name but is known to science as *Gymnarchus niloticus*. This same fish has been facing me for the past 12 years, ever since I brought it from Africa. By observation and experiment I have tried to understand its behavior in response to stimuli from its environment. I am now convinced that *Gymnarchus* lives in a world totally alien to man: its most important

sense is an electric one, different from any we possess.

From time to time over the past century investigators have examined and dissected this curious animal. The literature describes its locomotive apparatus, central nervous system, skin and electric organs, its habitat and its family relation to the "elephant-trunk fishes," or mormyrids, of Africa. But the parts have not been fitted together into a functional pattern, comprehending the design of the animal as a whole and the history of its development. In this line of biological research one must resist the temptation to be deflected by details, to follow the fashion of putting the pieces too early under the electron microscope. The magnitude of a scientific revelation is not always paralleled by the degree of magnification employed. It is easier to select the points on which attention should be concentrated once the plan is understood. In the case of *Gymnarchus*, I think, this can now be attempted.

A casual observer is at once impressed by the grace with which *Gymnarchus* swims. It does not lash its tail from side to side, as most other fishes do, but keeps its spine straight. A beautiful undulating fin along its back propels its body through the water—forward or backward with equal ease. *Gymnarchus* can maintain its rigid posture even when turning, with complex wave forms running hither and thither over different regions of the dorsal fin at one and the same time.

Closer observation leaves no doubt that the movements are executed with great precision. When *Gymnarchus* darts after the small fish on which it feeds, it never bumps into the walls of its tank, and it clearly takes evasive action at some distance from obstacles placed in

its aquarium. Such maneuvers are not surprising in a fish swimming forward, but *Gymnarchus* performs them equally well swimming backward. As a matter of fact it should be handicapped even when it is moving forward: its rather degenerate eyes seem to react only to excessively bright light.

Still another unusual aspect of this fish and, it turns out, the key to all the puzzles it poses, is its tail, a slender, pointed process bare of any fin ("gymnarchus" means "naked tail"). The tail was first dissected by Michael Pius Erdl of the University of Munich in 1847. He found tissue resembling a small electric organ, consisting of four thin spindles running up each side to somewhere beyond the middle of the body. Electric organs constructed rather differently, once thought to be "pseudoelectric," are also found at the hind end of the related mormyrids.

Such small electric organs have been an enigma for a long time. Like the powerful electric organs of electric eels and some other fishes, they are derived from muscle tissue. Apparently in the course of evolution the tissue lost its power to contract and became specialized in various ways to produce electric discharges [see "Electric Fishes," by Harry Grundfest; SCIENTIFIC AMERICAN, October, 1960]. In the strongly electric fishes this adaptation serves to deter predators and to paralyze prey. But the powerful electric organs must have evolved from weak ones. The original swimming muscles would therefore seem to have possessed or have acquired at some stage a subsidiary electric function that had survival value. Until recently no one had found a function for weak electric organs. This was one of the questions on my mind when I began to study *Gymnarchus*.

I noticed quite early, when I placed a

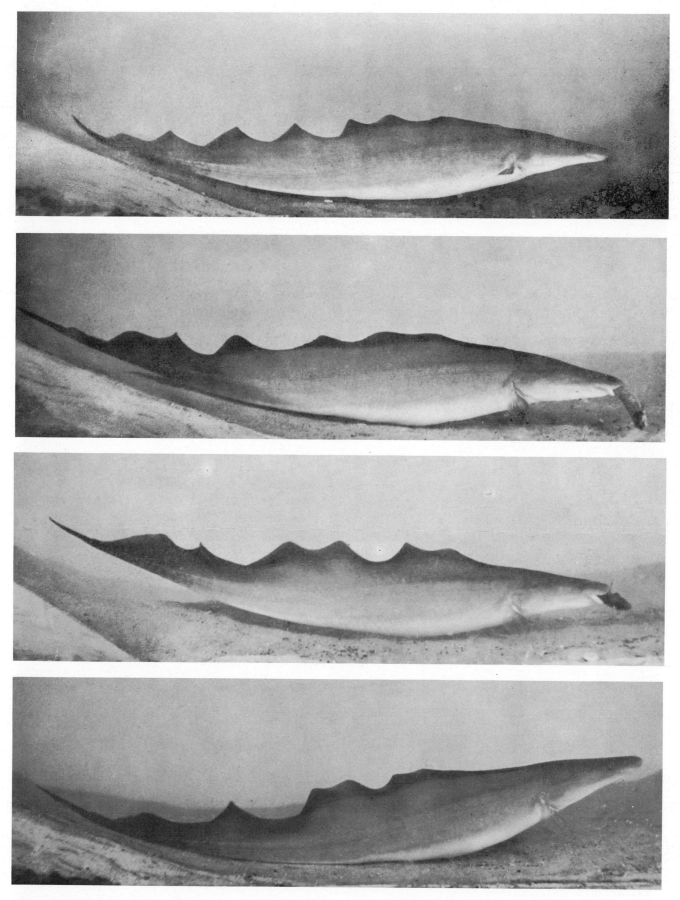

ELECTRIC FISH *Gymnarchus niloticus*, from Africa, generates weak discharges that enable it to detect objects. In this sequence the fish catches a smaller fish. *Gymnarchus* takes its name, which means "naked tail," from the fact that its pointed tail has no fin.

new object in the aquarium of a well-established *Gymnarchus*, that the fish would approach it with some caution, making what appeared to be exploratory movements with the tip of its tail. It occurred to me that the supposed electric organ in the tail might be a detecting mechanism. Accordingly I put into the water a pair of electrodes, connected to an amplifier and an oscilloscope. The result was a surprise. I had expected to find sporadic discharges co-ordinated with the swimming or exploratory motions of the animal. Instead the apparatus recorded a continuous stream of electric discharges at a constant frequency of about 300 per second, waxing and waning in amplitude as the fish changed position in relation to the stationary electrodes. Even when the fish was completely motionless, the electric activity remained unchanged.

This was the first electric fish found to behave in such a manner. After a brief search I discovered two other kinds that emit an uninterrupted stream of weak discharges. One is a mormyrid relative of *Gymnarchus;* the other is a gymnotid, a small, fresh-water South American relative of the electric eel, belonging to a group of fish rather far removed from *Gymnarchus* and the mormyrids.

It had been known for some time that the electric eel generates not only strong discharges but also irregular series of weaker discharges. Various functions had been ascribed to these weak discharges of the eel. Christopher W. Coates, director of the New York Aquarium, had suggested that they might serve in navigation, postulating that the eel somehow measured the time delay between the output of a pulse and its reflection from an object. This idea was untenable on physical as well as physiological grounds. The eel does not, in the first place, produce electromagnetic waves; if it did, they would travel too fast to be timed at the close range at which such a mechanism might be useful, and in any case they would hardly penetrate water. Electric current, which the eel does produce, is not reflected from objects in the surrounding environment.

Observation of *Gymnarchus* suggested another mechanism. During each discharge the tip of its tail becomes momentarily negative with respect to the head. The electric current may thus be pictured as spreading out into the surrounding water in the pattern of lines that describes a dipole field [*see illustration on opposite page*]. The exact configuration of this electric field depends on the conductivity of the water and on the distortions introduced in the field by objects with electrical conductivity different from that of the water. In a large volume of water containing no objects the field is symmetrical. When objects are present, the lines of current will converge on those that have better conductivity and diverge from the poor conductors [*see top illustration on page 210*]. Such objects alter the distribution of electric potential over the surface of the fish. If the fish could register these changes, it would have a means of detecting the objects.

Calculations showed that *Gymnarchus* would have to be much more sensitive electrically than any fish was known to be if this mechanism were to work. I had observed, however, that *Gymnarchus* was sensitive to extremely small external electrical disturbances. It responded violently when a small magnet or an electrified insulator (such as a comb that had just been drawn through a person's hair) was moved near the aquarium. The electric fields produced in the water by such objects must be very small indeed, in the range of fractions of a millionth of one volt per centimeter. This crude observation was enough to justify a series of experiments under more stringent conditions.

In the most significant of these experiments Kenneth E. Machin and I trained the fish to distinguish between objects that could be recognized only by an electric sense. These were enclosed in porous ceramic pots or tubes with thick walls. When they were soaked in water, the ceramic material alone had little effect on the shape of the electric field. The pots excluded the possibility of discrimination by vision or, because each test lasted only a short time, by a chemical sense such as taste or smell.

The fish quickly learned to choose between two pots when one contained aquarium water or tap water and the other paraffin wax (a nonconductor). After training, the fish came regularly to pick a piece of food from a thread suspended behind a pot filled with aquarium or tap water and ignored the pot filled with wax [*see bottom illustration on page 210*]. Without further conditioning it also avoided pots filled with air, with distilled water, with a close-fitting glass tube or with another nonconductor. On the other hand, when the electrical conductivity of the distilled water was matched to that of tap or aquarium water by the addition of salts or acids, the fish would go to the pot for food.

A more prolonged series of trials showed that *Gymnarchus* could distinguish mixtures in different proportions of tap water and distilled water and perform other remarkable feats of discrimination. The limits of this performance can best be illustrated by the fact that the fish could detect the presence of a glass rod two millimeters in diameter and would fail to respond to a glass rod .8 millimeter in diameter, each hidden in a

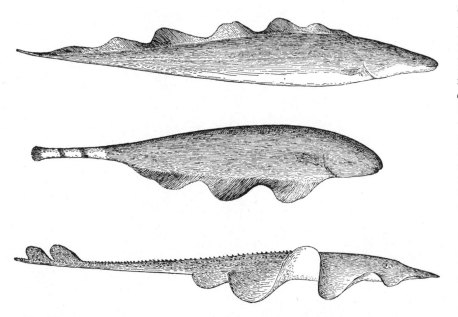

UNUSUAL FINS characterize *Gymnarchus* (top), a gymnotid from South America (*middle*) and sea-dwelling skate (*bottom*). All swim with spine rigid, probably in order to keep electric generating and detecting organs aligned. *Gymnarchus* is propelled by undulating dorsal fin, gymnotid by similar fin underneath and skate by lateral fins resembling wings.

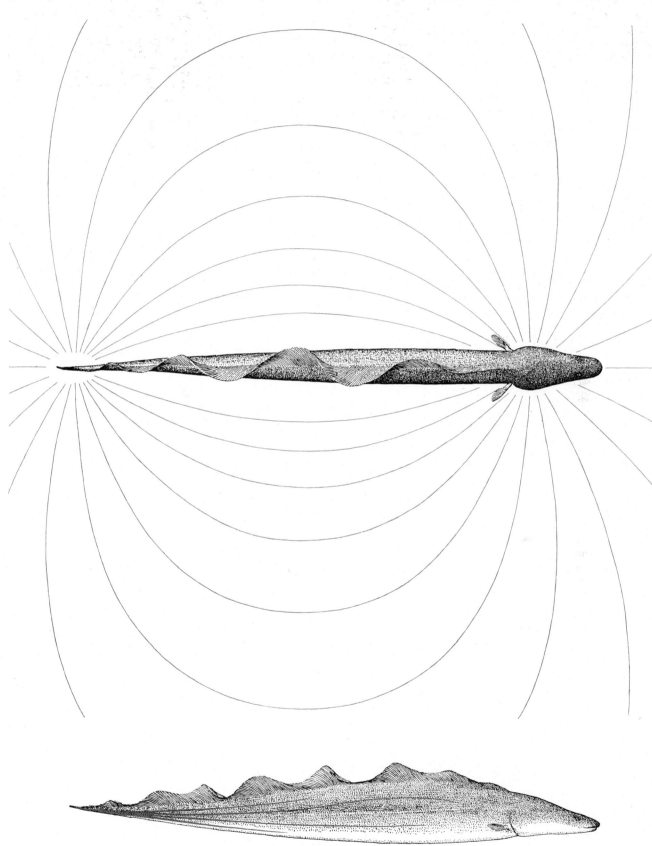

ELECTRIC FIELD of *Gymnarchus* and location of electric generating organs are diagramed. Each electric discharge from organs in rear portion of body (*color in side view*) makes tail negative with respect to head. Most of the electric sensory pores or organs are in head region. Undisturbed electric field resembles a dipole field, as shown, but is more complex. The fish responds to changes in the distribution of electric potential over the surface of its body. The conductivity of objects affects distribution of potential.

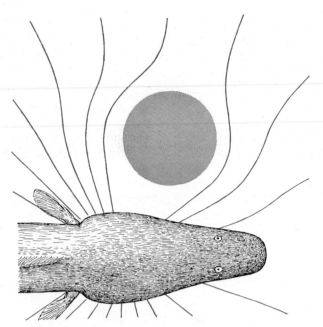

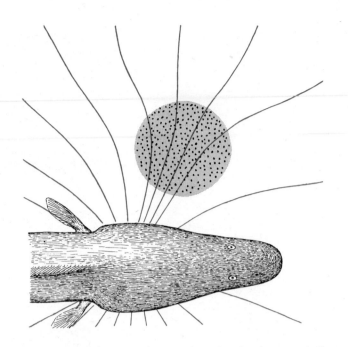

OBJECTS IN ELECTRIC FIELD of *Gymnarchus* distort the lines of current flow. The lines diverge from a poor conductor (*left*) and converge toward a good conductor (*right*). Sensory pores in the head region detect the effect and inform the fish about the object.

pot of the same dimensions. The threshold of its electric sense must lie somewhere between these two values.

These experiments seemed to establish beyond reasonable doubt that *Gymnarchus* detects objects by an electrical mechanism. The next step was to seek the possible channels through which the electrical information may reach the brain. It is generally accepted that the tissues and fluids of a fresh-water fish are relatively good electrical conductors enclosed in a skin that conducts poorly. The skin of *Gymnarchus* and of many mormyrids is exceptionally thick, with layers of platelike cells sometimes arrayed in a remarkable hexagonal pattern [*see top illustration on page 213*]. It can therefore be assumed that natural selection has provided these fishes with better-than-average exterior insulation.

In some places, particularly on and around the head, the skin is closely perforated. The pores lead into tubes often filled with a jelly-like substance or a loose aggregation of cells. If this jelly is a good electrical conductor, the arrangement would suggest that the lines of electric current from the water into the body of the fish are made to converge at these pores, as if focused by a lens. Each jelly-filled tube widens at the base into

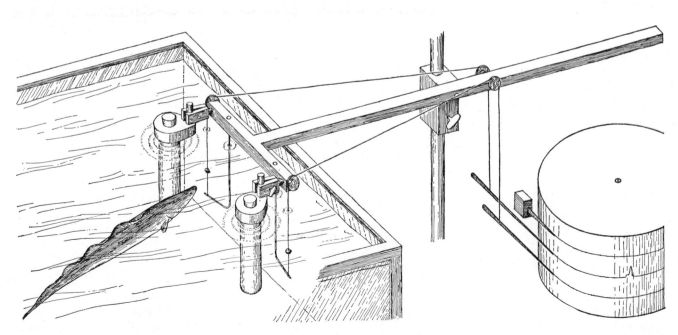

EXPERIMENTAL ARRANGEMENT for conditioned-reflex training of *Gymnarchus* includes two porous pots or tubes and recording mechanism. The fish learns to discriminate between objects of different electrical conductivity placed in the pots and to seek bait tied to string behind the pot holding the object that conducts best. *Gymnarchus* displays a remarkable ability to discriminate.

a small round capsule that contains a group of cells long known to histologists by such names as "multicellular glands," "mormyromasts" and "snout organs." These, I believe, are the electric sense organs.

The supporting evidence appears fairly strong: The structures in the capsule at the base of a tube receive sensory nerve fibers that unite to form the stoutest of all the nerves leading into the brain. Electrical recording of the impulse traffic in such nerves has shown that they lead away from organs highly sensitive to electric stimuli. The brain centers into which these nerves run are remarkably large and complex in *Gymnarchus*, and in some mormyrids they completely cover the remaining portions of the brain [*see illustration on next page*].

If this evidence for the plan as well as the existence of an electric sense does not seem sufficiently persuasive, corroboration is supplied by other weakly electric fishes. Except for the electric eel, all species of gymnotids investigated so far emit continuous electric pulses. They are also highly sensitive to electric fields. Dissection of these fishes reveals the expected histological counterparts of the structures found in the mormyrids: similar sense organs embedded in a similar skin, and the corresponding regions of the brain much enlarged.

Skates also have a weak electric organ in the tail. They are cartilaginous fishes, not bony fishes, or teleosts, as are the mormyrids and gymnotids. This means that they are far removed on the family line. Moreover, they live in the sea, which conducts electricity much better than fresh water does. It is almost too much to expect structural resemblances to the fresh-water bony fishes, or an electrical mechanism operating along similar lines. Yet skates possess sense organs, known as the ampullae of Lorenzini, that consist of long jelly-filled tubes opening to the water at one end and terminating in a sensory vesicle at the other. Recently Richard W. Murray of the University of Birmingham has found that these organs respond to very delicate electrical stimulation. Unfortunately, either skates are rather uncooperative animals or we have not mastered the trick of training them; we have been unable to repeat with them the experiments in discrimination in which *Gymnarchus* performs so well.

Gymnarchus, the gymnotids and skates all share one obvious feature: they swim in an unusual way. *Gymnarchus* swims with the aid of a fin on its back; the gymnotids have a similar fin on their

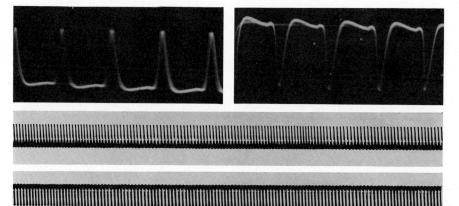

ELECTRIC DISCHARGES of *Gymnarchus* show reversal of polarity when detecting electrodes are rotated 180 degrees (*enlarged records at top*). The discharges, at rate of 300 per second, are remarkably regular even when fish is resting, as seen in lower records.

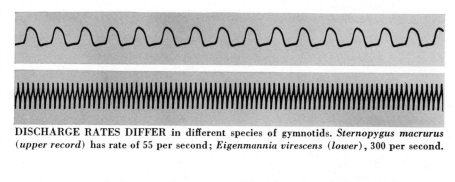

DISCHARGE RATES DIFFER in different species of gymnotids. *Sternopygus macrurus* (*upper record*) has rate of 55 per second; *Eigenmannia virescens* (*lower*), 300 per second.

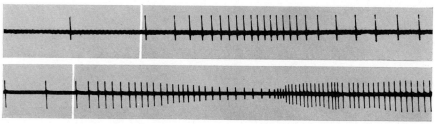

VARIABLE DISCHARGE RATE is seen in some species. Tap on tank (*white line in upper record*) caused mormyrid to increase rate. Tap on fish (*lower record*) had greater effect.

underside; skates swim with pectoral fins stuck out sideways like wings [*see illustration on page 208*]. They all keep the spine rigid as they move. It would be rash to suggest that such deviations from the basic fish plan could be attributed to an accident of nature. In biology it always seems safer to assume that any redesign has arisen for some reason, even if the reason obstinately eludes the investigator. Since few fishes swim in this way or have electric organs, and since the fishes that combine these features are not related, a mere coincidence would appear most unlikely.

A good reason for the rigid swimming posture emerged when we built a model to simulate the discharge mecha-

nism and the sensory-perception system. We placed a pair of electrodes in a large tank of water; to represent the electric organ they were made to emit repetitive electric pulses. A second pair of electrodes, representing the electric sense organ, was placed some distance away to pick up the pulses. We rotated the second pair of electrodes until they were on a line of equipotential, where they ceased to record signals from the sending electrodes. With all the electrodes clamped in this position, we showed that the introduction of either a conductor or a nonconductor into the electric field could cause sufficient distortion of the field for the signals to reappear in the detectors.

In a prolonged series of readings the

slightest displacement of either pair of electrodes would produce great variations in the received signal. These could be smoothed to some extent by recording not the change of potential but the change in the potential gradient over the "surface" of our model fish. It is probable that the real fish uses this principle, but to make it work the electrode system must be kept more or less constantly aligned. Even though a few cubic centimeters of fish brain may in some respects put many electronic computers in the shade, the fish brain might be unable to obtain any sensible information if the fish's electrodes were to be misaligned by the tail-thrashing that propels an ordinary fish. A mode of swimming that keeps the electric field symmetrical with respect to the body most of the time would therefore offer obvious advantages. It seems logical to assume that *Gymnarchus*, or its ancestors, acquired the rigid mode of swimming along with the electric sensory apparatus and subsequently lost the broad, oarlike tail fin.

Our experiments with models also showed that objects could be detected only at a relatively short distance, in spite of high amplification in the receiving system. As an object was moved farther and farther away, a point was soon reached where the signals arriving at the oscilloscope became submerged in the general "noise" inherent in every detector system. Now, it is known that minute amounts of energy can stimulate a sense organ: one quantum of light registers on a visual sense cell; vibrations of subatomic dimensions excite the ear; a single molecule in a chemical sense organ can produce a sensation, and so on. Just how such small external signals can be picked out from the general noise in and around a metabolizing cell represents one of the central questions of sensory physiology. Considered in connection with the electric sense of fishes, this question is complicated further by the high frequency of the discharges from the electric organ that excite the sensory apparatus.

In general, a stimulus from the environment acting on a sense organ produces a sequence of repetitive impulses in the sensory nerve. A decrease in the strength of the stimulus causes a lower frequency of impulses in the nerve. Conversely, as the stimulus grows stronger, the frequency of impulses rises, up to a certain limit. This limit may vary from one sense organ to another, but 500 impulses per second is a common upper limit, although 1,000 per second have been recorded over brief intervals.

In the case of the electric sense organ of a fish the stimulus energy is provided by the discharges of the animal's electric organ. *Gymnarchus* discharges at the rate of 300 pulses per second. A change in the amplitude—not the rate—of these pulses, caused by the presence of an object in the field, constitutes the effective stimulus at the sense organ. Assuming that the reception of a single discharge of small amplitude excites one impulse in a sensory nerve, a discharge of larger amplitude that excited two impulses would probably reach and exceed the upper limit at which the nerve can generate impulses, since the nerve would now be firing 600 times a second (twice the rate of discharge of the electric organ). This would leave no room

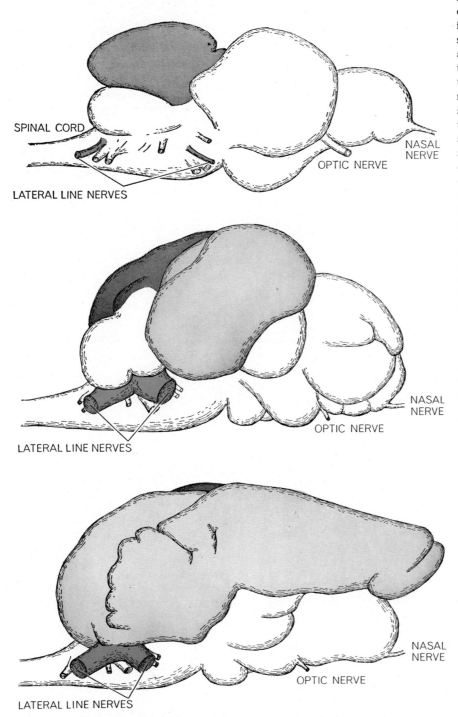

BRAIN AND NERVE ADAPTATIONS of electric fish are readily apparent. Brain of typical nonelectric fish (*top*) has prominent cerebellum (*gray*). Regions associated with electric sense (*color*) are quite large in *Gymnarchus* (*middle*) and even larger in the mormyrid (*bottom*). Lateral-line nerves of electric fishes are larger, nerves of nose and eyes smaller.

to convey information about gradual changes in the amplitude of incoming stimuli. Moreover, the electric organs of some gymnotids discharge at a much higher rate; 1,600 impulses per second have been recorded. It therefore appears unlikely that each individual discharge is communicated to the sense organs as a discrete stimulus.

We also hit on the alternative idea that the frequency of impulses from the sensory nerve might be determined by the mean value of electric current transmitted to the sense organ over a unit of time; in other words, that the significant messages from the environment are averaged out and so discriminated from the background of noise. We tested this idea on *Gymnarchus* by applying trains of rectangular electric pulses of varying voltage, duration and frequency across the aquarium. Again using the conditioned-reflex technique, we determined the threshold of perception for the different pulse trains. We found that the fish is in fact as sensitive to high-frequency pulses of short duration as it is to low-frequency pulses of identical voltage but correspondingly longer duration. For any given pulse train, reduction in voltage could be compensated either by an increase in frequency of stimulus or an increase in the duration of the pulse. Conversely, reduction in the frequency required an increase in the voltage or in the duration of the pulse to reach the threshold. The threshold would therefore appear to be determined by the product of voltage times duration times frequency.

Since the frequency and the duration of discharges are fixed by the output of the electric organ, the critical variable at the sensory organ is voltage. Threshold determinations of the fish's response to single pulses, compared with quantitative data on its response to trains of pulses, made it possible to calculate the time over which the fish averages out the necessarily blurred information carried within a single discharge of its own. This time proved to be 25 milliseconds, sufficient for the electric organ to emit seven or eight discharges.

The averaging out of information in this manner is a familiar technique for improving the signal-to-noise ratio; it has been found useful in various branches of technology for dealing with barely perceptible signals. In view of the very low signal energy that *Gymnarchus* can detect, such refinements in information processing, including the ability to average out information picked up by a large number of separate sense organs,

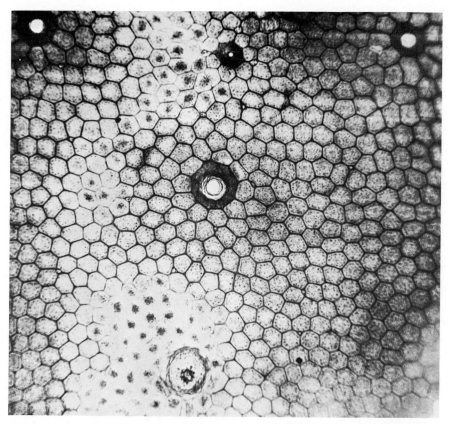

SKIN OF MORMYRID is made up of many layers of platelike cells having remarkable hexagonal structure. The pores contain tubes leading to electric sense organs. This photomicrograph by the author shows a horizontal section through the skin, enlarged 100 diameters.

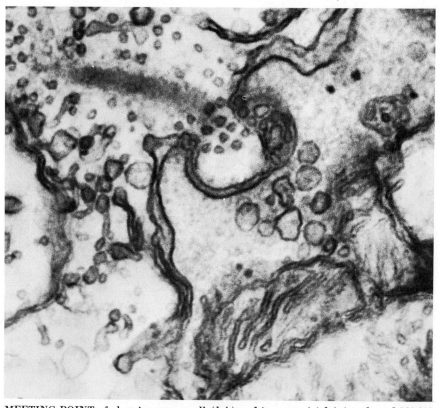

MEETING POINT of electric sensory cell (*left*) and its nerve (*right*) is enlarged 120,000 diameters in this electron micrograph by the author and Ann M. Mullinger. Bulge of sensory cell into nerve ending displays the characteristic dense streak surrounded by vesicles.

appear to be essential. We have found that *Gymnarchus* can respond to a continuous direct-current electric stimulus of about .15 microvolt per centimeter, a value that agrees reasonably well with the calculated sensitivity required to recognize a glass rod two millimeters in diameter. This means that an individual sense organ should be able to convey information about a current change as small as .003 micromicroampere. Extended over the integration time of 25 milliseconds, this tiny current corresponds to a movement of some 1,000 univalent, or singly charged, ions.

The intimate mechanism of the single sensory cell of these organs is still a complete mystery. In structure the sense organs differ somewhat from species to species and different types are also found in an individual fish. The fine structure of the sensory cells, their nerves and associated elements, which Ann M. Mullinger and I have studied with both the light microscope and the electron microscope, shows many interesting details. Along specialized areas of the boundary between the sensory cell and the nerve fiber there are sites of intimate contact where the sensory cell bulges into the fiber. A dense streak extends from the cell into this bulge, and the vesicles alongside it seem to penetrate the intercellular space. The integrating system of the sensory cell may be here.

These findings, however, apply only to *Gymnarchus* and to about half of the species of gymnotids investigated to date. The electric organs of these fishes emit pulses of constant frequency. In the other gymnotids and all the mormyrids the discharge frequency changes with the state of excitation of the fish. There is therefore no constant mean value of current transmitted in a unit of time; the integration of information in these species may perhaps be carried out in the brain. Nevertheless, it is interesting that both types of sensory system should have evolved independently in the two different families, one in Africa and one in South America.

The experiments with *Gymnarchus*, which indicate that no information is carried by the pulse nature of the discharges, leave us with a still unsolved problem. If the pulses are "smoothed out," it is difficult to see how any one fish can receive information in its own frequency range without interference from its neighbors. In this connection Akira Watanabe and Kimihisa Takeda at the University of Tokyo have made the potentially significant finding that the gymnotids respond to electric oscillations close in frequency to their own by shifting their frequency away from the applied frequency. Two fish might thus react to each other's presence.

For reasons that are perhaps associated with the evolutionary origin of their electric sense, the electric fishes are elusive subjects for study in the field. I have visited Africa and South America in order to observe them in their natural habitat. Although some respectable specimens were caught, it was only on rare occasions that I actually saw a *Gymnarchus*, a mormyrid or a gymnotid in the turbid waters in which they live. While such waters must have favored the evolution of an electric sense, it could not have been the only factor. The same waters contain a large number of

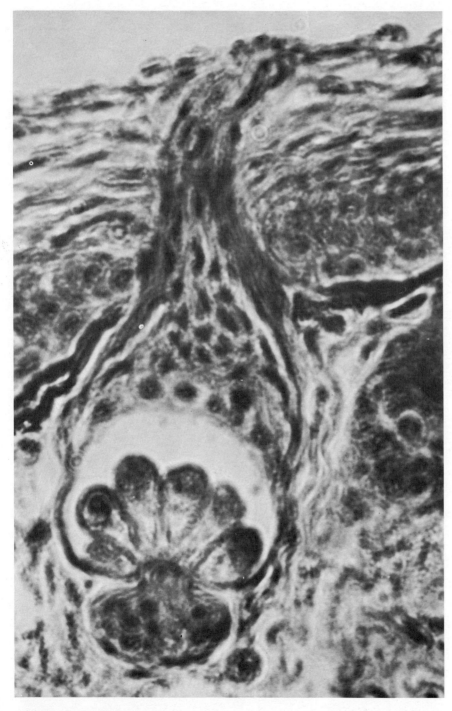

VERTICAL SECTION through skin and electric sense organ of a gymnotid shows tube containing jelly-like substance widening at base into a capsule, known as multicellular gland, that holds a group of special cells. Enlargement of this photomicrograph is 1,000 diameters.

other fishes that apparently have no electric organs.

Although electric fishes cannot be seen in their natural habitat, it is still possible to detect and follow them by picking up their discharges from the water. In South America I have found that the gymnotids are all active during the night. Darkness and the turbidity of the water offer good protection to these fishes, which rely on their eyes only for the knowledge that it is day or night. At night most of the predatory fishes, which have well-developed eyes, sleep on the bottom of rivers, ponds and lakes. Early in the morning, before the predators wake up, the gymnotids return from their nightly excursions and occupy inaccessible hiding places, where they often collect in vast numbers. In the rocks and vegetation along the shore the ticking, rattling, humming and whistling can be heard in bewildering profusion when the electrodes are connected to a loudspeaker. With a little practice one can begin to distinguish the various species by these sounds.

When one observes life in this highly competitive environment, it becomes clear what advantages the electric sense confers on these fishes and why they have evolved their curiously specialized sense organs, skin, brain, electric organs and peculiar mode of swimming. Such well-established specialists must have originated, however, from ordinary fishes in which the characteristics of the specialists are found in their primitive state: the electric organs as locomotive muscles and the sense organs as mechanoreceptors along the lateral line of the body that signal displacement of water. Somewhere there must be intermediate forms in which the contraction of a muscle, with its accompanying change in electric potential, interacts with these sense organs. For survival it may be important to be able to distinguish water movements caused by animate or inanimate objects. This may have started the evolutionary trend toward an electric sense.

Already we know some supposedly nonelectric fishes from which, nevertheless, we can pick up signals having many characteristics of the discharges of electric fishes. We know of sense organs that appear to be structurally intermediate between ordinary lateral-line receptors and electroreceptors. Furthermore, fishes that have both of these characteristics are also electrically very sensitive. We may hope one day to piece the whole evolutionary line together and express, at least in physical terms, what it is like to live in an electric world.

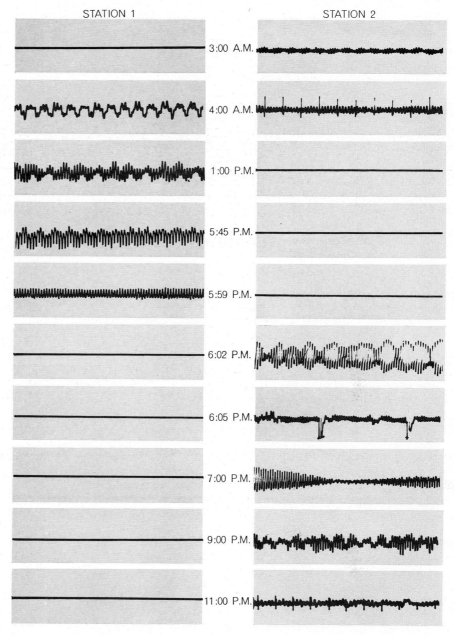

TRACKING ELECTRIC FISH in nature involves placing electrodes in water they inhabit. Records at left were made in South American stream near daytime hiding place of gymnotids, those at right out in main channel of stream, where they seek food at night.

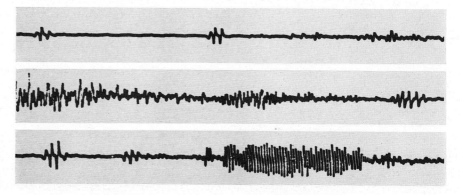

AFRICAN CATFISH, supposedly nonelectric, produced the discharges shown here. Normal action potentials of muscles are seen, along with odd regular blips and still other oscillations of higher frequency. Such fish may be evolving an electric sense or may already have one.

22

MOTHS AND ULTRASOUND

KENNETH D. ROEDER   April 1965

If an animal is to survive, it must be able to perceive and react to predators or prey. What nerve mechanisms are used when one animal reacts to the presence of another? Those animals that have a central nervous system perceive the outer world through an array of sense organs connected with the brain by many thousands of nerve fibers. Their reactions are expressed as critically timed sequences of nerve impulses traveling along motor nerve fibers to specific muscles. Exactly how the nervous system converts a particular pattern of sensory input into a specific pattern of motor output remains a subject of investigation in many branches of zoology, physiology and psychology.

Even with the best available techniques one can simultaneously follow the traffic of nerve impulses in only five or perhaps 10 of the many thousands of separate nerve fibers connecting a mammalian sense organ with the brain. Trying to learn how information is encoded and reported among all the fibers by following the activity of so few is akin to basing a public opinion poll on one or two interviews. (Following the activity of all the fibers would of course be like sampling public opinion by having the members of the population give their different answers in chorus.) Advances in technique may eventually make it possible to follow the traffic in thousands of fibers; in the meantime much can be learned by studying animals with less profusely innervated sense organs.

With several colleagues and students at Tufts University I have for some time been trying to decode the sensory patterns connecting the ear and central nervous system of certain nocturnal moths that have only two sense cells in each ear. Much of the behavior of these simple invertebrates is built in, not learned, and therefore is quite stereotyped and stable under experimental conditions. Working with these moths offers another advantage: because they depend on their ears to detect their principal predators, insect-eating bats, we are able to discern in a few cells the nervous mechanisms on which the moth's survival depends.

Insectivorous bats are able to find their prey while flying in complete darkness by emitting a series of ultrasonic cries and locating the direction and distance of sources of echoes. So highly sophisticated is this sonar that it enables the bats to find and capture flying insects smaller than mosquitoes. Some night-flying moths—notably members of the families Noctuidae, Geometridae and Arctiidae—have ears that can detect the bats' ultrasonic cries. When they hear the approach of a bat, these moths take evasive action, abandoning their usual cruising flight to go into sharp dives or erratic loops or to fly at top speed directly away from the source of ultrasound. Asher E. Treat of the College of the City of New York has demonstrated that moths taking evasive action on a bat's approach have a significantly higher chance of survival than those that continue on course.

A moth's ears are located on the sides of the rear part of its thorax and are directed outward and backward into the constriction that separates the thorax and the abdomen [see top illustration on page 218]. Each ear is extremely visible as a small cavity, and within the cavity is a transparent eardrum. Behind the eardrum is the tympanic air sac; a fine strand of tissue containing the sensory apparatus extends across the air sac from the center of the eardrum to a skeletal support. Two acoustic cells, known as A cells, are located within this strand. Each A cell sends a fine sensory strand outward to the eardrum and a nerve fiber inward to the skeletal support. The two A fibers pass close to a large nonacoustic cell, the B cell, and are joined by its nerve fiber. The three fibers continue as the tympanic nerve into the central nervous system of the moth. From the two A fibers, then, it is possible—and well within our technical means—to obtain all the information about ultrasound that is transmitted from the moth's ear to its central nervous system.

Nerve impulses in single nerve fibers can be detected as "action potentials," or self-propagating electrical transients, that have a magnitude of a few millivolts and at any one point on the fiber last less than a millisecond. In the moth's A fibers action potentials travel from the sense cells to the central nervous system in less than two milliseconds. Action potentials are normally an all-or-nothing phenomenon; once initiated by the sense cell, they travel to the end of the nerve fiber. They can be detected on the outside of the fiber by means of fine electrodes, and they are displayed as "spikes" on the screen of an oscilloscope.

Tympanic-nerve signals are demonstrated in the following way. A moth, for example the adult insect of one of the common cutworms or armyworms, is immobilized on the stage of a microscope. Some of its muscles are dissected away to expose the tympanic nerves at a point outside the central nervous system. Fine silver hooks are placed under one or both nerves, and the pattern of passing action potentials is observed on the oscilloscope. With moths thus prepared we have spent much time in impromptu outdoor laboratories, where the cries of passing bats provided the necessary stimuli.

In order to make precise measurements we needed a controllable source of ultrasonic pulses for purposes of comparison. Such pulses can be generated by electronic gear to approximate natural bat cries in frequency and duration. The natural cries are frequency-modulated: their frequency drops from about 70 kilocycles per second at the beginning of each cry to some 35 kilocycles at the end. Their duration ranges from one to 10 milliseconds, and they are repeated from 10 to 100 times a second. Our artificial stimulus is a facsimile of an average series of bat cries; it is not frequency-modulated, but such modulation is not detected by the moth's ear. Our sound pulses can be accurately graded in intensity by decibel steps; in the sonic range a decibel is roughly equivalent to the barely noticeable difference to human ears in the intensity of two sounds.

By using electronic apparatus to elicit and follow the responses of the A cells we have been able to define the amount of acoustic information avail-

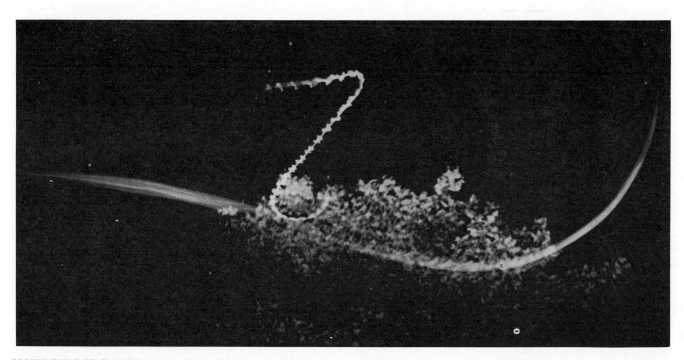

MOTH EVADED BAT by soaring upward just as the bat closed in to capture it. The bat entered the field at right; the path of its flight is the broad white streak across the photograph. The smaller white streak shows the flight of the moth. A tree is in background. The shutter of the camera was left open as contest began. Illumination came from continuous light source below field.

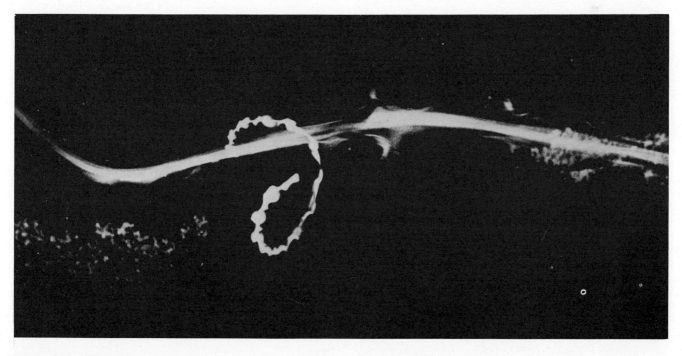

BAT CAPTURED MOTH at point where two white streaks intersect. Small streak shows the flight pattern of the moth. Broad streak shows the flight path of the bat. Both streak photographs were made by Frederic Webster of the Sensory Systems Laboratories.

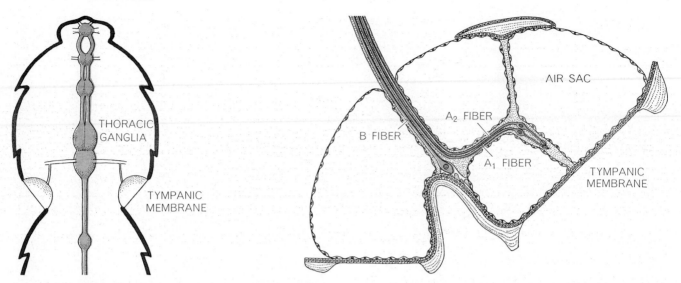

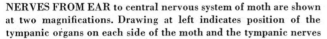

NERVES FROM EAR to central nervous system of moth are shown at two magnifications. Drawing at left indicates position of the tympanic organs on each side of the moth and the tympanic nerves connecting them with the thoracic ganglia. Central nervous system is colored. Drawing at right shows two nerve fibers of the acoustic cells joined by a nonacoustic fiber to form the tympanic nerve.

able to the moth by way of its tympanic nerve. It appears that the tympanic organ is not particularly sensitive; to elicit any response from the A cell requires ultrasound roughly 100 times more intense than sound that can just be heard by human ears. The ear of a moth can nonetheless pick up at distances of more than 100 feet ultrasonic bat cries we cannot hear at all. The reason it cannot detect frequency modulation is simply that it cannot discriminate one frequency from another; it is tone-deaf. It can, however, detect frequencies from 10 kilocycles to well over 100 kilocycles per second, which covers the range of bat cries. Its greatest talents are the detection of pulsed sound—short bursts of sound with intervening silence—and the discrimination of differences in the loudness of sound pulses.

When the ear of a moth is stimulated by the cry of a bat, real or artificial, spikes indicating the activity of the A cell appear on the oscilloscope in various configurations. As the stimulus increases in intensity several changes are apparent. First, the number of A spikes increases. Second, the time interval between the spikes decreases. Third, the spikes that had first appeared only on the record of one A fiber (the "A_1" fiber, which is about 20 decibels more sensitive than the A_2 fiber) now appear on the records of both fibers. Fourth, the greater the intensity of the stimulus, the sooner the A cell generates a spike in response.

The moth's ears transmit to the oscilloscope the same configuration of spikes they transmit normally to the central nervous system, and therein lies our interest. Which of the changes in auditory response to an increasingly in-

tense stimulus actually serve the moth as criteria for determining its behavior under natural conditions? Before we face up to this question let us speculate on the possible significance of these criteria from the viewpoint of the moth. For the moth to rely on the first kind of information—the number of A spikes—might lead it into a fatal error: the long, faint cry of a bat at a distance could be confused with the short, intense cry of a bat closing for the kill. This error could be avoided if the moth used the second kind of information—the interval between spikes—for estimating the loudness of the bat's cry. The third kind of information—the activity of the A_2 fiber—might serve to change an "early warning" message to a "take cover" message. The fourth kind of information—the length of time it takes for a spike to be generated—might provide the moth with

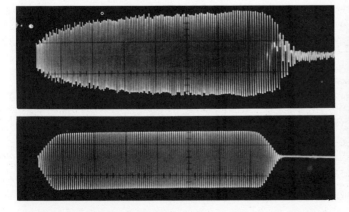

OSCILLOSCOPE TRACES of a real bat cry (top) and a pulse of sound generated electronically (bottom) are compared. The two ultrasonic pulses are of equal duration (length), 2.5 milliseconds, but differ in that the artificial pulse has a uniform frequency.

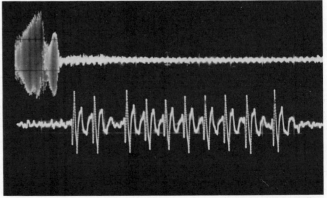

BAT CRY AND MOTH RESPONSE were traced on same oscilloscope from tape recording by Webster. The bat cry, detected by microphone, yielded the pattern at left in top trace. Reaction of the moth's acoustic cells produced the row of spikes at bottom.

the means for locating a cruising bat; for example, if the sound was louder in the moth's left ear than in its right, then *A* spikes would reach the left side of the central nervous system a fraction of a millisecond sooner than the right side.

Speculations of this sort are profitable only if they suggest experiments to prove or disprove them. Our tympanic-nerve studies led to field experiments designed to find out what moths do when they are exposed to batlike sounds from a loudspeaker. In the first such study moths were tracked by streak photography, a technique in which the shutter of a camera is left open as the subject passes by. As free-flying moths approached the area on which our camera was trained they were exposed to a series of ultrasonic pulses.

More than 1,000 tracks were recorded in this way. The moths were of many species; since they were free and going about their natural affairs most of them could not be captured and identified. This was an unavoidable disadvantage; earlier observations of moths captured, identified and then released in an enclosure revealed nothing. The moths were apparently "flying scared" from the beginning, and the ultrasound did not affect their behavior. Hence all comers were tracked in the field.

Because moths of some families lack ears, a certain percentage of the moths failed to react to the loudspeaker. The variety of maneuvers among the moths that did react was quite unpredictable and bewildering [*see illustrations at top of next page*]. Since the evasive behavior presumably evolved for the purpose of bewildering bats, it is hardly surprising that another mammal should find it confusing! The moths that flew close to the loudspeaker and encountered high-intensity ultrasound would maneuver toward the ground either by dropping passively with their wings closed, by power dives, by vertical and horizontal turns and loops or by various combinations of these evasive movements.

One important finding of this field work was that moths cruising at some distance from the loudspeaker would turn and fly at high speed directly away from it. This happened only if the sound the moths encountered was of low intensity. Moths closer to the loudspeaker could be induced to flee only if the signal was made weaker. Moths at about the height of the loudspeaker flew away in the horizontal plane; those above the loudspeaker were observed to turn directly upward

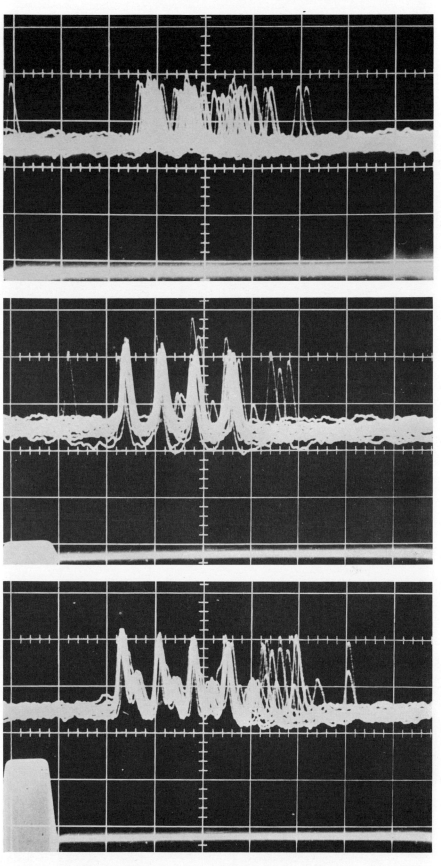

CHANGES ARE REPORTED by moth's tympanic nerve to the oscilloscope as pulses used to simulate bat cries gain intensity. Pulses (*lower trace in each frame*) were at five decibels (*top frame*), 20 (*middle*) and 35 (*bottom*). An increased number of tall spikes appear as intensity of stimulus rises. The time interval between spikes decreases slightly. Smaller spikes from the less sensitive nerve fiber appear at the higher intensities, and the higher the intensity of the stimulus, the sooner (*left on horizontal axis*) the first spike appears.

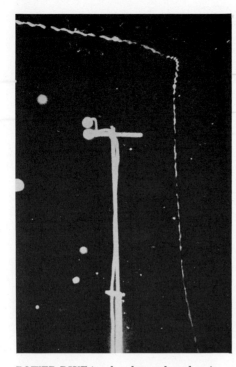

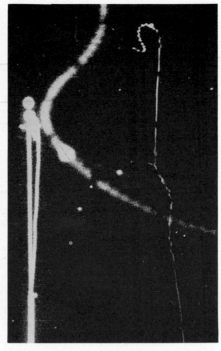

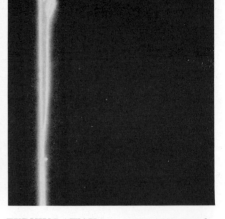

POWER DIVE is taken by moth on hearing simulated bat cry from loudspeaker mounted on thin tower (*left of moth's flight path*).

PASSIVE DROP was executed by another moth, which simply folded its wings. Blur at left and dots were made by other insects.

TURNING AWAY, an evasive action involving directional change, is illustrated. These streak photographs were made by author.

or at other sharp angles. To make such directional responses with only four sensory cells is quite a feat. A horizontal response could be explained on the basis that one ear of the moth detected the sound a bit earlier than the other. It is harder to account for a vertical response, although experiments I shall describe provide a hint.

Our second series of field experiments was conducted in another outdoor laboratory—my backyard. They were designed to determine which of the criteria of intensity encoded in the pattern of A-fiber spikes play an important part in determining evasive behavior. The percentage of moths showing "no re-

action," "diving," "looping" and "turning away" was noted when a 50-kilocycle signal was pulsed at different rates and when it was produced as a continuous tone. The continuous tone delivers more A impulses in a given fraction of a second and therefore should be a more effective stimulus if the number of A impulses is important. On the other hand, because the A cells, like many other sensory cells, become progressively less sensitive with continued stimulation, the interspike interval lengthens rapidly as continuous-tone stimulation proceeds. When the sound is pulsed, the interspike interval remains short because the A cells have had time to regain their sensitivity during the

brief "off" periods. If the spike-generation time—which is associated with difference in the time at which the A spike arrives at the nerve centers for each ear—plays an important part in evasive behavior, then continuous tones should be less effective. The difference in arrival time would be detected only once at the beginning of the stimulus; with pulsed sound it would be reiterated with each pulse.

The second series of experiments occupied many lovely but mosquito-ridden summer nights in my garden and provided many thousands of observations. Tabulation of the figures showed that continuous ultrasonic tones were much less effective in producing evasive

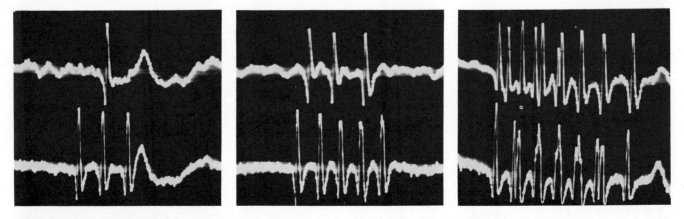

RESPONSE BY BOTH EARS of a moth to an approaching bat was recorded on the oscilloscope and photographed by the author. In trace at left the tympanic nerve from one ear transmits only one

spike (*upper curve*) while the nerve from the other ear sends three. As the bat advances, the ratio becomes three to five (*middle*), then 10 to 10 (*right*), suggesting that the bat has flown overhead.

behavior than pulses. The number of nonreacting moths increased threefold, diving occurred only at higher sound intensities and turning away was essentially absent. Only looping seemed to increase slightly.

Ultrasound pulsed between 10 and 30 times a second proved to be more effective than ultrasound pulsed at higher or lower rates. This suggests that diving, and possibly other forms of nondirectional evasive behavior, are triggered in the moth's central nervous system not so much by the number of A impulses delivered over a given period as by short intervals (less than 2.5 milliseconds) between consecutive A impulses. Turning away from the sound source when it is operating at low intensity levels seems to be set off by the reiterated difference in arrival time of the first A impulse in the right and left tympanic nerves.

These conclusions were broad but left unanswered the question: How can a moth equipped only with four A cells orient itself with respect to a sound source in planes that are both vertical and horizontal to its body axis? The search for an answer was undertaken by Roger Payne of Tufts University, assisted by Joshua Wallman, a Harvard undergraduate. They set out to plot the directional capacities of the tympanic organ by moving a loudspeaker at various angles with respect to a captive moth's body axis and registering (through the A_1 fiber) the organ's relative sensitivity to ultrasonic pulses coming from various directions. They took precautions to control acoustic shadows and reflections by mounting the moth and the recording electrodes on a thin steel tower in the center of an echo-free chamber; the effect of the moth's wings on the reception of sound was tested by systematically changing their position during the course of many experiments. A small loudspeaker emitted ultrasonic pulses 10 times a second at a distance of one meter. These sounds were presented to the moths from 36 latitude lines 10 degrees apart.

The response of the A fibers to the ultrasonic pulses was continuously recorded as the loudspeaker was moved. At the same time the intensity of ultrasound emitted by the loudspeaker was regulated so that at any angle it gave rise to the same response. Thus the intensity of the sound pulses was a measure of the moth's acoustic sensitivity. A pen recorder continuously graphed the changing intensity of the ultrasonic pulses against the angle from which

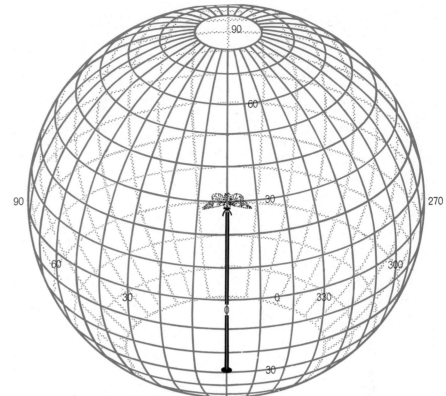

SPHERE OF SENSITIVITY, the range in which a moth with wings in a given position can hear ultrasound coming from various angles, was the subject of a study by Roger Payne of Tufts University and Joshua Wallman, a Harvard undergraduate. Moths with wings in given positions were mounted on a tower in an echo-free chamber. Data were compiled on the moths' sensitivity to ultrasound presented from 36 latitude lines 10 degrees apart.

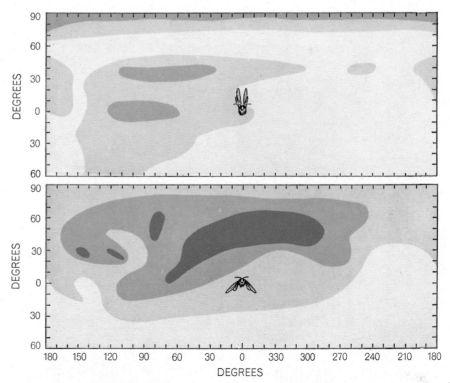

MERCATORIAL PROJECTIONS represent auditory environment of a moth with wings at end of upstroke (top) and near end of downstroke (bottom). Vertical scale shows rotation of loudspeaker around moth's body in vertical plane; horizontal scale shows rotation in horizontal plane. At top the loudspeaker is above moth; at far right and left, behind it. In Mercatorial projections, distortions are greatest at poles. The lighter the shading at a given angle of incidence, the more sensitive the moth to sound from that angle.

they were presented to the moth. Each chart provided a profile of sensitivity in a certain plane, and the data from it were assembled with those from others to provide a "sphere of sensitivity" for the moth at a given wing position.

This ingenious method made it possible to assemble a large amount of data in a short time. In the case of one moth it was possible to obtain the data for nine spheres of sensitivity (about 5,000 readings), each at a different wing position, before the tympanic nerve of the moth finally stopped transmitting impulses. Two of these spheres, taken from one moth at different wing positions, are presented as Mercatorial projections in the bottom illustration on the preceding page.

It is likely that much of the information contained in the fine detail of such

projections is disregarded by a moth flapping its way through the night. Certain general patterns do seem related, however, to the moth's ability to escape a marauding bat. For instance, when the moth's wings are in the upper half of their beat, its acoustic sensitivity is 100 times less at a given point on its side facing away from the source of the sound than at the corresponding point on the side facing toward the source. When flight movements bring the wings below the horizontal plane, sound coming from each side above the moth is in acoustic shadow, and the left-right acoustic asymmetry largely disappears. Moths commonly flap their wings from 30 to 40 times a second. Therefore left-right acoustic asymmetry must alternate with up-down asymmetry at this frequency. A left-right difference in the

A-fiber discharge when the wings are up might give the moth a rough horizontal bearing on the position of a bat with respect to its own line of flight. The absence of a left-right difference and the presence of a similar fluctuation in both left and right tympanic nerves at wingbeat frequency might inform the moth that the bat was above it. If neither variation occurred at the regular wingbeat frequency, it would mean that the bat was below or behind the moth.

This analysis uses terms of precise directionality that idealize the natural situation. A moth certainly does not zoom along on an even keel and a straight course like an airliner. Its flapping progress—even when no threat is imminent—is marked by minor yawing and pitching; its overall course is rare-

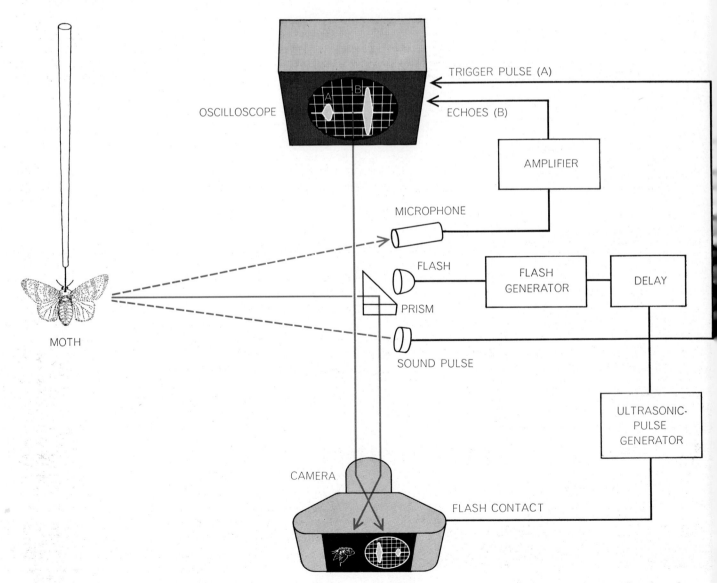

ARTIFICIAL BAT, the electronic device depicted schematically at right, was built by the author to determine at what position with respect to a bat a moth casts its greatest echo. As a moth supported by a wire flapped its wings in stationary flight, a film was made by means of a prism of its motions and of an oscilloscope that showed the pulse generated by the loudspeaker and the echo picked up by the microphone. Each frame of film thus resembled the composite picture of moth and two pulses shown inverted at bottom.

ly straight and commonly consists of large loops and figure eights. Even so, the localization experiments of Payne and Wallman suggest the ways in which a moth receives information that enables it to orient itself in three dimensions with respect to the source of an ultrasonic pulse.

The ability of a moth to perceive and react to a bat is not greatly superior or inferior to the ability of a bat to perceive and react to a moth. Proof of this lies in the evolutionary equality of their natural contest and in the observation of a number of bat-moth confrontations. Donald R. Griffin of Harvard University and Frederic Webster of the Sensory Systems Laboratories have studied in detail the almost unbelievable ability of bats to locate, track and intercept small flying targets, all on the basis of a string of echoes thrown back from ultrasonic cries. Speaking acoustically, what does a moth "look like" to a bat? Does the prey cast different echoes under different circumstances?

To answer this question I set up a crude artificial bat to pick up echoes from a live moth. The moth was attached to a wire support and induced to flap its wings in stationary flight. A movie camera was pointed at a prism so that half of each frame of film showed an image of the moth and the other half the screen of an oscilloscope. Mounted closely around the prism and directed at the moth from one meter away were a stroboscopic-flash lamp, an ultrasonic loudspeaker and a microphone. Each time the camera shutter opened and exposed a frame of film a short ultrasonic pulse was sent out by the loudspeaker and the oscilloscope began its sweep. The flash lamp was controlled through a delay circuit to go off the instant the ultrasonic pulse hit the moth, whose visible attitude was thereby frozen on the film. Meanwhile the echo thrown back by the moth while it was in this attitude was picked up by the microphone and finally displayed as a pulse of a certain height on the oscilloscope. All this took place before the camera shutter closed and the film moved on to the next frame. Thus each frame shows the optical and acoustic profiles of the moth from approximately the same angle and at the same instant of its flight. The camera was run at speeds close to the wingbeat frequency of the moth, so that the resulting film presents a regular series of wing positions and the echoes cast by them.

Films made of the same moth flying at different angles to the camera and the sound source show that by far the strongest echo is returned when the moth's wings are at right angles to the recording array [see illustrations at left]. The echo from a moth with its wings in this position is perhaps 100 times stronger than one from a moth with its wings at other angles. Apparently if a bat and a moth were flying horizontal courses at the same altitude, the moth would be in greatest danger of detection if it crossed the path of the approaching bat at right angles. From the bat's viewpoint at this instant the moth must appear to flicker acoustically at its wingbeat frequency. Since the rate at which the bat emits its ultrasonic cries is independent of the moth's wingbeat frequency, the actual sequence of echoes the bat receives must be complicated by the interaction of the two frequencies. Perhaps this enables the bat to discriminate a flapping target, likely to be prey, from inert objects floating in its acoustic field.

The moth has one advantage over the bat: it can detect the bat at a greater range than the bat can detect it. The bat, however, has the advantage of greater speed. This creates a nice problem for a moth that has picked up a bat's cries. If a moth immediately turns and flies directly away from a source of ultrasound, it has a good chance of disappearing from the sonar system of a still-distant bat. If the bat has also detected the moth, and is near enough to receive a continuous signal from its target, turning away on a straight course is a bad tactic because the moth is not likely to outdistance its pursuer. It is then to the moth's advantage to

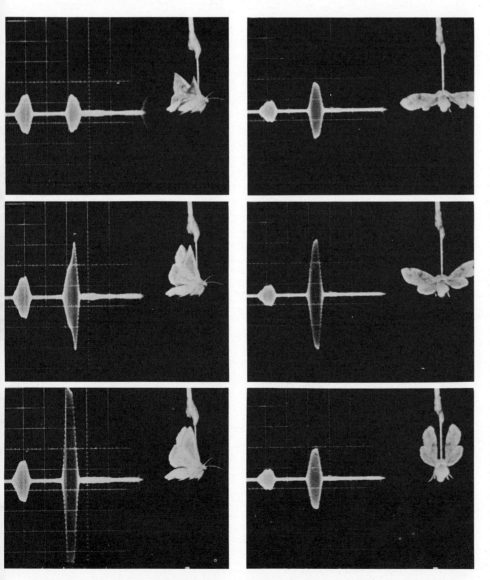

COMPOSITE PHOTOGRAPHS each show an artificial bat's cry (left) and the echo thrown back (middle) by a moth (right). The series of photographs at left is of a moth in stationary flight at right angles to the artificial bat. Those at right are of a moth oriented in flight parallel to the bat. The echo produced in the series of photographs at left is much the larger.

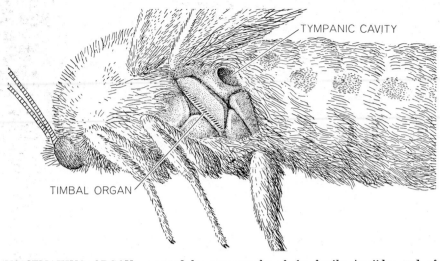

TYMPANIC CAVITY

TIMBAL ORGAN

NOISEMAKING ORGAN possessed by many moths of the family Arctiidae and of other families is a row of fine parallel ridges of cuticle that bend and unbend when a leg muscle contracts and relaxes. This produces a rapid sequence of high-pitched clicks.

go into tight turns, loops and dives, some of which may even take it toward the bat.

In this contest of hide-and-seek it seems much to a moth's advantage to remain as quiet as possible. The sensitive ears of a bat would soon locate a noisy target. It is therefore surprising to find that many members of the moth family Arctiidae (which includes the moths whose caterpillars are known as woolly bears) are capable of generating trains of ultrasonic clicks. David Blest and David Pye of University College London have demonstrated the working of the organ that arctiids use for this purpose.

In noisemaking arctiids the basal joint of the third pair of legs (which roughly corresponds to the hip) bulges outward and overlies an air-filled cavi-

ty. The stiff cuticle of this region has a series of fine parallel ridges [see illustration above]. Each ridge serves as a timbal that works rather like the familiar toy incorporating a thin strip of spring steel that clicks when it is pressed by the thumb. When one of the moth's leg muscles contracts and relaxes in rapid sequence, it bends and unbends the overlying cuticle, causing the row of timbals to produce rapid sequences of high-pitched clicks. Blest and Pye found that such moths would click when they were handled or poked, that the clicks occurred in short bursts of 1,000 or more per second and that each click contained ultrasonic frequencies within the range of hearing of bats.

My colleagues and I found that certain arctiids common in New England could also be induced to click if they were exposed to a string of ultrasonic

pulses while they were suspended in stationary flight. In free flight these moths showed the evasive tactics I have already described. The clicking seems almost equivalent to telling the bat, "Here I am, come and get me." Since such altruism is not characteristic of the relation between predators and prey, there must be another answer.

Dorothy C. Dunning, a graduate student at Tufts, is at present trying to find it. She has already shown that partly tamed bats, trained to catch mealworms that are tossed into the air by a mechanical device, will commonly swerve away from their target if they hear tape-recorded arctiid clicks just before the moment of contact. Other ultrasounds, such as tape-recorded bat cries and "white" noise (noise of all frequencies), have relatively little effect on the bats' feeding behavior; the tossed mealworms are caught in midair and eaten. Thus the clicks made by arctiids seem to be heeded by bats as a warning rather than as an invitation. But a warning against what?

One of the pleasant things about scientific investigation is that the last logbook entry always ends with a question. In fact, the questions proliferate more rapidly than the answers and often carry one along unexpected paths. I suggested at the beginning of this article that it is my intention to trace the nervous mechanisms involved in the evasive behavior of moths. By defining the information conveyed by the acoustic cells I have only solved the least complex half of that broad problem. As I embark on the second half of the investigation, I hope it will lead up as many diverting side alleys as the study of the moth's acoustic system has.

23

THE REPRODUCTIVE BEHAVIOR OF RING DOVES

by DANIEL S. LEHRMAN   November 1964

In recent years the study of animal behavior has proceeded along two different lines, with two groups of investigators formulating problems in different ways and indeed approaching the problems from different points of view. The comparative psychologist traditionally tends first to ask a question and then to attack it by way of animal experimentation. The ethologist, on the other hand, usually begins by observing the normal activity of an animal and then seeks to identify and analyze specific behavior patterns characteristic of the species.

The two attitudes can be combined. The psychologist can begin, like the ethologist, by watching an animal do what it does naturally, and only then ask questions that flow from his observations. He can go on to manipulate experimental conditions in an effort to discover the psychological and biological events that give rise to the behavior under study and perhaps to that of other animals as well. At the Institute of Animal Behavior at Rutgers University we have taken this approach to study in detail the reproductive-behavior cycle of the ring dove (*Streptopelia risoria*). The highly specific changes in behavior that occur in the course of the cycle, we find, are governed by complex psycho-

REPRODUCTIVE-BEHAVIOR CYCLE begins soon after a male and a female ring dove are introduced into a cage containing nesting material (hay in this case) and an empty glass nest bowl (*1*). Courtship activity, on the first day, is characterized by the "bowing

CYCLE CONTINUES as the adult birds take turns incubating the eggs (*6*), which hatch after about 14 days (*7*). The newly hatched squabs are fed "crop-milk," a liquid secreted in the gullets of the adults (*8*). The parents continue to feed them, albeit reluctantly,

biological interactions of the birds' inner and outer environments.

The ring dove, a small relative of the domestic pigeon, has a light gray back, creamy underparts and a black semicircle (the "ring") around the back of its neck. The male and female look alike and can only be distinguished by surgical exploration. If we place a male and a female ring dove with previous breeding experience in a cage containing an empty glass bowl and a supply of nesting material, the birds invariably enter on their normal behavioral cycle, which follows a predictable course and a fairly regular time schedule. During the first day the principal activity is courtship: the male struts around, bowing and cooing at the female. After several hours the birds announce their selection of a nest site (which in nature would be a concave place and in our cages is the glass bowl) by crouching in it and uttering a distinctive coo. Both birds participate in building the nest, the male usually gathering material and carrying it to the female, who stands in the bowl and constructs the nest. After a week or more of nest-building, in the course of which the birds copulate, the female be-

comes noticeably more attached to the nest and difficult to dislodge; if one attempts to lift her off the nest, she may grasp it with her claws and take it along. This behavior usually indicates that the female is about to lay her eggs. Between seven and 11 days after the beginning of the courtship she produces her first egg, usually at about five o'clock in the afternoon. The female dove sits on the egg and then lays a second one, usually at about nine o'clock in the morning two days later. Sometime that day the male takes a turn sitting; thereafter the two birds alternate, the male sitting for about six hours in the middle of each day, the female for the remaining 18 hours a day.

In about 14 days the eggs hatch and the parents begin to feed their young "crop-milk," a liquid secreted at this stage of the cycle by the lining of the adult dove's crop, a pouch in the bird's gullet. When they are 10 or 12 days old, the squabs leave the cage, but they continue to beg for and to receive food from the parents. This continues until the squabs are about two weeks old, when the parents become less and less willing to feed them as the young birds

gradually develop the ability to peck for grain on the floor of the cage. When the young are about 15 to 25 days old, the adult male begins once again to bow and coo; nest-building is resumed, a new clutch of eggs is laid and the cycle is repeated. The entire cycle lasts about six or seven weeks and—at least in our laboratory, where it is always spring because of controlled light and temperature conditions—it can continue throughout the year.

The variations in behavior that constitute the cycle are not merely casual or superficial changes in the birds' preoccupations; they represent striking changes in the overall pattern of activity and in the atmosphere of the breeding cage. At its appropriate stage each of the kinds of behavior I have described represents the predominant activity of the animals at the time. Furthermore, these changes in behavior are not just responses to changes in the external situation. The birds do not build the nest merely because the nesting material is available; even if nesting material is in the cage throughout the cycle, nest-building behavior is concentrated,

coo" of the male (2). The male and then the female utter a distinctive "nest call" to indicate their selection of a nesting site (3).

There follows a week or more of cooperation in nest-building (4), culminating in the laying of two eggs at precise times of day (5).

as the young birds learn to peck for grain themselves (9). When the squabs are between two and three weeks old, the adults ignore

them and start to court once again, and a new cycle begins (10). Physical changes during the cycle are shown on the next page.

as described, at one stage. Similarly, the birds react to the eggs and to the young only at appropriate stages in the cycle.

These cyclic changes in behavior therefore represent, at least in part, changes in the internal condition of the animals rather than merely changes in their external situation. Furthermore, the changes in behavior are associated with equally striking and equally pervasive changes in the anatomy and the physiological state of the birds. For example, when the female dove is first introduced into the cage, her oviduct weighs some 800 milligrams. Eight or nine days later, when she lays her first egg, the oviduct may weigh 4,000 milligrams. The crops of both the male and the female weigh some 900 milligrams when the birds are placed in the cage, and when they start to sit on the eggs some 10 days later they still weigh about the same. But two weeks afterward, when the eggs hatch, the parents' crops may weigh as much as 3,000 milligrams. Equally striking changes in the condition of the ovary, the weight of the testes, the length of the gut, the weight of the liver, the microscopic structure of the pituitary gland and other physiological indices are correlated with the behavioral cycle.

Now, if a male or a female dove is placed alone in a cage with nesting material, no such cycle of behavioral or anatomical changes takes place. Far from producing two eggs every six or seven weeks, a female alone in a cage lays no eggs at all. A male alone shows no interest when we offer it nesting material, eggs or young. The cycle of psychobiological changes I have described is, then, one that occurs more or less synchronously in each member of a pair of doves living together but that will not occur independently in either of the pair living alone.

In a normal breeding cycle both the male and the female sit on the eggs almost immediately after they are laid. The first question we asked ourselves was whether this is because the birds are always ready to sit on eggs or because they come into some special condition of readiness to incubate at about the time the eggs are produced.

We kept male and female doves in isolation for several weeks and then placed male-female pairs in test cages, each supplied with a nest bowl containing a normal dove nest with two eggs. The birds did not sit; they acted almost as if the eggs were not there. They courted, then built their own nest (usually on top of the planted nest and its eggs, which we had to keep fishing out to keep the stimulus situation constant!), then finally sat on the eggs—five to seven days after they had first encountered each other.

This clearly indicated that the doves are not always ready to sit on eggs; under the experimental conditions they changed from birds that did not want to incubate to birds that did want to incubate in five to seven days. What had induced this change? It could not have been merely the passage of time since their last breeding experience, because this had varied from four to six or more weeks in different pairs, whereas the variation in time spent in the test cage before sitting was only a couple of days.

Could the delay of five to seven days represent the time required for the birds to get over the stress of being handled and become accustomed to the strange cage? To test this possibility we placed pairs of doves in cages without any nest bowls or nesting material and separated each male and female by an opaque partition. After seven days we removed the partition and introduced nesting material and a formed nest with eggs. If the birds had merely needed time to recover from being handled and become acclimated to the cage, they should now have sat on the eggs immediately. They did not do so; they sat only after five to seven days, just as if they had been introduced into the cage only when the opaque partition was removed.

The next possibility we considered was that in this artificial situation stimulation from the eggs might induce the change from a nonsitting to a sitting "mood" but that this effect required five to seven days to reach a threshold value at which the behavior would change.

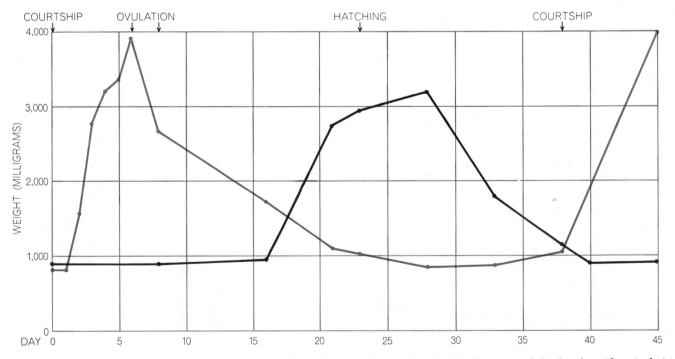

ANATOMICAL AND PHYSIOLOGICAL changes are associated with the behavioral changes of the cycle. The chart gives average weights of the crop (*black curve*) and the female oviduct (*color*) at various stages measured in days after the beginning of courtship.

We therefore placed pairs of birds in test cages with empty nest bowls and a supply of nesting material but no eggs. The birds courted and built nests. After seven days we removed the nest bowl and its nest and replaced it with a fresh bowl containing a nest and eggs. All these birds sat within two hours.

It was now apparent that some combination of influences arising from the presence of the mate and the availability of the nest bowl and nesting material induced the change from nonreadiness to incubate to readiness. In order to distinguish between these influences we put a new group of pairs of doves in test cages without any nest bowl or nesting material. When, seven days later, we offered these birds nesting material and nests with eggs, most of them did not sit immediately. Nor did they wait the full five to seven days to do so; they sat after one day, during which they engaged in intensive nest-building. A final group, placed singly in cages with nests and eggs, failed to incubate at all, even after weeks in the cages.

In summary, the doves do not build nests as soon as they are introduced into a cage containing nesting material, but they will do so immediately if the nesting material is introduced for the first time after they have spent a while together; they will not sit immediately on eggs offered after the birds have been in a bare cage together for some days, but they will do so if they were able to do some nest-building during the end of their period together. From these experiments it is apparent that there are two kinds of change induced in these birds: first, they are changed from birds primarily interested in courtship to birds primarily interested in nest-building, and this change is brought about by stimulation arising from association with a mate; second, under these conditions they are further changed from birds primarily interested in nest-building to birds interested in sitting on eggs, and this change is encouraged by participation in nest-building.

The course of development of readiness to incubate is shown graphically by the results of another experiment, which Philip N. Brody, Rochelle Wortis and I undertook shortly after the ones just described. We placed pairs of birds in test cages for varying numbers of days, in some cases with and in others without a nest bowl and nesting material. Then we introduced a nest and eggs into the cage. If neither bird sat within three hours, the test was scored as nega-

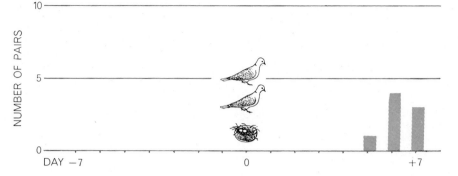

READINESS TO INCUBATE was tested with four groups of eight pairs of doves. Birds of the first group were placed in a cage containing a nest and eggs. They went through courtship and nest-building behavior before finally sitting after between five and seven days.

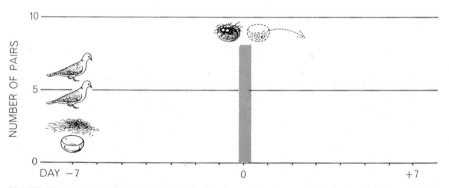

EFFECT OF HABITUATION was tested by keeping two birds separated for seven days in the cage before introducing nest and eggs. They still sat only after five to seven days.

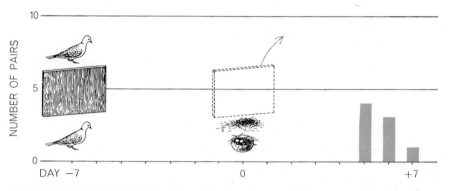

MATE AND NESTING MATERIAL had a dramatic effect on incubation-readiness. Pairs that had spent seven days in courtship and nest-building sat as soon as eggs were offered.

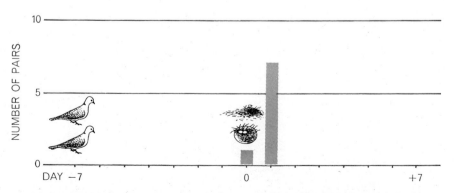

PRESENCE OF MATE without nesting activity had less effect. Birds that spent a week in cages with no nest bowls or hay took a day to sit after nests with eggs were introduced.

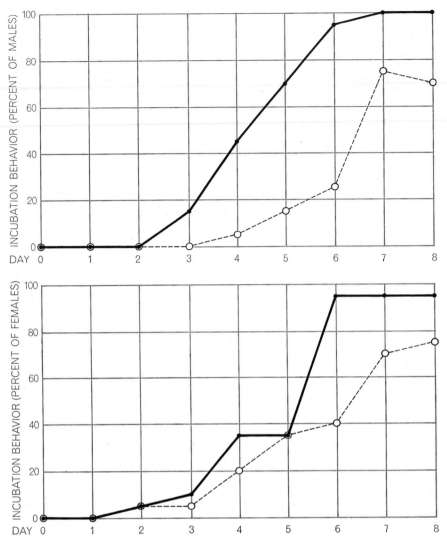

DURATION OF ASSOCIATION with mate and nesting material affects incubation behavior. The abscissas give the length of the association for different groups of birds. The plotted points show what percentage of each group sat within three hours of being offered eggs. The percentage increases for males (*top*) and females (*bottom*) as a function of time previously spent with mate (*open circles*) or with mate and nesting material (*solid dots*).

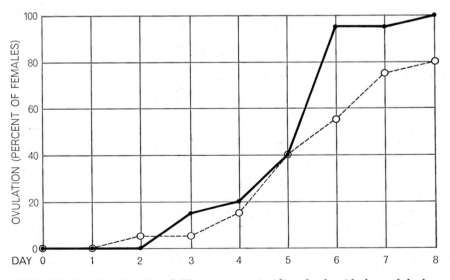

OVULATION is similarly affected. These curves, coinciding closely with those of the bottom chart above, show the occurrence of ovulation in the same birds represented there.

tive and both birds were removed for autopsy. If either bird sat within three hours, that bird was removed and the other bird was given an additional three hours to sit. The experiment therefore tested—independently for the male and the female—the development of readiness to incubate as a function of the number of days spent with the mate, with or without the opportunity to build a nest.

It is apparent [*see top illustration at left*] that association with the mate gradually brings the birds into a condition of readiness to incubate and that this effect is greatly enhanced by the presence of nesting material. Exposure to the nesting situation does not stimulate the onset of readiness to incubate in an all-or-nothing way; rather, its effect is additive with the effect of stimulation provided by the mate. Other experiments show, moreover, that the stimulation from the mate and nesting material is sustained. If either is removed, the incidence of incubation behavior decreases.

The experiments described so far made it clear that external stimuli normally associated with the breeding situation play an important role in inducing a state of readiness to incubate. We next asked what this state consists of physiologically. As a first approach to this problem we attempted to induce incubation behavior by injecting hormones into the birds instead of by manipulating the external stimulation. We treated birds just as we had in the first experiment but injected some of the birds with hormones while they were in isolation, starting one week before they were due to be placed in pairs in the test cages. When both members of the pair had been injected with the ovarian hormone progesterone, more than 90 percent of the eggs were covered by one of the birds within three hours after their introduction into the cage instead of five to seven days later. When the injected substance was another ovarian hormone—estrogen—the effect on most birds was to make them incubate after a latent period of one to three days, during which they engaged in nest-building behavior. The male hormone testosterone had no effect on incubation behavior.

During the 14 days when the doves are sitting on the eggs, their crops increase enormously in weight. Crop growth is a reliable indicator of the secretion of the hormone prolactin by the birds' pituitary glands. Since this growth coincides with the development of incubation behavior and culminates in the secretion of the crop-milk the birds feed to their young after the eggs hatch, Brody and I have recently examined the effect of injected prolactin on incubation behavior. We find that prolactin is not so effective as progesterone in inducing incubation behavior, even at dosage levels that induce full development of the crop. For example, a total prolactin dose of 400 international units induced only 40 percent of the birds to sit on eggs early, even though their average crop weight was about 3,000 milligrams, or more than three times the normal weight. Injection of 10 units of the hormone induced significant increases in crop weight (to 1,200 milligrams) but no increase in the frequency of incubation behavior. These results, together with the fact that in a normal breeding cycle the crop begins to increase in weight only after incubation begins, make it unlikely that prolactin plays an important role in the initiation of normal incubation behavior in this species. It does, however, seem to help to maintain such behavior until the eggs hatch.

Prolactin is much more effective in inducing ring doves to show regurgitation-feeding responses to squabs. When 12 adult doves with previous breeding experience were each injected with 450 units of prolactin over a seven-day period and placed, one bird at a time, in cages with squabs, 10 of the 12 fed the squabs from their engorged crops, whereas none of 12 uninjected controls did so or even made any parental approaches to the squabs.

This experiment showed that prolactin, which is normally present in considerable quantities in the parents when the eggs hatch, does contribute to the doves' ability to show parental feeding behavior. I originally interpreted it to mean that the prolactin-induced engorgement of the crop was necessary in order for any regurgitation feeding to take place, but E. Klinghammer and E. H. Hess of the University of Chicago have correctly pointed out that this was an error, that ring doves are capable of feeding young if presented with them rather early in the incubation period. They do so even though they have no crop-milk, feeding a mixture of regurgitated seeds and a liquid. We are now studying the question of how early the birds can do this and how this ability is related to the onset of prolactin secretion.

The work with gonad-stimulating hormones and prolactin demonstrates that the various hormones successively produced by the birds' glands during their reproductive cycle are capable of inducing the successive behavioral changes that characterize the cycle.

Up to this point I have described two main groups of experiments. One group demonstrates that external stimuli induce changes in behavioral status of a kind normally associated with the progress of the reproductive cycle; the second shows that these behavioral changes can also be induced by hormone administration, provided that the choice of hormones is guided by knowledge of the succession of hormone secretions during a normal reproductive cycle. An obvious—and challenging—implication of these results is that external stimuli may induce changes in hormone secretion, and that environment-induced hormone secretion may constitute an integral part of the mechanism of the reproductive behavior cycle. We have attacked the problem of the environmental stimulation of hormone secretion in a series of experiments in which, in addition to examining the effects of external stimuli on the birds' behavioral status, we have examined their effects on well-established anatomical indicators of the presence of various hormones.

Background for this work was provided by two classic experiments with the domestic pigeon, published during the 1930's, which we have verified in the ring dove. At the London Zoo, L. H. Matthews found that a female pigeon would lay eggs as a result of being placed in a cage with a male from whom she was separated by a glass plate. This was an unequivocal demonstration that visual and/or auditory stimulation provided by the male induces ovarian development in the female. (Birds are quite insensitive to olfactory stimulation.) And M. D. Patel of the University of Wisconsin found that the crops of breeding pigeons, which develop strikingly during the incubation period, would regress to their resting state if the incubating birds were removed from their nests and would fail to develop at all if the birds were removed before crop growth had begun. If, however, a male pigeon, after being removed from his nest, was placed in an adjacent cage from which he could see his mate still sitting on the eggs, his crop would develop just as if he were himself incubating! Clearly stimuli arising from participation in incubation, including visual stimuli, cause the doves' pituitary glands to secrete prolactin.

Our autopsies showed that the incidence of ovulation in females that had associated with males for various periods coincided closely with the incidence of incubation behavior [see bottom illustration on opposite page]; statistical analysis reveals a very high degree of association. The process by which the dove's ovary develops to the point of ovulation includes a period of estrogen secretion followed by one of progesterone secretion, both induced by appropriate ovary-stimulating hormones from the pituitary gland. We therefore conclude that stimuli provided by the male, augmented by the presence of the nest bowl and nesting material, induce the secretion of gonad-stimulating hormones by the female's pituitary, and that the onset of readiness to incubate is a result of this process.

As I have indicated, ovarian development, culminating in ovulation and egg-laying, can be induced in a female dove merely as a result of her seeing a male through a glass plate. Is this the result of the mere presence of another bird or of something the male does because he is a male? Carl Erickson and I have begun to deal with this question. We placed 40 female doves in separate cages, each separated from a male by a glass plate. Twenty of the stimulus animals were normal, intact males, whereas the remaining 20 had been castrated several weeks before. The intact males all exhibited vigorous bow-cooing immediately on being placed in the cage, whereas none of the castrates did so. Thirteen of the 20 females with intact males ovulated during the next seven days, whereas only two of those with the castrates did so. Clearly ovarian development in the female is not induced merely by seeing another bird but by seeing or hearing it act like a male as the result of the effects of its own male hormone on its nervous system.

Although crop growth, which begins early in the incubation period, is apparently stimulated by participation in incubation, the crop continues to be large and actively secreting for quite some time after the hatching of the eggs. This suggests that stimuli provided by the squabs may also stimulate prolactin secretion. In our laboratory Ernst Hansen substituted three-day-old squabs for eggs in various stages of incubation and after four days compared the adults' crop weights with those of birds that had continued to sit on their eggs dur-

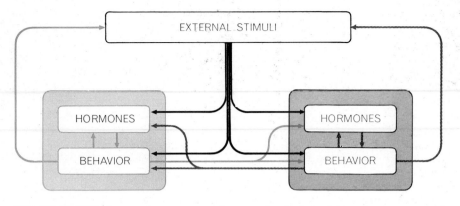

INTERACTIONS that appear to govern the reproductive-behavior cycle are suggested here. Hormones regulate behavior and are themselves affected by behavioral and other stimuli. And the behavior of each bird affects the hormones and the behavior of its mate.

ing the four days. He found that the crops grow even faster when squabs are in the nest than when the adults are under the influence of the eggs; the presence of squabs can stimulate a dove's pituitary glands to secrete more prolactin even before the stage in the cycle when the squabs normally appear.

This does not mean, however, that any of the stimuli we have used can induce hormone secretion at *any* time, regardless of the bird's physiological condition. If we place a pair of ring doves in a cage and allow them to go through the normal cycle until they have been sitting on eggs for, say, six days and we then place a glass partition in the cage to separate the male from the female and the nest, the female will continue to sit on the eggs and the male's crop will continue to develop just as if he were himself incubating. This is a simple replication of one of Patel's experiments. Miriam Friedman and I have found, however, that if the male and female are separated from the beginning, so that the female must build the nest by herself and sit alone from the beginning, the crop of the male does

not grow. By inserting the glass plate at various times during the cycle in different groups of birds, we have found that the crop of the male develops fully only if he is not separated from the female until 72 hours or more after the second egg is laid. This means that the sight of the female incubating induces prolactin secretion in the male only if he is in the physiological condition to which participation in nest-building brings him. External stimuli associated with the breeding situation do indeed induce changes in hormone secretion.

The experiments summarized here point to the conclusion that changes in the activity of the endocrine system are induced or facilitated by stimuli coming from various aspects of the environment at different stages of the breeding cycle, and that these changes in hormone secretion induce changes in behavior that may themselves be a source of further stimulation.

The regulation of the reproductive cycle of the ring dove appears to depend, at least in part, on a double set of reciprocal interrelations. First, there

is an interaction of the effects of hormones on behavior and the effects of external stimuli—including those that arise from the behavior of the animal and its mate—on the secretion of hormones. Second, there is a complicated reciprocal relation between the effects of the presence and behavior of one mate on the endocrine system of the other and the effects of the presence and behavior of the second bird (including those aspects of its behavior induced by these endocrine effects) back on the endocrine system of the first. The occurrence in each member of the pair of a cycle found in neither bird in isolation, and the synchronization of the cycles in the two mates, can now readily be understood as consequences of this interaction of the inner and outer environments.

The physiological explanation of these phenomena lies partly in the fact that the activity of the pituitary gland, which secretes prolactin and the gonad-stimulating hormones, is largely controlled by the nervous system through the hypothalamus. The precise neural mechanisms for any complex response are still deeply mysterious, but physiological knowledge of the brain-pituitary link is sufficiently detailed and definite so that the occurrence of a specific hormonal response to a specific external stimulus is at least no more mysterious than any other stimulus-response relation. We are currently exploring these responses in more detail, seeking to learn, among other things, the precise sites at which the various hormones act. And we have begun to investigate another aspect of the problem: the effect of previous experience on a bird's reproductive behavior and the interactions between these experiential influences and the hormonal effects.

24

PHEROMONES

EDWARD O. WILSON   May 1963

It is conceivable that somewhere on other worlds civilizations exist that communicate entirely by the exchange of chemical substances that are smelled or tasted. Unlikely as this may seem, the theoretical possibility cannot be ruled out. It is not difficult to design, on paper at least, a chemical communication system that can transmit a large amount of information with rather good efficiency. The notion of such a communication system is of course strange because our outlook is shaped so strongly by our own peculiar auditory and visual conventions. This limitation of outlook is found even among students of animal behavior; they have favored species whose communication methods are similar to our own and therefore more accessible to analysis. It is becoming increasingly clear, however, that chemical systems provide the dominant means of communication in many animal species, perhaps even in most. In the past several years animal behaviorists and organic chemists, working together, have made a start at deciphering some of these systems and have discovered a number of surprising new biological phenomena.

In earlier literature on the subject, chemicals used in communication were usually referred to as "ectohormones." Since 1959 the less awkward and etymologically more accurate term "pheromones" has been widely adopted. It is used to describe substances exchanged among members of the same animal species. Unlike true hormones, which are secreted internally to regulate the organism's own physiology, or internal environment, pheromones are secreted externally and help to regulate the organism's external environment by influencing other animals. The mode of influence can take either of two general forms. If the pheromone produces a more or less immediate and reversible change

in the behavior of the recipient, it is said to have a "releaser" effect. In this case the chemical substance seems to act directly on the recipient's central nervous system. If the principal function of the pheromone is to trigger a chain of physiological events in the recipient, it has what we have recently labeled a "primer" effect. The physiological changes, in turn, equip the organism with a new behavioral repertory, the components of which are thenceforth evoked by appropriate stimuli. In termites, for example, the reproductive and soldier castes prevent other termites from developing into

their own castes by secreting substances that are ingested and act through the *corpus allatum,* an endocrine gland controlling differentiation [see "The Termite and the Cell," by Martin Lüscher; SCIENTIFIC AMERICAN, May, 1953].

These indirect primer pheromones do not always act by physiological inhibition. They can have the opposite effect. Adult males of the migratory locust *Schistocerca gregaria* secrete a volatile substance from their skin surface that accelerates the growth of young locusts. When the nymphs detect this substance with their antennae, their hind legs,

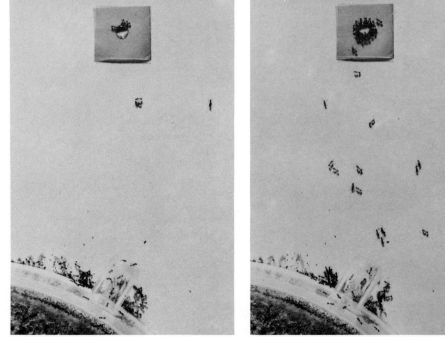

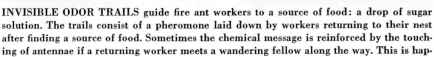

INVISIBLE ODOR TRAILS guide fire ant workers to a source of food: a drop of sugar solution. The trails consist of a pheromone laid down by workers returning to their nest after finding a source of food. Sometimes the chemical message is reinforced by the touching of antennae if a returning worker meets a wandering fellow along the way. This is hap-

some of their mouth parts and the antennae themselves vibrate. The secretion, in conjunction with tactile and visual signals, plays an important role in the formation of migratory locust swarms.

A striking feature of some primer pheromones is that they cause important physiological change without an immediate accompanying behavioral response, at least none that can be said to be peculiar to the pheromone. Beginning in 1955 with the work of S. van der Lee and L. M. Boot in the Netherlands, mammalian endocrinologists have discovered several unexpected effects on the female mouse that are produced by odors of other members of the same species. These changes are not marked by any immediate distinctive behavioral patterns. In the "Lee-Boot effect" females placed in groups of four show an increase in the percentage of pseudopregnancies. A completely normal reproductive pattern can be restored by removing the olfactory bulbs of the mice or by housing the mice separately. When more and more female mice are forced to live together, their oestrous cycles become highly irregular and in most of the mice the cycle stops completely for long periods. Recently W. K. Whitten of the Australian National University has discovered that the odor of a male mouse can initiate and synchronize the oestrous cycles of female mice. The male odor also reduces the frequency of reproductive abnormalities arising when female mice are forced to live under crowded conditions.

A still more surprising primer effect has been found by Helen Bruce of the National Institute for Medical Research in London. She observed that the odor of a strange male mouse will block the pregnancy of a newly impregnated female mouse. The odor of the original stud male, of course, leaves pregnancy undisturbed. The mouse reproductive pheromones have not yet been identified chemically, and their mode of action is only partly understood. There is evidence that the odor of the strange male suppresses the secretion of the hormone prolactin, with the result that the *corpus luteum* (a ductless ovarian gland) fails to develop and normal oestrus is restored. The pheromones are probably part of the complex set of control mechanisms that regulate the population density of animals [see "Population Density and Social Pathology," by John B. Calhoun; SCIENTIFIC AMERICAN, February, 1962].

Pheromones that produce a simple releaser effect—a single specific response mediated directly by the central nervous system—are widespread in the animal kingdom and serve a great many functions. Sex attractants constitute a large and important category. The chemical structures of six attractants are shown on page 241. Although two of the six—the mammalian scents muskone and civetone—have been known for some 40 years and are generally assumed to serve a sexual function, their exact role has never been rigorously established by experiments with living animals. In fact, mammals seem to employ musklike compounds, alone or in combination with other substances, to serve several functions: to mark home ranges, to assist in territorial defense and to identify the sexes.

The nature and role of the four insect sex attractants are much better understood. The identification of each represents a technical feat of considerable magnitude. To obtain 12 milligrams of esters of bombykol, the sex attractant of the female silkworm moth, Adolf F. J. Butenandt and his associates at the Max Planck Institute of Biochemistry in Munich had to extract material from 250,000 moths. Martin Jacobson, Morton Beroza and William Jones of the U.S. Department of Agriculture processed 500,000 female gypsy moths to get 20 milligrams of the gypsy-moth attractant gyplure. Each moth yielded only about .01 microgram (millionth of a gram) of

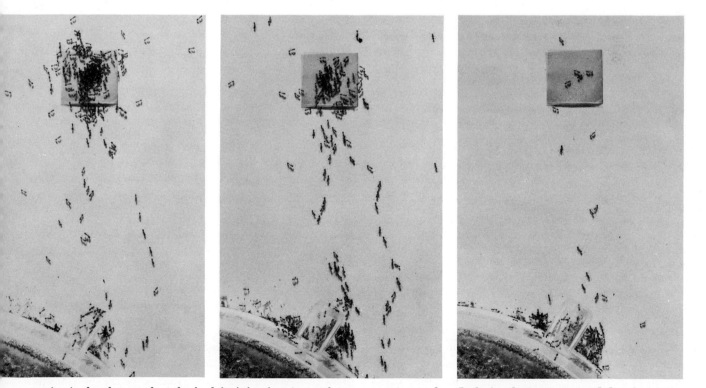

pening in the photograph at the far left. A few foraging workers have just found the sugar drop and a returning trail-layer is communicating the news to another ant. In the next two pictures the trail has been completed and workers stream from the nest in increasing numbers. In the fourth picture unrewarded workers return to the nest without laying trails and outward-bound traffic wanes. In the last picture most of the trails have evaporated completely and only a few stragglers remain at the site, eating the last bits of food.

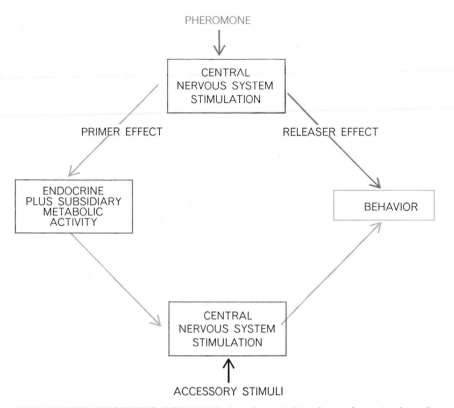

PHEROMONE

CENTRAL
NERVOUS SYSTEM
STIMULATION

PRIMER EFFECT   RELEASER EFFECT

ENDOCRINE
PLUS SUBSIDIARY
METABOLIC
ACTIVITY

BEHAVIOR

CENTRAL
NERVOUS SYSTEM
STIMULATION

ACCESSORY STIMULI

PHEROMONES INFLUENCE BEHAVIOR directly or indirectly, as shown in this schematic diagram. If a pheromone stimulates the recipient's central nervous system into producing an immediate change in behavior, it is said to have a "releaser" effect. If it alters a set of long-term physiological conditions so that the recipient's behavior can subsequently be influenced by specific accessory stimuli, the pheromone is said to have a "primer" effect.

gyplure, or less than a millionth of its body weight. Bombykol and gyplure were obtained by killing the insects and subjecting crude extracts of material to chromatography, the separation technique in which compounds move at different rates through a column packed with a suitable adsorbent substance. Another technique has been more recently developed by Robert T. Yamamoto of the U.S. Department of Agriculture, in collaboration with Jacobson and Beroza, to harvest the equally elusive sex attractant of the American cockroach. Virgin females were housed in metal cans and air was continuously drawn through the cans and passed through chilled containers to condense any vaporized materials. In this manner the equivalent of 10,000 females were "milked" over a nine-month period to yield 12.2 milligrams of what was considered to be the pure attractant.

The power of the insect attractants is almost unbelievable. If some 10,000 molecules of the most active form of bombykol are allowed to diffuse from a source one centimeter from the antennae of a male silkworm moth, a characteristic sexual response is obtained in most cases. If volatility and diffusion rate are taken into account, it can be estimated that the threshold concentration is no more than a few hundred molecules per cubic centimeter, and the actual number required to stimulate the male is probably even smaller. From this one can calculate that .01 microgram of gyplure, the minimum average content of a single female moth, would be theoretically adequate, if distributed with maximum efficiency, to excite more than a billion male moths.

In nature the female uses her powerful pheromone to advertise her presence over a large area with a minimum expenditure of energy. With the aid of published data from field experiments and newly contrived mathematical models of the diffusion process, William H. Bossert, one of my associates in the Biological Laboratories at Harvard University, and I have deduced the shape and size of the ellipsoidal space within which male moths can be attracted under natural conditions [*see bottom illustration on opposite page*]. When a moderate wind is blowing, the active space has a long axis of thousands of meters and a transverse axis parallel to the ground of more than 200 meters at the widest point. The 19th-century

French naturalist Jean Henri Fabre, speculating on sex attraction in insects, could not bring himself to believe that the female moth could communicate over such great distances by odor alone, since "one might as well expect to tint a lake with a drop of carmine." We now know that Fabre's conclusion was wrong but that his analogy was exact: to the male moth's powerful chemoreceptors the lake is indeed tinted.

One must now ask how the male moth, smelling the faintly tinted air, knows which way to fly to find the source of the tinting. He cannot simply fly in the direction of increasing scent; it can be shown mathematically that the attractant is distributed almost uniformly after it has drifted more than a few meters from the female. Recent experiments by Ilse Schwinck of the University of Munich have revealed what is probably the alternative procedure used. When male moths are activated by the pheromone, they simply fly upwind and thus inevitably move toward the female. If by accident they pass out of the active zone, they either abandon the search or fly about at random until they pick up the scent again. Eventually, as they approach the female, there is a slight increase in the concentration of the chemical attractant and this can serve as a guide for the remaining distance.

If one is looking for the most highly developed chemical communication systems in nature, it is reasonable to study the behavior of the social insects, particularly the social wasps, bees, termites and ants, all of which communicate mostly in the dark interiors of their nests and are known to have advanced chemoreceptive powers. In recent years experimental techniques have been developed to separate and identify the pheromones of these insects, and rapid progress has been made in deciphering the hitherto intractable codes, particularly those of the ants. The most successful procedure has been to dissect out single glandular reservoirs and see what effect their contents have on the behavior of the worker caste, which is the most numerous and presumably the most in need of continuing guidance. Other pheromones, not present in distinct reservoirs, are identified in chromatographic fractions of crude extracts.

Ants of all castes are constructed with an exceptionally well-developed exocrine glandular system. Many of the most prominent of these glands, whose function has long been a mystery to entomologists, have now been identified as the source of pheromones [*see illustra-*

tion on page 239]. The analysis of the gland-pheromone complex has led to the beginnings of a new and deeper understanding of how ant societies are organized.

Consider the chemical trail. According to the traditional view, trail secretions served as only a limited guide for worker ants and had to be augmented by other kinds of signals exchanged inside the nest. Now it is known that the trail substance is extraordinarily versatile. In the fire ant (*Solenopsis saevissima*), for instance, it functions both to activate and to guide foraging workers in search of food and new nest sites. It also contributes as one of the alarm signals emitted by workers in distress. The trail of the fire ant consists of a substance secreted in minute amounts by Dufour's gland; the substance leaves the ant's body by way of the extruded sting, which is touched intermittently to the ground much like a moving pen dispensing ink. The trail pheromone, which has not yet been chemically identified, acts primarily to attract the fire ant workers. Upon encountering the attractant the workers move automatically up the gradient to the source of emission. When the substance is drawn out in a line, the workers run along the direction of the line away from the nest. This simple response brings them to the food source or new nest site from which the trail is laid. In our laboratory we have extracted the pheromone from the Dufour's glands of freshly killed workers and have used it to create artificial trails. Groups of workers will follow these trails away from the nest and along arbitrary routes (including circles leading back to the nest) for considerable periods of time. When the pheromone is presented to whole colonies in massive doses, a large portion of the colony, including the queen, can be drawn out in a close simulation of the emigration process.

The trail substance is rather volatile, and a natural trail laid by one worker diffuses to below the threshold concentration within two minutes. Consequently outward-bound workers are able to follow it only for the distance they can travel in this time, which is about 40 centimeters. Although this strictly limits the distance over which the ants can communicate, it provides at least two important compensatory advantages. The more obvious advantage is that old, useless trails do not linger to confuse the hunting workers. In addition, the intensity of the trail laid by many workers provides a sensitive index of the amount of food at a given site and the rate of its depletion. As workers move to and from

ANTENNAE OF GYPSY MOTHS differ radically in structure according to their function. In the male (*left*) they are broad and finely divided to detect minute quantities of sex attractant released by the female (*right*). The antennae of the female are much less developed.

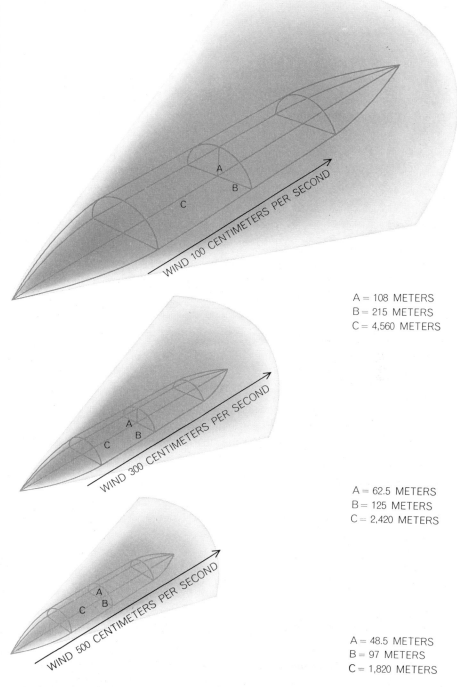

A = 108 METERS
B = 215 METERS
C = 4,560 METERS

A = 62.5 METERS
B = 125 METERS
C = 2,420 METERS

A = 48.5 METERS
B = 97 METERS
C = 1,820 METERS

ACTIVE SPACE of gyplure, the gypsy moth sex attractant, is the space within which this pheromone is sufficiently dense to attract males to a single, continuously emitting female. The actual dimensions, deduced from linear measurements and general gas-diffusion models, are given at right. Height (*A*) and width (*B*) are exaggerated in the drawing. As wind shifts from moderate to strong, increased turbulence contracts the active space.

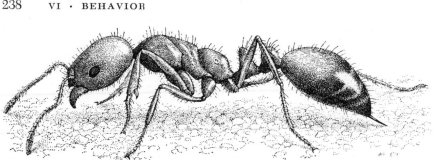

FIRE ANT WORKER lays an odor trail by exuding a pheromone along its extended sting. The sting is touched to the ground periodically, breaking the trail into a series of streaks.

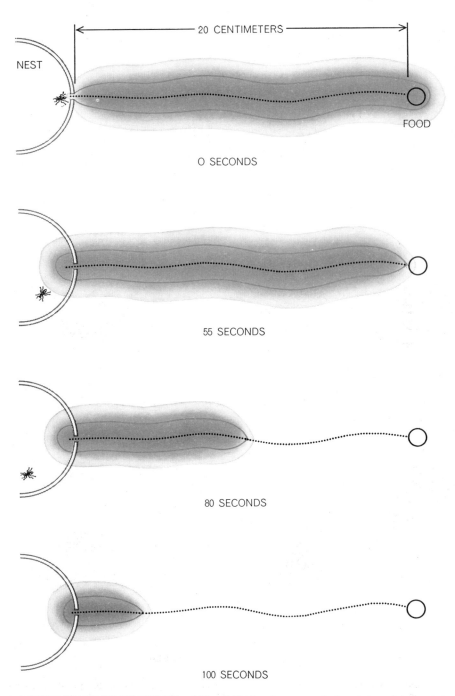

ACTIVE SPACE OF ANT TRAIL, within which the pheromone is dense enough to be perceived by other workers, is narrow and nearly constant in shape with the maximum gradient situated near its outer surface. The rapidity with which the trail evaporates is indicated.

the food finds (consisting mostly of dead insects and sugar sources) they continuously add their own secretions to the trail produced by the original discoverers of the food. Only if an ant is rewarded by food does it lay a trail on its trip back to the nest; therefore the more food encountered at the end of the trail, the more workers that can be rewarded and the heavier the trail. The heavier the trail, the more workers that are drawn from the nest and arrive at the end of the trail. As the food is consumed, the number of workers laying trail substance drops, and the old trail fades by evaporation and diffusion, gradually constricting the outward flow of workers.

The fire ant odor trail shows other evidences of being efficiently designed. The active space within which the pheromone is dense enough to be perceived by workers remains narrow and nearly constant in shape over most of the length of the trail. It has been further deduced from diffusion models that the maximum gradient must be situated near the outer surface of the active space. Thus workers are informed of the space boundary in a highly efficient way. Together these features ensure that the following workers keep in close formation with a minimum chance of losing the trail.

The fire ant trail is one of the few animal communication systems whose information content can be measured with fair precision. Unlike many communicating animals, the ants have a distinct goal in space—the food find or nest site—the direction and distance of which must both be communicated. It is possible by a simple technique to measure how close trail-followers come to the trail end, and, by making use of a standard equation from information theory, one can translate the accuracy of their response into the "bits" of information received. A similar procedure can be applied (as first suggested by the British biologist J. B. S. Haldane) to the "waggle dance" of the honeybee, a radically different form of communication system from the ant trail [for further information, see "Dialects in the Language of Bees," by Karl von Frisch, Offprint #130]. Surprisingly, it turns out that the two systems, although of wholly different evolutionary origin, transmit about the same amount of information with reference to distance (two bits) and direction (four bits in the honeybee, and four or possibly five in the ant). Four bits of information will direct an ant or a bee into one of 16 equally probable sectors of a circle and two bits will identify one of four equally probable distances. It is conceivable that these information values represent the maximum that can be achieved with the insect brain and sensory apparatus.

Not all kinds of ants lay chemical trails. Among those that do, however, the pheromones are highly species-specific in their action. In experiments in which artificial trails extracted from one species were directed to living colonies of other species, the results have almost always been negative, even among related species. It is as if each species had its own private language. As a result there is little or no confusion when the trails of two or more species cross.

Another important class of ant pheromone is composed of alarm substances. A simple backyard experiment will show that if a worker ant is disturbed by a clean instrument, it will, for a short time, excite other workers with whom it comes in contact. Until recently most students of ant behavior thought that

the alarm was spread by touch, that one worker simply jostled another in its excitement or drummed on its neighbor with its antennae in some peculiar way. Now it is known that disturbed workers discharge chemicals, stored in special glandular reservoirs, that can produce all the characteristic alarm responses solely by themselves. The chemical structure of four alarm substances is shown on page 243. Nothing could illustrate more clearly the wide differences between the human perceptual world and that of chemically communicating animals. To the human nose the alarm substances are mild or even pleasant, but to the ant they represent an urgent tocsin that can propel a colony into violent and instant action.

As in the case of the trail substances, the employment of the alarm substances appears to be ideally designed for the purpose it serves. When the contents of the mandibular glands of a worker of the harvesting ant (*Pogonomyrmex badius*)

are discharged into still air, the volatile material forms a rapidly expanding sphere, which attains a radius of about six centimeters in 13 seconds. Then it contracts until the signal fades out completely some 35 seconds after the moment of discharge. The outer shell of the active space contains a low concentration of pheromone, which is actually attractive to harvester workers. This serves to draw them toward the point of disturbance. The central region of the active space, however, contains a concentration high enough to evoke the characteristic frenzy of alarm. The "alarm sphere" expands to a radius of about three centimeters in eight seconds and, as might be expected, fades out more quickly than the "attraction sphere."

The advantage to the ants of an alarm signal that is both local and short-lived becomes obvious when a *Pogonomyrmex* colony is observed under natural conditions. The ant nest is subject to almost innumerable minor disturbances. If the

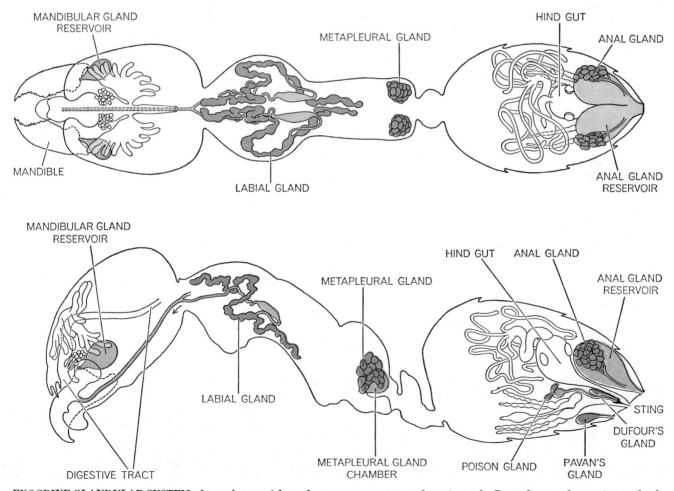

EXOCRINE GLANDULAR SYSTEM of a worker ant (*shown here in top and side cutaway views*) is specially adapted for the production of chemical communication substances. Some pheromones are stored in reservoirs and released in bursts only when needed; others are secreted continuously. Depending on the species, trail substances are produced by Dufour's gland, Pavan's gland or the poison glands; alarm substances are produced by the anal and mandibular glands. The glandular sources of other pheromones are unknown.

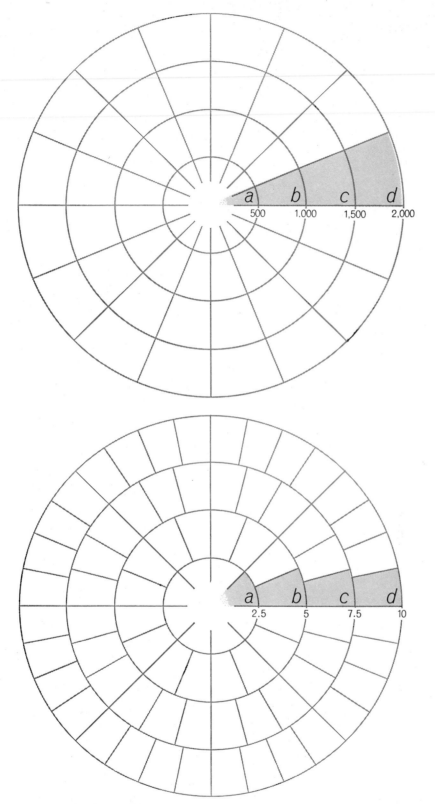

FORAGING INFORMATION conveyed by two different insect communication systems can be represented on two similar "compass" diagrams. The honeybee "waggle dance" (*top*) transmits about four bits of information with respect to direction, enabling a honeybee worker to pinpoint a target within one of 16 equally probable angular sectors. The number of "bits" in this case remains independent of distance, given in meters. The pheromone system used by trail-laying fire ants (*bottom*) is superior in that the amount of directional information increases with distance, given in centimeters. At distances *c* and *d*, the probable sector in which the target lies is smaller for ants than for bees. (For ants, directional information actually increases gradually and not by jumps.) Both insects transmit two bits of distance information, specifying one of four equally probable distance ranges.

alarm spheres generated by individual ant workers were much wider and more durable, the colony would be kept in ceaseless and futile turmoil. As it is, local disturbances such as intrusions by foreign insects are dealt with quickly and efficiently by small groups of workers, and the excitement soon dies away.

The trail and alarm substances are only part of the ants' chemical vocabulary. There is evidence for the existence of other secretions that induce gathering and settling of workers, acts of grooming, food exchange, and other operations fundamental to the care of the queen and immature ants. Even dead ants produce a pheromone of sorts. An ant that has just died will be groomed by other workers as if it were still alive. Its complete immobility and crumpled posture by themselves cause no new response. But in a day or two chemical decomposition products accumulate and stimulate the workers to bear the corpse to the refuse pile outside the nest. Only a few decomposition products trigger this funereal response; they include certain long-chain fatty acids and their esters. When other objects, including living workers, are experimentally daubed with these substances, they are dutifully carried to the refuse pile. After being dumped on the refuse the "living dead" scramble to their feet and promptly return to the nest, only to be carried out again. The hapless creatures are thrown back on the refuse pile time and again until most of the scent of death has been worn off their bodies by the ritual.

Our observation of ant colonies over long periods has led us to believe that as few as 10 pheromones, transmitted singly or in simple combinations, might suffice for the total organization of ant society. The task of separating and characterizing these substances, as well as judging the roles of other kinds of stimuli such as sound, is a job largely for the future.

Even in animal species where other kinds of communication devices are prominently developed, deeper investigation usually reveals the existence of pheromonal communication as well. I have mentioned the auxiliary roles of primer pheromones in the lives of mice and migratory locusts. A more striking example is the communication system of the honeybee. The insect is celebrated for its employment of the "round" and "waggle" dances (augmented, perhaps, by auditory signals) to designate the location of food and new nest sites. It is not so widely known that chemical signals play equally important roles in other aspects of honeybee life. The mother queen regulates the reproductive cycle of the colony by secreting from her mandibular glands a substance recently identified as 9-ketodecanoic acid. When this pheromone is ingested by the worker bees, it inhibits development of their ovaries and also their ability to manufacture the royal cells in which new queens are reared. The same pheromone serves as a sex attractant in the queen's nuptial flights.

Under certain conditions, including the discovery of new food sources, worker bees release geraniol, a pleasant-smelling alcohol, from the abdominal Nassanoff glands. As the geraniol diffuses through the air it attracts other workers and so supplements information contained in the waggle dance. When a worker stings an intruder, it discharges, in addition to the venom, tiny amounts of a secretion from clusters of unicellular

glands located next to the basal plates of the sting. This secretion is responsible for the tendency, well known to beekeepers, of angry swarms of workers to sting at the same spot. One component, which acts as a simple attractant, has been identified as isoamyl acetate, a compound that has a banana-like odor. It is possible that the stinging response is evoked by at least one unidentified alarm substance secreted along with the attractant.

Knowledge of pheromones has advanced to the point where one can make some tentative generalizations about their chemistry. In the first place, there appear to be good reasons why sex attractants should be compounds that contain between 10 and 17 carbon atoms and that have molecular weights between about 180 and 300—the range actually observed in attractants so far identified. (For comparison, the weight of a single

carbon atom is 12.) Only compounds of roughly this size or greater can meet the two known requirements of a sex attractant: narrow specificity, so that only members of one species will respond to it, and high potency. Compounds that contain fewer than five or so carbon atoms and that have a molecular weight of less than about 100 cannot be assembled in enough different ways to provide a distinctive molecule for all the insects that want to advertise their presence.

It also seems to be a rule, at least with insects, that attraction potency increases with molecular weight. In one series of esters tested on flies, for instance, a doubling of molecular weight resulted in as much as a thousandfold increase in efficiency. On the other hand, the molecule cannot be too large and complex or it will be prohibitively difficult for the insect to synthesize. An equally important limitation on size is

BOMBYKOL (SILKWORM MOTH)

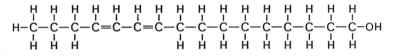

GYPLURE (GYPSY MOTH)

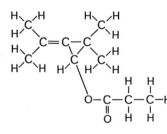

2,2-DIMETHYL-3-ISOPROPYLIDENECYCLOPROPYL PROPIONATE (AMERICAN COCKROACH)

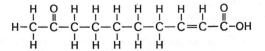

HONEYBEE QUEEN SUBSTANCE

CIVETONE (CIVET)

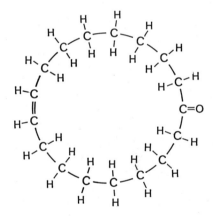

MUSKONE (MUSK DEER)

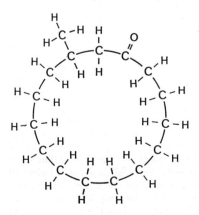

SIX SEX PHEROMONES include the identified sex attractants of four insect species as well as two mammalian musks generally believed to be sex attractants. The high molecular weight of most sex pheromones accounts for their narrow specificity and high potency.

the fact that volatility—and, as a result, diffusibility—declines with increasing molecular weight.

One can also predict from first principles that the molecular weight of alarm substances will tend to be less than those of the sex attractants. Among the ants there is little specificity; each species responds strongly to the alarm substances of other species. Furthermore, an alarm substance, which is used primarily within the confines of the nest, does not need the stimulative potency of a sex attractant, which must carry its message for long distances. For these reasons small molecules will suffice for alarm purposes. Of seven alarm substances known in the social insects, six have 10 or fewer carbon atoms and one (dendrolasin) has 15. It will be interesting to see if future discoveries bear out these early generalizations.

Do human pheromones exist? Primer pheromones might be difficult to detect, since they can affect the endocrine system without producing overt specific behavioral responses. About all that can be said at present is that striking sexual differences have been observed in the ability of humans to smell certain

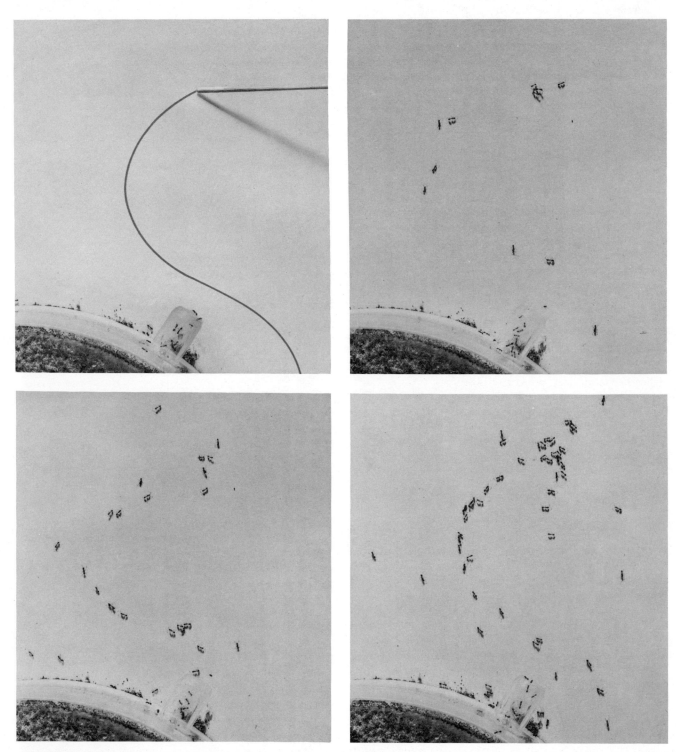

ARTIFICIAL TRAIL can be laid down by drawing a line (*colored curve in frame at top left*) with a stick that has been treated with the contents of a single Dufour's gland. In the remaining three frames, workers are attracted from the nest, follow the artificial route in close formation and mill about in confusion at its arbitrary terminus. Such a trail is not renewed by the unrewarded workers.

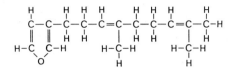

DENDROLASIN (*LASIUS FULIGINOSUS*)

CITRAL (*ATTA SEXDENS*)

CITRONELLAL (*ACANTHOMYOPS CLAVIGER*)

2-HEPTANONE (*IRIDOMYRMEX PRUINOSUS*)

FOUR ALARM PHEROMONES, given off by the workers of the ant species indicated, have so far been identified. Disturbing stimuli trigger the release of these substances from various glandular reservoirs.

substances. The French biologist J. Le-Magnen has reported that the odor of Exaltolide, the synthetic lactone of 14-hydroxytetradecanoic acid, is perceived clearly only by sexually mature females and is perceived most sharply at about the time of ovulation. Males and young girls were found to be relatively insensitive, but a male subject became more sensitive following an injection of estrogen. Exaltolide is used commercially as a perfume fixative. LeMagnen also reported that the ability of his subjects to detect the odor of certain steroids paralleled that of their ability to smell Exaltolide. These observations hardly represent a case for the existence of human pheromones, but they do suggest that the relation of odors to human physiology can bear further examination.

It is apparent that knowledge of chemical communication is still at an early stage. Students of the subject are in the position of linguists who have learned the meaning of a few words of a nearly indecipherable language. There is almost certainly a large chemical vocabulary still to be discovered. Conceiv-

ably some pheromone "languages" will be found to have a syntax. It may be found, in other words, that pheromones can be combined in mixtures to form new meanings for the animals employing them. One would also like to know if some animals can modulate the intensity or pulse frequency of pheromone emission to create new messages. The solution of these and other interesting problems will require new techniques in analytical organic chemistry combined with ever more perceptive studies of animal behavior.

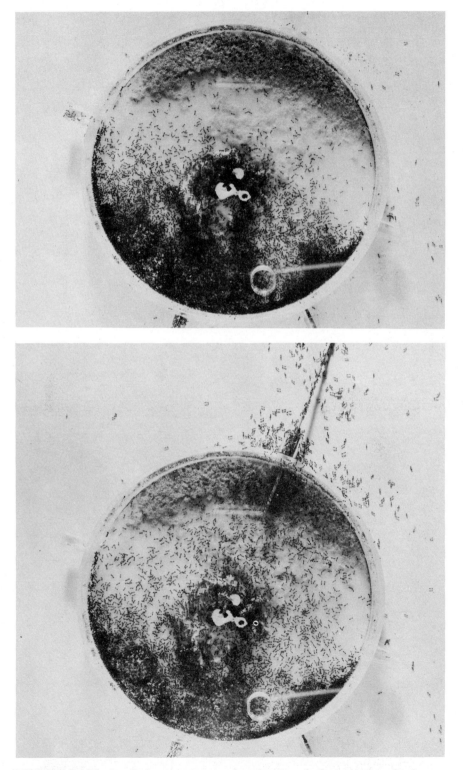

MASSIVE DOSE of trail pheromone causes the migration of a large portion of a fire ant colony from one side of a nest to another. The pheromone is administered on a stick that has been dipped in a solution extracted from the Dufour's glands of freshly killed workers.

Biographical Notes and Bibliographies

The biographical and bibliographical data are those that were available at the time the articles appeared in SCIENTIFIC AMERICAN.

PART I: DEVELOPMENTAL INTEGRATION

General References

Austin, C. R. FERTILIZATION. Prentice-Hall, 1965.

Balinsky, B. I. AN INTRODUCTION TO EMBRYOLOGY. Saunders, 1960.

Berrill, N. J. GROWTH, DEVELOPMENT, AND PATTERN. W. H. Freeman and Company, 1961.

Ebert, J. D. INTERACTING SYSTEMS IN DEVELOPMENT. Holt, Rinehart, and Winston, 1965.

Hillman, W. S. THE PHYSIOLOGY OF FLOWERING. Holt, Rinehart, and Winston, 1964.

Sinnott, E. W. PLANT MORPHOGENESIS. McGraw-Hill, 1960.

1. Differentiation in Social Amoebae

The Author

JOHN TYLER BONNER ("Differentiation in Social Amoebae") is professor of biology at Princeton University. He was born in New York City in 1920 and took his degrees at Harvard University. During World War II he did research in the Aero Medical Laboratory at Wright Field, and afterward was a junior fellow at Harvard. Bonner joined the faculty at Princeton in 1947. He started his research on social amoebae as an undergraduate under William H. Weston, and has been studying these life forms ever since. This is his fifth article for SCIENTIFIC AMERICAN.

Bibliography

THE CELLULAR SLIME MOLDS. John Tyler Bonner. Princeton University Press, 1959.

EVIDENCE FOR THE SORTING OUT OF CELLS IN THE DEVELOPMENT OF THE CELLULAR SLIME MOLDS. John Tyler Bonner in *Proceedings of the National Academy of Sciences*, Vol. 45, No. 3, pages 379-384; March, 1959.

2. "The Organizer"

The Author

GEORGE W. GRAY was born in Texas and, at the University of Texas, began to prepare for a career in physiology. Within a year, however, he left the University because of illness. After his recovery he worked for the Houston *Post* and then went to Harvard University. He graduated in 1912 and joined the staff of the New York *World*. During the 1920s he turned freelance and began to concentrate on the natural sciences, contributing many articles to magazines, as well as writing a number of books. For most of his career, Gray enjoyed a unique appointment on the staff of the Rockefeller Foundation, visiting the scientists and institutions supported by the Foundation and interpreting their work in reports for other members of the staff and the board of directors. Gray passed away in February, 1962.

Bibliography

ANALYSIS OF DEVELOPMENT. Edited by Benjamin H. Willier, Paul A. Weiss and Viktor Hamburger. W. B. Saunders Company, 1955.

ASPECTS OF SYNTHESIS AND ORDER IN GROWTH. Edited by Dorothea Rudnick. Princeton University Press, 1954.

DYNAMICS OF GROWTH PROCESSES. Edited by Edgar J. Boell. Princeton University Press, 1954.

EMBRYONIC TRANSPLANTATION AND THE DEVELOPMENT OF THE NERVOUS SYSTEM. Ross G. Harrison in *The Harvey Lectures*, pages 199–222; 1907-08.

EMBRYONIC DEVELOPMENT AND INDUCTION. Hans Spemann. Yale University Press, 1938.

3. The Embryological Origin of Muscle

The Author

IRWIN R. KONIGSBERG is on the faculty of the Department of Biology at the University of Virginia. A graduate of Brooklyn College, Konigsberg acquired a Ph.D. in biology from Johns Hopkins University in 1952. From 1952 to 1958 he did research in pediatrics in the Laboratory of Chemical Embryology of the University of Colorado School of Medicine. In 1958 he joined the gerontology branch of the National Heart Institute. From 1961 to 1966 he did research in the Department of Embryology of the Carnegie Institution of Washington.

Bibliography

CELLULAR DIFFERENTIATION IN COLONIES DERIVED FROM SINGLE CELL PLATINGS OF FRESHLY ISOLATED CHICK EMBRYO MUSCLE CELLS. Irwin R. Konigsberg in *Proceedings of the National Academy of Sciences*, Vol. 47, No. 11, pages 1868–1872; November, 1961.

CLONAL ANALYSIS OF MYOGENESIS. Irwin R. Konigsberg in *Science*, Vol. 140, No. 3573, pages 1273–1284; June, 1963.

GROWTH AND GENETICS OF SOMATIC MAMMALIAN CELLS IN VITRO. Theodore T. Puck in *Journal of Cellular and Comparative Physiology*, Vol. 52, Supplement 1; December, 1958.

4. Light and Plant Development

The Authors

W. L. BUTLER and ROBERT J. DOWNS, when this article was written, were both employed by the U.S. Department of Agriculture, the former as a biophysicist at the Instrumentation Research Laboratory, the latter as a plant physiologist at the Plant Industry Station. Butler acquired a B.A. in physics at Reed College in 1949. He did graduate work in photosynthesis at the University of Chicago under the Nobel laureate James Franck, receiving a Ph.D. in biophysics in 1955. He joined the Instrumentation Laboratory in 1956, and in 1964 went to the University of California at San Diego as professor of biology. Downs served in the Navy from 1941 to 1947, first as a repair machinist and later as an optical repairman and instructor. He took three degrees at George Washington University, receiving his Ph.D. in botany two years after he joined the Plant Industry Station in 1952. He is now at North Carolina State University. This article describes the results of a research program that was carried out by a group of workers which included the authors.

Bibliography

CONTROL OF GROWTH AND REPRODUCTION BY LIGHT AND DARKNESS. Sterling B. Hendricks in *American Scientist*, Vol. 44, No. 3, pages 229–247; July, 1956.

DETECTION, ASSAY, AND PRELIMINARY PURIFICATION OF THE PIGMENT CONTROLLING PHOTORESPONSIVE DEVELOPMENT OF PLANTS. W. L. Butler, K. H. Norris, H. W. Siegelman and S. B. Hendricks in *Proceedings of the National Academy of Sciences*, Vol. 45, No. 12, pages 1703–1708; December, 1959.

INFLUENCE OF LIGHT ON PLANT GROWTH. M. W. Parker and H. A. Borthwick in *Annual Review of Plant Physiology*, Vol. 1, pages 43–58; 1950.

PHYSIOLOGY OF SEED GERMINATION. E. H. Toole, S. B. Hendricks, H. A. Borthwick and Vivian K. Toole in *Annual Review of Plant Physiology*, Vol. 7, pages 299–324; 1956.

PART II: EXCHANGE AND TRANSPORT

General References

Florey, E. INTRODUCTION TO GENERAL AND COMPARATIVE ANIMAL PHYSIOLOGY. Saunders, 1966 (Chapters 5, 9, 10, 13).

Hughes, G. M. THE COMPARATIVE PHYSIOLOGY OF VERTEBRATE RESPIRATION. Harvard University Press, 1963.

Krogh, A. THE COMPARATIVE PHYSIOLOGY OF RESPIRATION MECHANISMS. University of Pennsylvania Press, 1941.

Potts, W. T. W., and G. Parry. OSMOTIC AND IONIC REGULATION IN ANIMALS. Pergamon, 1964.

Prosser, C. L., and F. A. Brown, Jr. COMPARATIVE ANIMAL PHYSIOLOGY. 2nd Edition, Saunders, 1961 (Chapters 1–8, 13).

5. The Kidney

The Author

HOMER W. SMITH was, for most of his career, professor of physiology at New York University. Before his death in 1962, he was widely esteemed as the foremost authority on the kidney, and was, without doubt, this organ's foremost publicist. A native of Denver, Colorado, and a graduate of the University of Denver, he received his doctorate in renal physiology from Johns Hopkins University in 1921. Apart from his numerous research papers and his definitive books on the kidney, his writings embrace a wide range of concern with the philosophical and ethical concerns of scientific knowledge. In *Kamongo* (1932), *The End of Illusion* (1935), *Man and His Gods* (1952) and *From Fish to Philosopher* (1953), Smith deployed his free-swinging scholarship in robust assault on ancient and modern superstitions, shibboleths, myths, conventions and sacred cows.

Bibliography

THE KIDNEY: STRUCTURE AND FUNCTION IN HEALTH AND DISEASE. Homer W. Smith. Oxford University Press, 1951.

PRINCIPLES OF RENAL PHYSIOLOGY. Homer W. Smith. Oxford University Press, 1956.

6. The Heart

The Author

CARL J. WIGGERS retired in 1953 as professor of physiology at Western Reserve University, where he had taught since 1913. He is now honorary professor at the Bunts Educational Institute in Cleveland and editor of *Circulation Research*, a journal which he founded in 1949. Wiggers received his M.D. from the University of Michigan, taught there for several years, and did advanced work at the Physiological Institute in Munich before going to Western Reserve. He has written books titled *The Physiology of Shock* and *Circulatory Dynamics*, served as president of the American Physiological Society, and won the Gold Heart Medal of the American Heart Association and the Lasker Award. More than 24 of his former students now hold university professorships in the U.S. and abroad.

Bibliography

PHYSIOLOGY IN HEALTH AND DISEASE. Carl J. Wiggers. Lea & Febiger, 1949.

THE MOTION OF THE HEART: THE STORY OF CARDIOVASCULAR RESEARCH. Blake Cabot. Harper & Brothers, 1954.

7. The Microcirculation of the Blood

The Author

BENJAMIN W. ZWEIFACH says that he is a "complete product of New York City's public-school system." Upon graduating from the College of the City of New York in 1931, at the height of the depression, he found himself jobless, and so decided to enroll for a year in the New York University Graduate School. There he came under the influence of Robert Chambers, a pioneer in the microsurgery of cells. Chamber's work on the permeability of cell membranes led Zweifach immediately into the study of how the cells are nourished by the blood capillaries. Zweifach took his Ph.D. in 1936, was a research fellow at Tufts University for two years, then returned to N.Y.U., where (except for a period of a few years at the Cornell University Medical College) he has worked ever since.

Bibliography

THE ANATOMY AND PHYSIOLOGY OF CAPILLARIES. August Krogh. Yale University Press, 1929.

GENERAL PRINCIPLES GOVERNING THE BEHAVIOR OF THE MICROCIRCULA-TION. B. W. Zweifach in *The American Journal of Medicine*, Vol. 23, No. 5, pages 684–696; November, 1957.

THE ROLE OF MEDIATORS IN THE INFLAMMATORY TISSUE RESPONSE. W. Feldberg in *International Archives of Allergy and Applied Immunology*, Vol. 8, No. 1–2, pages 15–31; 1956.

8. How Sap Moves in Trees

The Author

MARTIN H. ZIMMERMANN is a forest physiologist at Harvard University who does his research under the auspices of the Maria Moors Cabot Foundation for Botanical Research. He also holds the title of Lecturer on Forest Physiology at Harvard. Zimmermann, who was born and educated in Switzerland, studied botany at the Swiss Federal Institute of Technology. During that time he became particularly interested in trees as a result of several summers spent as a lumberjack. In 1951 Zimmermann became assistant plant physiologist at the Federal Institute and acquired a D.Sc. in 1953. He went to Harvard the following year.

Bibliography

ASCENT OF SAP. K. N. H. Greenidge in *Annual Review of Plant Physiology*, Vol. 8, pages 237–256; 1957.

COHESIVE LIFT OF SAP IN THE RATTAN VINE. P. F. Scholander, E. Hemmingsen and W. Garey in *Science*, Vol. 134, No. 3493, pages 1835–1838; December, 1961.

TRANSPORT IN THE PHLOEM. Martin H. Zimmermann in *Annual Review of Plant Physiology*, Vol. 11, pages 167–190; 1960.

TRANSPORT IN THE XYLEM. E. G. Bollard in *Annual Review of Plant Physiology*, Vol. 11, pages 141–166; 1960.

PART III: METABOLIC REGULATION

General References

Gorbman, A., and H. A. Bern. A TEXTBOOK OF COMPARATIVE ENDOCRINOLOGY. Wiley, 1962.

Schmidt-Nielsen, K. ANIMAL PHYSIOLOGY. 2nd Edition, Prentice-Hall, 1964.

Telfer, W. H., and D. Kennedy. THE BIOLOGY OF ORGANISMS. Wiley, 1965 (Chapters 6, 7).

Turner, C. D. GENERAL ENDOCRINOLOGY. 4th Edition, Saunders, 1966.

9. The Human Thermostat

The Author

T. H. BENZINGER is head of the Calorimetry Branch and the Bio-Energetics Division of the Naval Medical Research Institute in Bethesda, Md. He was born in Germany in 1905, and studied at the universities of Tübingen and Munich and at the Stuttgart Institute of Technology. From 1927 to 1933 he studied medicine at Tübingen and the universities of Berlin and Freiburg, receiving his M.D. degree from the latter institution. During World War II Benzinger was head of both the Medical Division of the German Air Force Testing Center and the Medical Branch of the Research Division in the Air Ministry. After two years at the U.S. Air Force Aero-Medical Center in Heidelberg, he went to the Naval Medical Research Institute in 1947.

Bibliography

ACTIVATION OF HEAT LOSS MECHANISMS BY LOCAL HEATING OF THE BRAIN. H. W. Magoun, F. Harrison, J. R. Brobeck and S. W. Ranson in *Journal of Neurophysiology*, Vol. 1, No. 2, pages 101–114; March, 1938.

THE RELATION OF THE NERVOUS SYSTEM TO THE TEMPERATURE OF THE BODY. Isaac Ott in *The Journal of Nervous and Mental Disease*, Vol. XI, No. 2, pages 141–152; April, 1884.

THE ROLE OF THE ANTERIOR HYPOTHALAMUS IN TEMPERATURE REGULATION. R. S. Teague and S. W. Ranson in *The American Journal of Physiology*, Vol. 117, No. 3, pages 562–570; November 1, 1936.

10. The Master Switch of Life

The Author

P. F. SCHOLANDER is professor of physiology and director of the newly established Physiological Research Laboratory at the Scripps Institution of Oceanography. Scholander was born in Örebro, Sweden, in 1905 and received an M.D. from the University of Oslo in 1932. After two years as instructor in anatomy at Oslo he acquired a Ph.D. in botany there in 1934. He did research in botany and physiology in Norway until 1939, when he came to this country to join the department of zoology at Swarthmore College. During World War II Scholander served as an aviation physiologist in the U.S. Army Air Force, returning to Swarthmore in 1946 to continue his studies on the comparative physiology of diving. He investigated climatic adaptations of arctic and tropical animals and plants at the Arctic Research Laboratory in Point Barrow, Alaska, and in Panama from 1947 until 1949, when he became a research fellow in the department of biological chemistry at the

Harvard Medical School. In 1952 he joined the staff of the Woods Hole Oceanographic Institution, and in 1955 he was appointed professor of physiology and director of the Institute of Zoophysiology at the University of Oslo. He joined the Scripps Institution in 1958.

Bibliography

CIRCULATORY ADJUSTMENT IN PEARL DIVERS. P. F. Scholander, H. T. Hammel, H. LeMessurier, E. Hemmingsen and W. Garey in *Journal of Applied Physiology*, Vol. 17, No. 2, pages 184–190; March, 1962.

MESENTERIC VASCULAR INSUFFICIENCY: INTESTINAL ISCHEMIA INDUCED BY REMOTE CIRCULATORY DISTURBANCES. Eliot Corday, David W. Irving, Herbert Gold, Harold Bernstein and Robert B. T. Skelton in *The American Journal of Medicine*, Vol. 33, No. 3, pages 365–376; September, 1962.

RESPIRATION IN DIVING MAMMALS. Laurence Irving in *Physiological Reviews*, Vol. 19, No. 1, pages 112–134; January, 1939.

SELECTIVE ISCHEMIA IN DIVING MAN. R. W. Elsner, W. F. Garey and P. F. Scholander in *American Heart Journal*, Vol. 65, No. 4, pages 571–572; April, 1963.

11. The Desert Rat

The Authors

KNUT and BODIL SCHMIDT-NIELSEN ("The Desert Rat") have collaborated in investigating the physiology of desert mammals. Knut was born in Trondheim, Norway, in 1915 and educated at the universities of Oslo and Copenhagen. At Copenhagen he studied with August Krogh, Nobel laureate in physiology and medicine, and received a doctorate of philosophy in 1946. Bodil Schmidt-Nielsen was born in Copenhagen in 1918, holds a dental degree and Ph.D., and has done much work in general physiology. Both Schmidt-Nielsens came to the U.S. in 1946 and have successively been on the faculties of Swarthmore College, Stanford University, the University of Cincinnati, and Duke University, where Knut Schmidt-Nielsen is now a professor of zoology. Bodil Schmidt-Nielsen is a member of the Department of Biology at Western Reserve University.

Bibliography

THE WATER ECONOMY OF DESERT MAMMALS. Bodil and Knut Schmidt-Nielsen in *The Scientific Monthly*, Vol. 69, No. 3; September, 1949.

12. Hormones and Genes

The Author

ERIC H. DAVIDSON is a research associate at the Rockefeller Institute, working in cell biology. As a high school student in Nyack, N.Y., he worked summers at the Marine Biological Laboratory in Woods Hole, Mass., and was one of the national winners of the Westinghouse Science Talent Search. Davidson was graduated from the University of Pennsylvania in 1958, having majored in zoology. For the next five years he was a graduate fellow at the Rockefeller Institute, obtaining a doctor's degree there in 1963. He writes that as a research fellow at the Institute he is "collaborating with Alfred E. Mirsky in the study of gene action in the initiation and control of embryological development."

Bibliography

EFFECT OF ACTINOMYCIN AND INSULIN ON THE METABOLISM OF ISOLATED RAT DIAPHRAGM. Ira G. Wool and Arthur N. Moyer in *Biochimica et Biophysica Acta*, Vol. 91, No. 2, pages 248–256; October 16, 1964.

ON THE MECHANISM OF ACTION OF ALDOSTERONE ON SODIUM TRANSPORT: THE ROLE OF RNA SYNTHESIS. George A. Porter, Rita Bogoroch and Isidore S. Edelman in *Proceedings of the National Academy of Sciences*, Vol. 52, No. 6, pages 1326–1333; December, 1964.

PREVENTION OF HORMONE ACTION BY LOCAL APPLICATION OF ACTINOMYCIN. D. G. P. Talwar and Sheldon J. Segal in *Proceedings of the National Academy of Sciences*, Vol. 50, No. 1, pages 226–230; July 15, 1963.

SELECTIVE ALTERATIONS OF MAMMALIAN MESSENGER-RNA SYNTHESIS: EVIDENCE FOR DIFFERENTIAL ACTION OF HORMONES ON GENE TRANSCRIPTION. Chev Kidson and K. S. Kirby in *Nature*, Vol. 203, No. 4945, pages 599–603; August 8, 1964.

TRANSFER RIBONUCLEIC ACIDS. E. N. Carlsen, G. J. Trelle and O. A. Schjeide in *Nature*, Vol. 202, No. 4936, pages 984–986; June 6, 1964.

PART IV: SENSORY RECEPTORS

General References

Beament, J. W. L. (Ed.). BIOLOGICAL RECEPTOR MECHANISMS. Symp. Soc. Experimental Biology XVI. Academic Press, 1962.

Case, James. SENSORY MECHANISMS. Macmillan, 1966.

Davson, H. (Ed). THE EYE. 2 vols. Academic Press, 1962.

Granit, R. RECEPTORS AND SENSORY PERCEPTION. Yale University Press, 1955.

Griffin, D. R. LISTENING IN THE DARK. Yale University Press, 1958.

13. Biological Transducers

The Author

WERNER R. LOEWENSTEIN is a member of the Faculty of Medicine at Columbia University. Born in Germany and raised in Chile, he decided at the age of 19 to become a neurophysiologist. He obtained his doctorate in physiological science at the University of Chile in 1950, and came to the U.S. in 1953 to work at Johns Hopkins University and at the University of California at Los Angeles. After further research at the University of Chile in 1955, he returned to take up his present position at Columbia, where he has worked principally on nerve-impulse initiation in receptors.

Bibliography

THE GENERATION OF ELECTRIC ACTIVITY IN A NERVE ENDING. Werner R.

Loewenstein in *Annals of the New York Academy of Sciences*, Vol. 81, Article 2, pages 367–387; August 28, 1959.

THE PHYSICAL BACKGROUND OF PERCEPTION. E. D. Adrian. Oxford University Press, 1947.

RECEPTORS AND SENSORY PERCEPTION. Ragnar Granit. Yale University Press, 1955.

14. The Ear

The Author

GEORG VON BÉKÉSY, winner of the 1961 Nobel Prize in medicine for the work described in this article, was for most of his career research fellow in psychophysics at Harvard University. Born in Hungary in 1899, he studied at the universities of Berne and Budapest and received a Ph.D. from the latter in 1923. Besides the teaching posts which he held at the University of Budapest, von Békésy was for many years a research scientist with the Hungarian telephone system. His work on adapting telephone receivers to the mechanics of the ear led eventually to a revision of the theory of hearing. After leaving Hungary in 1947 he was for several years a research professor at the Karolinska Institute in Stockholm as well as a fellow at Harvard. In 1955 he received the Howard Crosby Warren Medal of the Society of Experimental Psychologists.

Bibliography

THE EARLY HISTORY OF HEARING—OBSERVATIONS AND THEORIES. Georg v. Békésy and Walter A. Rosenblith in *The Journal of the Acoustical Society of America*, Vol. 20, No. 6, pages 727–748; November, 1948.

HEARING: ITS PSYCHOLOGY AND PHYSIOLOGY. Stanley Smith Stevens and Hallowell Davis. John Wiley & Sons, Inc., 1938.

PHYSIOLOGICAL ACOUSTICS. Ernest Glen Wever and Merle Lawrence. Princeton University Press, 1954.

15. Eye and Camera

The Author

GEORGE WALD is Harvard University's well-known authority on the chemistry of vision. Born in New York, he graduated from New York University in 1927, then did graduate work in zoology at Columbia University under Selig Hecht. After receiving his Ph.D. in 1932, he traveled to Germany on a National Research Council fellowship. While studying in Otto Warburg's laboratory at the Kaiser Wilhelm Institute in Berlin, Wald made his first notable contribution to knowledge of the eye—his discovery of vitamin A in the retina. After another year of postdoctoral study at the University of Chicago, he went to Harvard, where he is now professor of biology.

Bibliography

VISION AND THE EYE. M. H. Pirenne. The Pilot Press, Ltd., 1948.

THE RETINA. S. Polyak. University of Chicago Press, 1941.

THE LIGHT REACTION IN THE BLEACHING OF RHODOPSIN. George Wald, Jack Durell and C. C. St. George in *Science*, Vol. 3, No. 2,877, pages 179–181; February 17, 1950.

THE PHOTORECEPTOR PROCESS IN VISION. George Wald in *Handbook of Physiology: Neurophysiology*, Vol. 1, American Physiological Society, pages 671–692, 1959.

VISUAL PIGMENTS IN HUMAN AND MONKEY RETINAS. Paul Brown and George Wald in *Nature*, Vol. 200, No. 4901, pages 37–43, 1963.

16. Visual Pigments in Man

The Author

W. A. H. RUSHTON is a Fellow and director of medical studies at Trinity College of the University of Cambridge. The son of a London dental surgeon, Rushton numbered among his schoolmates at Gresham's School the poets W. H. Auden and Stephen Spender, the composer, Benjamin Britten and the physiologist A. L. Hodgkin. Rushton studied medicine at Cambridge and did research under the direction of the physiologist E. D. Adrian. Rushton came to the U.S. in 1929 as one of the original members of the Johnson Foundation of the University of Pennsylvania. He returned to Cambridge in 1931. Since 1948, when he was elected a Fellow of the Royal Society for his work on nerve excitation, Rushton has become increasingly interested in the role of pigments in vision.

Bibliography

CHEMICAL BASIS OF HUMAN COLOUR VISION. W. A. H. Rushton in *Research*, Vol. 11, No. 12, pages 478–483; December, 1958.

THE VISUAL PIGMENTS. H. J. A. Dartnall. John Wiley & Sons, Inc., 1957.

VISUAL PIGMENTS IN MAN. W. A. H. Rushton. Liverpool University Press, 1962.

PART V: NERVOUS INTEGRATION

General References

Bullock, Theodore Holmes, and G. Adrian Horridge. STRUCTURE AND FUNCTION IN THE NERVOUS SYSTEMS OF INVERTEBRATES. 2 vols. W. H. Freeman and Company, 1965.

Hodgkin, A. CONDUCTION OF THE NERVOUS IMPULSE. Thomas, 1964.

Katz, B. NERVE, MUSCLE AND SYNAPSE. McGraw-Hill, 1966.

Ochs, S. ELEMENTS OF NEUROPHYSIOLOGY. Wiley, 1965.

Ruch, T. C., H. D. Patton, J. W. Woodbury, and A. L. Towe. NEUROPHYSIOLOGY. Saunders, 1961.

17. The Synapse

The Author

SIR JOHN ECCLES is currently at Northwestern University. He shared the Nobel prize for physiology and medicine in 1963; in 1958 he was knighted by Queen Elizabeth. Sir John writes:

"My interest in nerve physiology dates from my medical-student days in Melbourne, when I became interested in philosophical problems relating to all the events experienced in consciousness. I became dissatisfied with the explanations given by psychologists and philosophers and decided to try and investigate for myself the basic phenomenon of nerve action. For that purpose I went to Oxford to work under Sir Charles Sherrington, who was then the acknowledged world leader in this field." Eccles was a Victorian Rhodes scholar at Oxford, receiving a D.Phil. there in 1929. He held fellowships at Exeter and Magdalen colleges at Oxford until 1937, when he returned to Australia as director of the Kanematsu Research Institute of Sydney Hospital. In 1944 he became professor of physiology at the University of Otago in New Zealand, and in 1951 he joined the faculty at the Australian National University, where he remained until 1966, when he assumed his present position at Northwestern University. Sir John is the author of the article entitled "The Physiology of Imagination," which appeared in the September 1958 issue of SCIENTIFIC AMERICAN.

Bibliography

EXCITATION AND INHIBITION IN SINGLE NERVE CELLS. Stephen W. Kuffler in *The Harvey Lectures, Series 54.* Academic Press, 1960.

PHYSIOLOGY OF NERVE CELLS. John C. Eccles. Johns Hopkins Press, 1957.

THE PHYSIOLOGY OF SYNAPSES. John C. Eccles. Academic Press, 1964.

THE TRANSMISSION OF IMPULSES FROM NERVE TO MUSCLE, AND THE SUBCELLULAR UNIT OF SYNAPTIC ACTION. B. Katz in *Proceedings of the Royal Society,* Vol. 155, No. 961, Series B, pages 455–477; April, 1962.

18. The Visual Cortex of the Brain

The Author

DAVID H. HUBEL is professor of neurophysiology at the Harvard Medical School. Born in Windsor, Ontario, in 1926, Hubel received a B.Sc. and an M.D. from McGill University in 1947 and 1951 respectively. He studied clinical neurology for three years at the Montreal Neurological Institute before coming to this country in 1954 to spend a year's residency in neurology at the Johns Hopkins Hospital. In 1955 he began neurophysiological research at the Walter Reed Army Institute of Research in Washington, and in 1960 he joined the Harvard faculty.

Bibliography

DISCHARGE PATTERNS AND FUNCTIONAL ORGANIZATION OF MAMMALIAN RETINA. Stephen W. Kuffler in *Journal of Neurophysiology,* Vol. 16, No. 1, pages 37–68; January, 1953.

INTEGRATIVE PROCESSES IN CENTRAL VISUAL PATHWAYS OF THE CAT. David M. Hubel in *Journal of the Optical Society of America,* Vol. 53, No. 1, pages 58–66; January, 1963.

RECEPTIVE FIELDS, BINOCULAR INTERACTION AND FUNCTIONAL ARCHITECTURE IN THE CAT'S VISUAL CORTEX. D. H. Hubel and T. N. Wiesel in *Journal of Physiology,* Vol. 160, No. 1, pages 106–154; January, 1962.

THE VISUAL PATHWAY. Ragnar Granit in *The Eye, Volume II: The Visual Process,* edited by Hugh Davson. Academic Press, 1962.

19. The Growth of Nerve Circuits

The Author

R. W. SPERRY is Hixon Professor of Psychobiology at the California Institute of Technology. He graduated from Oberlin College in 1935 and took his Ph.D. at the University of Chicago in 1941. Then he was a National Research Fellow at Harvard University and a research associate at the Yerkes Laboratories of Primate Biology. In 1946 he returned to Chicago, serving first as assistant professor of anatomy and later as associate professor of psychology. He spent a year at the National Institute of Health as chief of the section on developmental neurology, and in 1954 went to his present position at Cal Tech.

Bibliography

DEVELOPMENTAL BASIS OF BEHAVIOR. R. W. Sperry in *Behavior and Evolution,* edited by Anne Roe and George Gaylord Simpson, pages 128–138. Yale University Press, 1958.

MECHANISMS OF NEURAL MATURATION. R. W. Sperry in *Handbook of Experimental Psychology,* edited by S. S. Stevens, pages 236–275. John Wiley & Sons, Inc., 1951.

PHYSIOLOGICAL PLASTICITY AND BRAIN CIRCUIT THEORY. R. W. Sperry in *Biological and Biochemical Bases of Behavior,* edited by Harry H. Harlow and Clinton N. Woolsey, pages 401–421. University of Wisconsin Press, 1958.

20. Learning in the Octopus

The Author

BRIAN B. BOYCOTT is reader in zoology at University College London. He is a graduate of Birkbeck College London, which he attended while employed as a laboratory technician at the National Institute for Medical Research. After a short time lecturing in zoology at University College he spent five years as a research assistant in the anatomy department, returning to the zoology department in 1952. Boycott writes that in addition to his investigations of the octopus he is "currently working on the vertebrate retina and changes in the structure of the brain of hibernating mammals." Although he bears a surname that has entered the language as both a common noun and a verb, he is not related to the Captain Charles Boycott whose activities were responsible for the origin of the word.

Bibliography

BRAIN AND BEHAVIOR IN CEPHALOPODS. M. J. Wells. Stanford University Press, 1962.

THE FUNCTIONAL ORGANIZATION OF THE BRAIN OF THE CUTTLEFISH SEPIA OFFICINALIS. B. B. Boycott in *Proceedings of the Royal Society of London,* Series B, Vol. 153, No. 953, pages 503–534; February, 1961.

IN SEARCH OF THE ENGRAM. K. S. Lashley in *Physiological Mechanisms in Animal Behaviour.* Symposia of the Society for Experimental Biology, No. 4, 1950.

A MODEL OF THE BRAIN. J. Z. Young. Oxford University Press, 1964.

SOME ESSENTIALS OF NEURAL MEMORY SYSTEMS: PAIRED CENTRES THAT REGULATE AND ADDRESS THE SIGNALS OF THE RESULTS OF ACTION. J. Z. Young in *Nature,* Vol. 198, No. 4881, pages 626–632; May, 1963.

PART VI: BEHAVIOR

General References

Dethier, V. G., and Eliot Stellar. ANIMAL BEHAVIOR: ITS EVOLUTIONARY AND NEUROLOGICAL BASIS. 2nd Edition, Prentice-Hall, 1964.

Gray, James. HOW ANIMALS MOVE. Cambridge University Press, 1953.

Roeder, K. D. NERVE CELLS AND INSECT BEHAVIOR. Harvard University Press, 1963.

Thorpe, W. H. LEARNING AND INSTINCT IN ANIMALS. Methuen, 1956.

Tinbergen, N. THE STUDY OF INSTINCT. Oxford University Press, 1951.

21. Electric Location by Fishes

The Author

H. W. LISSMANN is lecturer in the department of zoology at the University of Cambridge. Lissmann, who was born in Russia, studied in Germany, where he obtained a Ph.D. at the University of Hamburg. He received an M.A. from Cambridge in 1947 and became assistant director of research there the same year. Lissmann was elected a Fellow of the Royal Society in 1954. He was made lecturer in zoology at the University of Cambridge the following year and was also elected a Fellow and Lecturer of Trinity College, Cambridge. Lissmann is interested mainly in the behavior of animals, their movements, sense organs and nervous systems. His studies of electric fishes, begun in 1950, are described in his article. Lissmann has found it "useful and illuminating" to collect his own fish in Africa and South America.

Bibliography

ECOLOGICAL STUDIES ON GYMNOTIDS. H. W. Lissmann in *Bioelectrogenesis: A Comparative Survey of its Mechanisms with Particular Emphasis on Electric Fishes.* American Elsevier Publishing Co., Inc., 1961.

ON THE FUNCTION AND EVOLUTION OF ELECTRIC ORGANS IN FISH. H. W. Lissmann in *Journal of Experimental Biology,* Vol. 35, No. 1, pages 156–191; March, 1958.

THE MECHANISM OF OBJECT LOCATION IN GYMNARCHUS NILOTICUS AND SIMILAR FISH. H. W. Lissmann and K. E. Machin in *Journal of Experimental Biology,* Vol. 35, No. 2, pages 451–486; June, 1958.

THE MODE OF OPERATION OF THE ELECTRIC RECEPTORS IN GYMNARCHUS NILOTICUS. K. E. Machin and H. W. Lissmann in *Journal of Experimental Biology,* Vol. 37, No. 4, pages 801–811; December, 1960.

22. Moths and Ultrasound

The Author

KENNETH ROEDER is professor of physiology at Tufts University, where he has served since 1931. He was born in England, was graduated from the University of Cambridge and did graduate work there and at the University of Toronto. "My lifelong interest in insects," he writes, "probably stems from a childhood enthusiasm for butterfly collecting." Roeder's research deals mainly with the biological aspects of insect behavior. In addition he is "an incurable tinkerer with mechanical and electronic gadgets," an activity that "led at one time to the construction of an electromechanical analogue of certain phases of cockroach behavior and has played a part in the work on moth hearing." He says he has "always felt that if one can make a subject clear and interesting to a nonspecialist, it becomes clearer and more interesting to oneself."

Bibliography

THE DETECTION AND EVASION OF BATS BY MOTHS. Kenneth D. Roeder and Asher E. Treat in *American Scientist,* Vol. 49, No. 2, pages 135–148; June, 1961.

MOTH SOUNDS AND THE INSECT-CATCHING BEHAVIOR OF BATS. Dorothy C. Dunning and Kenneth D. Roeder in *Science,* Vol. 147, No. 3654, pages 173–174; January 8, 1965.

NERVE CELLS AND INSECT BEHAVIOR. Kenneth D. Roeder. Harvard University Press, 1963.

23. The Reproductive Behavior of Ring Doves

The Author

DANIEL S. LEHRMAN is professor of psychology and director of the Institute of Animal Behavior at Rutgers University. Lehrman was graduated from the City College of New York in

1947 and obtained a Ph.D. in psychology from New York University in 1954. He has been a member of the Rutgers faculty since 1950.

Bibliography

CONTROL OF BEHAVIOR CYCLES IN REPRODUCTION. Daniel S. Lehrman in *Social Behavior and Organization among Vertebrates,* edited by William Etkin. The University of Chicago Press, 1964.

HORMONAL REGULATION OF PARENTAL BEHAVIOR IN BIRDS AND INFRAHUMAN MAMMALS. Daniel S. Lehrman in *Sex and Internal Secretions.* Edited by William C. Young. Williams & Wilkins Company, 1961.

INTERACTION OF HORMONAL AND EXPERIENTIAL INFLUENCES ON DEVELOPMENT OF BEHAVIOR. Daniel S. Lehrman in *Roots of Behavior,* edited by E. L. Bliss. Harper & Row, Publishers, 1962.

24. Pheromones

The Author

EDWARD O. WILSON is associate professor of zoology at Harvard University. As a native of Alabama, Wilson fairly early in life became acquainted with the Southern agricultural pest known as the fire ant, which he discussed in an article for SCIENTIFIC AMERICAN ("The Fire Ant," March, 1958). Wilson received B.S. and M.S. degrees from the University of Alabama in 1949 and 1950. He took a Ph.D. in biology at Harvard, where he held a National Science Foundation fellowship and a junior fellowship in the Society of Fellows. He joined the Harvard faculty in 1956.

Bibliography

OLFACTORY STIMULI IN MAMMALIAN REPRODUCTION. A. S. Parkes and H. M. Bruce in *Science,* Vol. 134, No. 3485, pages 1049–1054; October, 1961.

PHEROMONES (ECTOHORMONES) IN INSECTS. Peter Karlson and Adolf Butenandt in *Annual Review of Entomology,* Vol. 4, pages 39–58; 1959.

THE SOCIAL BIOLOGY OF ANTS. Edward O. Wilson in *Annual Review of Entomology,* Vol. 8, pages 345–368; 1963.

Index